Consuming mobility

Consuming mobility

A practice approach to sustainable mobility transitions

Jorrit O. Nijhuis

Environmental Policy Series – Volume 10

ISBN: 978-90-8686-242-9
e-ISBN: 978-90-8686-794-3
DOI: 10.3920/978-90-8686-794-3

First published, 2013

© Wageningen Academic Publishers
The Netherlands, 2013

This work is subject to copyright. All rights are reserved, whether the whole or part of the material is concerned. Nothing from this publication may be translated, reproduced, stored in a computerised system or published in any form or in any manner, including electronic, mechanical, reprographic or photographic, without prior written permission from the publisher:
Wageningen Academic Publishers
P.O. Box 220
6700 AE Wageningen
The Netherlands
www.WageningenAcademic.com
copyright@WageningenAcademic.com

The content of this publication and any liabilities arising from it remain the responsibility of the author.

The publisher is not responsible for possible damages, which could be a result of content derived from this publication.

Preface

Considering that this thesis is strongly inspired by Giddens' constitution of society, why not start with a very fitting quote from his work: 'this was not a particularly easy book to write'! All in all it has taken me over eight years to finish this thesis and during the last half its completion has balanced continuously on the edge of a knife. Lamenting on the hardships of the journey has been done more than enough though, so I will not bother you with that. Especially as my curiosity for the subject at hand, which, together with the aspiration to find educated work, prompted me start this research, has never abandoned me. Individual and collective human behaviour, especially in the fields of sustainable consumption and production and everyday mobility, is simply a fascinating world to explore. It continues to be a mind-boggling puzzle to connect the obvious necessity for radical change towards sustainability at the society level with the scope of everyday life interests of citizen- consumers, especially in times of economic crisis. Hopefully, this thesis has made a worthwhile contribution.

During my research I have discussed my work with many different experts working in science, policy, consultancy or business. From these experts I want to express my gratitude to Wendy Williams who selflessly ensured my way in to the private sector which resulted in two very useful case studies. Furthermore, for their contribution to my research I would like to thank Frank Versteege (marketing manager at Toyota) and Dick Bakker (commercial director at Arval) who showed me the meaning of service. Of course I also want to thank the 3Js (professors Johan Schot, Jan Rotmans, and John Grin), the three directors of the Knowledge network for System Innovations and Transitions (KSI) who made the Contrast research program possible. It has been a great honour to meet, study, learn, laugh and dance (or at least something that was aimed to look like making rhythmical movements) together. In the same light I want to thank my fellow PhD-students from the KSI mobility research group, especially Marc Dijk and Bonno Pel, who provided a safe and inspirational environment to discuss sustainable mobility transitions.

My research was conducted at the Environmental Policy Group at Wageningen University, a place which I have always felt to have an academically stimulating and atmospherically mellow work environment. I would like to thank the many national and international PhD-students and staff members who have made it such a nice location to work. Though many people come and go, the genuine ENP feeling always remains present. Bas van Vliet, my short educational 'career' was mainly conducted together with you and I must say that it was very pleasant and worthwhile. Astrid Hendriksen, thank you for guiding the focus group sessions with the car salesmen and car purchasers, and if I'm not mistaken, and I rarely am, I still owe you a bottle of whisky for a bet lost long ago. In addition, at the Veluweloop you both showed me year after year that running is definitely not one of my talents. Obviously, I want to consider the other members of the Contrast research programme as well, especially Elizabeth Sargant and Lenny Putman, my fellow PhD students at ENP, who ensured that this was never a solitary enterprise. Desirée, our life and PhD trajectories were and will always be strongly entwined. Typically, you have flown of towards far-away tourist destinations while I continue to focus on daily commutes. What else can I say but that it was like passing ships in the night.

As tradition goes, and I might be considered a traditional person, many credits go out to Corry whose collective and refreshingly down-to-earth spirit in the individualist world of academia far outweighs her secretarial and organisational skills. Thanks also go out to professor Arthur Mol, the chair of the Environmental Policy Group. I have the highest regard for his ability to perform at a world-class level both in academic publications and in academic management, and I'm sure he will turn out to be a supernatural being in disguise.

Naturally, special thanks are in order for my two promotors, professor Gert Spaargaren and professor Hans Mommaas. I can easily remember the effort I had to put in understanding the theoretical debates, especially in the first and second year of my PhD research. Often after an intense discussion I walked back to my room feeling I more or less understood every theoretical line of reasoning only to sense the knowledge acquired slipping rapidly through my fingers with every step taken to my room while vigorously trying to hold on to the grains of knowledge in my hands in order not to arrive empty-handed. Luckily, these theoretical discussions have become increasingly accessible to me over the years. Though sociology will never be an easy subject for me, it has greatly improved my understanding of social life. Furthermore, I need to thank Gert for his perseverance, patience, understanding, never wavering trust in my scientific capabilities, and the occasional fatherly advice throughout the whole endeavour.

Over the last four years I have been working at Rijkswaterstaat which has proven to be an interesting and enjoyable organisation to work for. I want to thank all the colleagues who 'make' the organisation work as it is, especially my direct colleagues of the 'team behaviour' whose youthful spirits and social science perspectives on traffic and mobility is very inspiring. I want to thank Marlies Emmen and Jolanda Vis for putting trust in a PhD-researcher at a time when knowledge was not unquestionably considered to be an asset at Rijkswaterstaat, Henk Pauwels and Gordon de Munck for teaching me so many things about the world of mobility management, and Loes Aarts for being the best colleague one could wish for and for continuously stimulating me to step out of my comfort zone.

Special thanks are in order for those who supported me when life turned towards the inevitable less enjoyable periods which every PhD researcher is hunted by. I feel fortunate to have spent much time with my (indoor-soccer-loving) friends from Feyenesse and the Charismatische Utrechtsche Mannschaft. Furthermore, I have the pleasure to be surrounded by the smartest group of friends that I know of, though it seems we invest much effort to reveal that fact. The fun we have had in old industrial English cities, at pop- and pub-quizzes, parties and midnight gaming sessions has been simply ridiculous. More specifically I want to thank Merlijn and Hannah for their warmth, Raoel and Jan-Willem for long-lasting friendships, Vincent – Zenmaster – van der Vlies for his never-ending advice, and Vibhi van Wersch who is like a brother I never had.

Naturally, I need to thank my father, who taught me to be respectful to the environment, and Corine for pushing me to continue working on my PhD. Also, I am glad that my sister and her family are here to share this moment of joy.

Maartje, we are only together for a relatively short time and luckily you have been spared the bulk of the unpleasantness of being a PhD-student's partner. Nevertheless, you understand better than all what this trajectory may take and you already know me better than I perhaps find comfortable. We are somewhere between the click of the light and the start of a dream. We don't know where we're going, but let's go.

Table of contents

Abbreviations

ACEA	Association des Constructeurs Européens d'Automobiles (European Automobile Manufacturers Association)
ADAC	Allgemeiner Deutscher Automobil-Club
AIDA	Awareness interest desire action
ANWB	Algemene Nederlandse Wielrijdersbond
CAFE	Corporate average fuel economy
CNG	Compressed natural gas
CO	Carbon monoxide
CO_2	Carbon dioxide
CONTRAST	Consumption transitions for sustainability
EKB model	Engel, Kollat, Blackwell model
EV	Electric vehicle
FFV	Flex-fuel vehicle
HEV	Hybrid electric vehicle
ICE	Internal combustion engine
IMF	International Monetary Fund
JAMA	Japan Automobile Manufacturers Association
KAMA	Korea Automobile Manufacturers Association
LTS	Large technical systems
MKB	Midden en klein bedrijf (small and medium sized enterprises)
MLP	Multi-level perspective
NCAP	New car assessment program
NGO	Non-governmental organisation
NOx	Nitrogen oxide
NVVP	Nationaal Verkeer- en Vervoersplan
OECD	Organisation for Economic Co-operation and Development
PM10	Particulate matter (with diameter of 10 micrometre or less)
P+R	Park and ride
PT	Public transport
SO_2	Sulphur dioxide
SPA	Social practices approach
RIVM	Rijksinstituut voor Volksgezondheid en Milieu
SUV	Sports utility vehicle
SVV2	Tweede Structuurschema Verkeer en Vervoer
TFMM	Taskforce Mobility Management
VNO NCW	Verbond van Nederlandse Ondernemingen-Nederlands Christelijk Werkgeversverbond (Dutch's employers federation)
VOS	Volatile organic compounds
WTO	World Trade Organization
WWF	World Wildlife Fund

Chapter 1. Introduction

So what if it only does three miles to the gallon, I'm a mom, not a conservationist.

(Fictitious) advertisement for the Maibatsu Monstrosity SUV in Grand Theft Auto III

1.1 Are you driving a flex-fuel vehicle?

Consumers play a crucial role in the transition to sustainable mobility. The infamous case of flex-fuel vehicles (FFV) in the United States provides an interesting example of the adverse effects of a neglect of citizen-consumer aspects in the diffusion and use of sustainable innovation in the domain of mobility. The United States has the second largest fleet of FFV in the world. From 1996 to the end of 2012 over fifteen million FFV have been sold throughout the years, of which 11 million flex-fuel cars and light trucks are still operational. Flexible fuel vehicles are specially designed vehicles that contain engines that allow a vehicle to operate on a blend of gasoline and ethanol which can vary between 0% (purely gasoline) and 85% (E85). The widespread diffusion of FFV was federally supported as bio-based fuels were seen by the US government to hold multiple benefits which included, next to a reduction of climate change emissions, a lowering of foreign petroleum dependency and an increase in job opportunities. Intended as an incentive to develop alternative fuel vehicles, the emission standard law[1] stated that vehicles running on alternative fuels such as bio-fuels are (partly) exempt from fuel efficiency standards. This prompted many car manufacturers to develop and sell FFV in the United States. Large automobile manufacturers such as GM, Ford and Chevrolet have developed dozens of vehicle types suitable for the use of bio fuels (in 2012 over sixty flexible fuel vehicle models were available on the US market). However, a study conducted in 2002 by the US Department of Transportation estimated that E85 constituted for only about 1% of the fuel consumed by FFV between 1996 and 2000! By 2010 approximately ten million flex-fuel vehicles were operational in the US, while only 600,000 of these FFV were actually believed to be using E85 as a fuel, most of which were part of the federal vehicle fleet (Davis *et al.*, 2013). Even more interesting, despite the large-scale diffusion of FFV in the US, many consumers were completely unaware of the flex-fuel capabilities of their vehicle! The reason for this lack of consumer awareness is that the sales price and the exterior of FFV are identical to 'regular' gasoline vehicles. While in a number of cases consumers were not able to purchase a certain model in anything *but* a flexible vehicle configuration, car manufacturers didn't feel the need to communicate this characteristic to its owners. In addition, these vehicles were sold nation-wide while E85 fuel stations were only available in a number of states, most notably in the Corn Belt. For example, in the state of Massachusetts in 2010 only three E85 stations were available to the more than 100,000 owners of FFV (Clean Fuels Foundation, 2011).

Only recently a coalition of major car companies, together with the US Environmental Protection Agency, the US Department of Agriculture, and E85 retailers have started to promote

[1] Corporate Average Fuel Economy requirements (CAFE).

the use of bio fuels in FFV in an attempt to meet the National Renewable Fuel Standard. A public education project was initiated under the heading of the National FFV Awareness Campaign whose mission is: 'to locate and educate the nation's nine million FFV owners and motivate them to use higher than 10% blends of ethanol' (www.ffv-awareness.org). Car drivers are prompted to check their vehicle manual, their fuel cap or to go to a special information website that lists all E85 compatible models. According to ethanol promoters most FFV owners were stunned to learn that their car could run on alternative fuel.

1.2 Mobility as a complex problem

Automobility provides a perfect example of the environmental challenges in a post-modern era. As the abovementioned case of flex-fuel vehicles in the United States illustrates, sustainable innovation in the mobility domain is not only dependent on the technological means available, but even more so on the developments in the preferences, opinions and actions of (car) consumers and citizens (ECN, 2009). Nevertheless, the discussion on large-scale sustainable transformation in the domain of mobility continues to focus predominantly on the (technological) characteristics of vehicles and fuels. This emphasis must also be seen in the light of the success of earlier generations of environmental policies.

Classic environmental policies of the 70s, 80s and 90s have contributed to a substantial improvement in the environmental quality in Western societies. Cleaner production processes, recycling programs, and eco-efficient products and services have reduced the environmental impact in various production-consumption chains. The development path of recycling (policies) of end-of-life vehicles can be seen as a case of a successful ecological modernization in the automotive industry (Sminke *et al.*, 2003). Increasing material costs, environmental policies (such as the EU guideline 2000/53/EC) and technological innovations in the handling of end-of-life vehicles have in a mutually interdependent way resulted in a situation that on average in the Netherlands over 95% of the vehicle weight is recycled (ARN, 2013). Simultaneously, in that same time period the excretion of polluting emissions of petrol and diesel cars (NO_x, CO, PM_{10}, SO_2) has been reduced substantially. Especially the traffic induced emission of sulphur dioxides has been dealt with almost completely. Since 1998 the EU-limit for SO_2 concentrations, implemented to protect public health and ecosystem quality, has not been exceeded anywhere in the Netherlands (CBS, PBL, Wageningen UR, 2013). Acidification of ecosystems, *the* environmental topic in the 1980s due to the forecasted disappearance of woodlands, is therefore not an issue anymore.

The abovementioned developments show how the cleaning of production processes can be seen as the 'low hanging fruit' of environmental policies in the domain of mobility. Grin *et al.* (2003) describe these approaches between the 1970s and the 1990s as the first and second generation of environmental policies. The focus of the first generation of environmental policies was on reducing the harmful effects (such as polluting emissions) via end-of-pipe measures in a specific domain. The second generation was characterised by a focus on reducing pollution at its source, and by a more regional approach. However, this (formerly) successful approach of government-led environmental policies with a strong reliance on a regulatory steering model is no longer sufficient to deal with current environmental challenges in the domain of mobility (PBL, 2013).

Contemporary mobility problems can be characterised by their complex and persistent nature[2]. These problems are not limited to environmental dimensions. From an economic, social and ecological perspective the current fossil fuel oriented system of mobility cannot be considered sustainable (Kahn Ribeiro *et al.*, 2007; Kemp *et al.*, 2012; WBCSD, 2004). Thus, mobility problems are multi-facetted, covering environmental, social and economic dimensions. Solutions devised for a problem in one dimension may have detrimental effects in another. Clearly, increasing road infrastructure and improving traffic flows through traffic management can reduce travel time loss for travellers but may have negative social and ecological consequences. Cleaner and more efficient car engines are beneficial for the environment but may weaken existing constraints on the growth of car travel and contribute to urban sprawl and car dependent lifestyles (Adams, 2005).

Furthermore, the effects of the problems generated by the car regime are not only local but also global in scope. This makes the negative impact of the car regime on environment and society not only less visible but also much more demanding to deal with as they surpass the authority and reach of single nations. In the economic sphere there are a number of global institutions such as the IMF, the WTO and the OECD which have the necessary authority and (political, economic and legal) means to have ensured global rules and regulations on international trade and currencies. In the ecological sphere comparable global institutions which have the mandate and power to enforce global environmental regulations are lacking. But even if these institutions were in place it is highly questionable whether these classical regulatory strategies applied by governmental agencies will be sufficient for the complexity of present-day mobility problems.

Another part of the complexity lies herein that over the years the role of citizen-consumers in dealing with sustainability problems has become a critical factor. Not in the least because cleaner production processes do not necessarily result into sustainable consumption practices. So far, technological improvements in production and end-of-life processes have been insufficient to compensate for the increasing environmental pressure due to consumption growth (MNP, 2006, 2007). People not only travel more kilometres in general, they also use more resource-intensive transport modes to move around. In the last decades for many people everyday mobility has increasingly become synonymous with automobility. Between 1950 and 1997 the world car fleet has grown from 50 million to 580 million vehicles, approximately five times more than the population growth in that time period (Kahn Ribeiro *et al.*, 2007). A generally accepted prognosis for daily passenger mobility is that until 2050 there will be a worldwide growth of 1.7% in passenger kilometres per year (WBCSD, 2004).

It is interesting to note that the growing role of citizen-consumers in environmental issues seemingly runs parallel with their decreasing involvement. In comparison to social, economic and political subjects, the interest for environmental problems has declined over the years (CBS, PBL, Wageningen UR, 2013). While most of the local (environmental) effects often have a tangible impact on the daily life of citizens, the global effects of the new generation of environmental challenges lack this concreteness and are therefore more difficult to distinguish (*ibid.*). Comparable to urban air quality and traffic noise, climate change may seem to be a very abstract phenomenon.

[2] Complex problems are problems that: 'occur on different levels of scale, have a variety of actors with different perspectives involved, are highly uncertain in terms of future developments, can only be dealt with on the long term and are hard to 'manage' in a traditional sense' (Loorbach, 2007, p. 14).

Climate change and other global environmental problems are to a large extent a blind spot in the everyday life of citizen-consumers. Also, rarely do consumers make the connection between the use of technologies, energy consumption, and climate change emissions (Goldblatt, 2002). That doesn't mean that citizen-consumers have no perception of climate change, it is just not a part of everyday life routines. Nevertheless, even though people have limited concrete awareness of the (unintended) implications of their actions, the cumulative effect of the combined local actions of human agents may be global in nature (Urry, 2003a).

1.3 The approach of system innovations and transitions

In the last decade there is a growing body of literature on system innovations and transitions which has as a common understanding that long-term transformative change is necessary to deal with the complexity and persistency of contemporary environmental problems. This systems approach in dealing with complex environmental problems can be considered the third generation of environmental policies (Grin *et al.*, 2003).

The shared point of view is that 'standard' short-term approaches to changes within societal subsystems, such as the system of mobility, will be insufficient to deal with the problems facing society because they will lead only to gradual changes instead of transformative change. For instance, innovation studies have shown that most innovations are of an incremental nature; the novel technologies introduced to the market fit in the existing social and technical infrastructure and do not require societal actors to substantially alter their routines (Elzen & Wieczorek, 2005). This means that the whole modus operandi remains unaltered, for instance the modus of how people travel and what it entails to be mobile. In contrast, a transition consists of a continuous process of structural change whereby a subsystem of society changes fundamentally (Loorbach, 2007; Rotmans *et al.*, 2001).

There is general agreement that transitions are defined by three main characteristics. First, a transition is a long-term process that covers one or more generations (at least 25 years). Second, transitions consist of reciprocal developments in policy, economy, science, technology, culture, markets and consumption patterns. Thirdly, transitions are the result of dynamics within and between three different levels of scale: niche, regime and landscape, generally referred to as the multi-level perspective (MLP) (Loorbach, 2007, Geels & Schot, 2007).

The general goal of transition studies is to understand how transitions in society develop. Over the years three branches of transition studies have emerged, each of which has a particular point of view on the analysis of transitions due to their differences in scientific origin. The first branch of transition studies conceptualises a transition as a shift from one socio-technical system to another. Derived primarily from the research domain of science and technology and evolutionary economics the perspective of socio-technical systems takes socio-technological configurations within a societal subsystem as the point of departure for analysing transitions (see Berkhout *et al.*, 2003; Elzen *et al.*, 2004; Geels, 2005c, Geels & Schot, 2007; Hoogma *et al.*, 2002; Schot *et al.*, 1994). The second branch of transition studies, the approach of transition management, is a form of action research as it is 'by definition (partly) applied and participatory' (Loorbach, 2007, p. 37). Grounded in theories of complex adaptive systems and complexity governance, scientists in the field of transition management, aside from analysing transitions, also actively contribute to sustainable transitions via Mode 2 knowledge production. By analysing historical and on-going transitions new methodologies and approaches for governance are developed to coordinate on-going and future transition processes (see Kemp & Loorbach, 2006; Loorbach, 2007; Rotmans *et al.*, 2001). Finally, the third branch of transition studies provides a more analytical perspective on transition governance. Based on political and sociological theories of reflexive modernisation this approach specifically focuses on the dynamic relationship between transitions, politics and power (Grin, 2012; Grin *et al.*, 2003).

1.3.1 The multi-level perspective

The socio-technical perspective in transition studies conceptualizes transport as a socio-technical system. The notion of socio-technical systems emphasizes that technologies are not stand-alone objects but are embedded in society through infrastructures, regulations, production networks, and consumption patterns. It is the alignment of these separate elements which result in a socio-technical system. While the socio-technical *system* refers to measurable elements (such as automobiles, road infrastructures, market shares), the notion of socio-technical *regime* refers to more intangible elements (such as the normative, cognitive and regulative rules) which actors draw upon in concrete actions (Geels & Kemp, 2012; Geels, 2004).The socio-technical regime refers therefore to the associations, rules, and thoughts carried by different social groups which are active in a socio-technical system.

As described above, transition studies claim that trajectories of transitions can best be understood as the alignment of developments at multiple levels (Geels & Schot, 2007). The multi-level perspective (MLP) is an analytical model which distinguishes three levels of increasing structuration: niches, socio-technical regime, and socio-technical landscapes. Radical innovations (unstable novel socio-technical configurations) 'emerge' at the niche-level where they are developed by networks of niche actors, often outside of the existing regime. The socio-technical landscape refers to the context in which niches and regimes are embedded but which are outside of the direct influence of niche and regime actors. They refer to aspects such as macro-economic and macro-political developments, climate change, and physical infrastructures. The socio-technical landscape may have a direct influence on the socio-technical regime and niche. For instance, climate change and peak oil put pressure on the car-regime, while the existing physical automotive infrastructures provide a barrier to non-automotive transport modes in technological niches.

The MLP is a helpful analytical tool in analysing and understanding stability and change in socio-technical systems such as the car-based system for transportation (Geels & Kemp, 2012). It is easy to understand how existing socio-technical regimes show a resistance to radical change due to investment costs (in production processes, infrastructures), interests by regime players (automotive industry, road construction) and established consumption patterns and lifestyles. Socio-technical systems are relatively stable in the sense that alterations and innovations are incremental and do not fundamentally alter the socio-technical system. The reduction of air polluting emissions by cars due to the introduction of technological innovations can be seen as a series of incremental innovations in the car-based regime which came about due to pressures at the landscape level (increasing global environmental awareness in the 1980s). As the regime alters due to developments within the regime, and responds to developments at the landscape and niche level, the regime is in a state of dynamic equilibrium.

Aside from analysing stability in regimes, the MLP also provides an analytical framework to describe how radical innovations may break through to the socio-technical regime. These transitions are considered to be the result of the alignment of three occurring processes: '(1) niche-innovations build internal momentum, (2) changes at the landscape level put pressure on the socio-technical regime, and (3) destabilization of the socio-technical regime provides windows of opportunities for niche-innovations' (Geels & Schot, 2007, p. 400; Schot & Geels, 2008, p. 545).

1.3.2 Criticism on the absence of citizen-consumers in the transition perspective

Transition studies have provided significant contributions to the understanding of innovation processes and the complex trajectories of systemic change in the system of mobility. Especially relevant is notion that systemic change is perceived more as a form of co-evolution between technology and society than as a diffusion of innovation (Geels *et al.*, 2012). This novel approach also provides a promising perspective on the governance of complex and persistent problems. As Paredis (2009) indicates, the discussion surrounding sustainable development has become increasingly concentrated on policy instruments and sustainable developments goals. A comprehensive vision on how the necessary change could take place has been lacking. The approach of transition provides this focal point for governance (*ibid.*).

However, when we go back over the case of the government-led technology push of flex-fuel vehicles described at the beginning of this chapter one can feel that something is amiss. What is needed most in the 'management' of complex and persistent mobility problems is more consideration for the social-cultural context in which socio-technical innovations must land (VROM-raad, 2005). This requires more attention for the role of citizen-consumers in transitions, both in their public role as voters and employees, but also in their private roles as consumers. In an evaluation of Dutch transition policy, Weterings (2010) concludes that end-users/consumers are barely involved in contemporary transition policies. The majority of the transition programmes are supply-oriented while there is little attention for the exploration of societal trends and needs within the transition programme. The systemic focus of transition management therefore brings with it the risk that end-users and consumers are seen as an external factor. In the debate on transitions policy-makers and corporate actors are presented as the key players while the vital role of citizen-consumers or practitioners in creating, sustaining, and changing of everyday life routines is neglected (Shove & Walker, 2010).

In the socio-technical approach to transitions the technological trajectory provides the starting point for the analysis. While this critical point has been recognized (see Geels & Schot, 2007; Grin *et al.*, 2011), knowledge on the role of citizen-consumers in sustainable transitions in general, and mobility transitions specifically, is still underdeveloped. The current transition policy in the field of mobility has no adequate answer to the question what a promising end-user perspective on sustainable mobility could or should be. A major unsolved challenge is to find a way in which the socialisation of sustainable mobility takes place in such a way that a connection is made with the everyday life experiences, concerns and actions of citizen-consumers. Clearly there is a challenge for transition studies to incorporate consumption patterns and the viewpoint of everyday life routines in theories and policies on sustainable mobility transitions.

1.4 Models of sustainable consumer behaviour

While the role of citizen-consumers in systemic changes in transportation regimes is of the essence, the question is whether or not conventional models on sustainable consumption may provide sufficient answers to the critical points raised above. Most disciplines in consumer behaviour see behavioural change predominantly as the result of changes in the individual values, beliefs and attitudes (Jackson, 2005; Spaargaren, 2003a).

The hypothesis of many social and cognitive psychology approaches is that consumption behaviour is the result of a rather linear and rational decision-making process (Hargreaves, 2008). A significant portion of these studies is based on adjusted expectancy-value models such as the Theory of Reasoned Action (Fishbein & Ajzen, 1975) and the Theory of Planned Behaviour (Ajzen & Madden, 1986) which see behavioural intention, that is the deliberate plan to perform the behaviour, as the prime predictor of that specific behaviour (see also Staats, 2003). More specifically, the intention to perform a specific behaviour is formed by beliefs about the outcomes of the available behavioural choices and an evaluation of these outcomes[3]. In ideal circumstances consumers will choose the outcome with the highest expected benefits. Based on expectancy-value models the logical consumer policy suggestion is to ensure that consumers have sufficient information to make an informed behavioural choice. However, in many circumstances the expected outcome of pro-environmental behaviour is evaluated as negative (the personal benefits do not outweigh the personal costs) (see Staats, 2003). This can be the case when people consider travelling by car or public transport. While the benefits of the choice for the car are on an individual level, the negative consequences are on a societal level. This is also framed as the social dilemma character of environmental problems.

Therefore, studies on pro-environmental behaviour are often combined with the perspective that environmentally relevant consumer behaviour is a form of moral behaviour[4]. In this perspective pro-environmental behaviour is primarily the result of value orientations of the individual (see Dunlap & Van Liere, 1978; Schwartz, 1977; Stern, 2000). As pro-environmental behaviour often has negative personal consequences, e.g. in terms of costs or effort, it must be seen as a form of altruistic or pro-social behaviour. Schwartz's norm-activation theory (1977) implies that consumers are more likely to conduct altruistic behaviour when they have a high problem-awareness, have

[3] While, according to these adjusted expectancy-value models, attitudes are the strongest variable in explaining behavioural intention, subjective norms and perceived behavioural control also directly or indirectly influence intention. Subjective norms are individual beliefs about what other (important) people think of a specific behaviour. It is therefore a perception of the existing normative rules in the social context of the individual. An important addition in the Theory of Planned Behaviour is the perceived behavioural control which basically is the perception of how easy or difficult it is to perform a specific behaviour.

[4] Though there is some disagreement how personal norms (performing pro-environmental behaviour out of personal motivation) relate to subjective norms (performing pro-environmental behaviour to comply with existing social norms) (Stern, 2005, p. 47). For instance, Schwartz (1977) sees pro-social behaviour primarily as the result of internalized personal norms. Failing to act in compliance with an internalized norm will lead to negative self-evaluations (Van Meegeren, 1996). Similarly, the personal benefits are also moral in nature.

insight into possible solutions, and feel personal responsible to contribute to the solution (Van Meegeren, 1996). Following the line of thought of these theories it is easy to explain the heavy emphasis placed on increasing environmental awareness and general environmental attitudes in society to promote pro-environmental behaviour. By means of environmental education and/or environmental communication, people will come to adopt more sustainable lifestyles and perform more pro-environmental behaviour. Nordlund and Garvill (2003) in a study to reduce car use conclude that 'strategies aimed to increase the willingness to reduce personal car use should emphasize self-transcendent and ecocentric values, clarify the negative environmental consequences of car use, and thereby accentuate the moral dilemma of personal car use' (p. 345).

The problem of consumer models based on (environmental) values and intentions is that generally speaking there is only a very weak relationship between the personal norms and the indicators of pro-environmental behaviour. On the basis of their research Vringer *et al.* (2007) conclude that a consumer-oriented energy policy solely based on a strategy of internalizing environmental responsibility will not be effective.

Stern (2000, p. 421) admits that his theory offers a good account of general predisposition toward pro-environmental behaviour, but that it may not be very useful for changing specific behaviour. In addition, Stern states that attitudinal factors have the highest explanatory value for behaviours which are not strongly constrained by context or require a high amount of effort. Steg (1999) concludes that generally speaking environmental awareness does not play an important role for environmentally relevant behaviour, especially when this behaviour requires more effort or time, or results in a reduction of comfort and freedom of movement.

1.5 A practice approach to sustainable mobility

Because of the criticisms on models of consumer behaviour and the perspective of socio-technical transitions a new paradigm is needed which can offer a new perspective on the analysis and management of sustainable mobility by recognizing the co-shaping influence of citizen-consumers on the one hand, and objects and socio-technological infrastructures on the other (Spaargaren, 2010). In (environmental) sociology, practice based approaches have been presented as the means to conceptualise the dynamics of the demand side in socio-technical transitions (Shove, 2012; Shove & Walker, 2010; Spaargaren, 2003). A social practice can be viewed as a routinized way of doing things, or a routinized type of behaviour (Giddens, 1984; Reckwitz, 2002). In this routinized type of behaviour several elements are interconnected with each other such activities, knowledge, skills, technologies, meanings.

The fundamental notion in analysing social practices is that individuals are no longer the prime subject of theories and policies on environmental behaviour and behaviour change. In contrast, in theories of social practice the processes of change and stability are primarily analysed at the level of practices themselves. Furthermore, novel in the approach of social practices is the acknowledgement of the active and dynamic relation between producers and consumers in a specific consumption domain. Mobility patterns are not only the result of individual decision-making but also depend on the socio-technical context in which the behaviour takes place. To understand why some socio-technical innovations are being picked up by specific groups of citizen-consumers and others not, lock-in factors in production-consumption chains have to be analysed

(Spaargaren, 2003). By adopting a practice approach, this thesis does not focus on one specific socio-technical system of mobility, or on the development pathways of specific niche innovations which may contribute to a regime change. The primal focus is on the human agents within these systems of mobility and the contextual factors influencing their actions in everyday routines.

While practice theories are gradually gaining more academic interest, in these approaches important questions remain with regard to environmentally relevant mobility behaviour. Which promises do practice theories entail for the study of everyday mobility patterns and transitions to sustainable mobility? What kind of factors triggers (environmentally relevant) changes in mobility practices and what factors may act as barriers to change (Spaargaren, 2010). What are the dynamics of change in practices of mobility? How can practice theories be used to develop and sustain policies for sustainable consumption in the domain of mobility? Addressing these questions is important for the study of complex and persistent problems in the domain of everyday mobility and may be helpful to develop a consumer-oriented transition policy in this domain.

1.6 Objective and research questions

The central objective of this thesis is to contribute to the academic and political discourse and debate on transitions to sustainable development in the domain of personal mobility. By adopting a practice based approach to study socio-technical transitions an attempt is made to develop a novel framework to analyse, understand and influence resource-intensive everyday mobility patterns. The main postulation is that a focus on social practices will provide a major asset to the study of innovation processes and the complex interaction between social-technical innovations and behaviour in the domain of mobility. The following research questions will be investigated in this book:

1. What are the characterizing elements in political and academic thought regarding the role of citizen-consumers in sustainable mobility transitions?
2. How can a practice approach help to transcend the limitations of the individualistic, voluntaristic perspective on consumer behaviour with the long-term, systemic perspective of socio-technical transitions?
3. When applied to everyday mobility routines, what can the social practices model contribute to the existing body of knowledge on transitions and system innovations?
4. By which means can environmental policy better incorporate citizen-consumers as agents of change in strategies towards sustainable mobility practices?

1.7 Methodology and outline of the book

Before drawing the outline of the book the specific context in which the research has taken place must be addressed. A thesis describing environmentally relevant phenomena from a citizen-consumer perspective is automatically region-specific. Though production-consumption chains are increasingly global in nature, a development which applies especially for the automotive sector, the actual performance of mobility on the level of everyday life is (also) locally construed. This clearly applies for the transport modes themselves. In the European Union as a whole, car use makes up around 78% of the share of passenger (non-aerial) transport kilometres (EEA, 2006). Globally, car ownership varies from 2 motor vehicles per thousand inhabitants in Bangladesh to 797 motor vehicles per thousand inhabitants in the United States, with the Netherlands located somewhere at the upper-end of the scale with 527 cars (World Bank, 2013). Aside from a manufactured object the car is representative of specific cultural meanings in the domain of everyday mobility (Urry, 2006). The automobile seen as an object of consumption embodies various meanings and identities, cultural elements which have been far from stable throughout the 20th century (Gartman, 2004). Similarly, car cultures and cultures of mobility (deliberately phrased in plural form) have a certain locality as they vary from country to country and within that country from community to community (Miller, 2001; Sheller, 2012; Vannini, 2009).

The focus in this thesis will be on the study of transition processes in Western countries, most notably European countries with densely populated nations, high levels of car use and, until recently, high economic development. More specifically, the case study research has taken place in the Netherlands making the results from the empirical chapters especially interesting for Dutch policy and research. The Dutch context in which the study has taken place obviously colours the outcomes to a certain extent[5]. Having said that, as this thesis addresses theories of socio-technical change and the contribution of practice theories to environmental policies, the theoretical and political outcomes apply for other Western countries as well.

The first research question will be investigated in the next chapter. Chapter 2 first describes the main trends and (sustainability) challenges in the domain of everyday mobility. Then, by reviewing from an historical perspective how policy-makers have addressed the sustainable development of everyday mobility, a closer look is taken at the prevailing paradigms in mobility governance. In this analysis we will focus both on the topics that have dominated the mobility policy agenda's and on the (sustainability) mobility strategies that have been pursued to reduce the negative impacts of mobility growth on society. We also provide initial answers to the first research question by portraying the fluctuating perspectives of policy actors on the possibilities of directly targeting the mobility practices of citizen-consumers.

Chapters 3 and 5 will provide the theoretical building blocks for this thesis. In Chapter 3, different approaches to sustainable development, such as stemming from social psychology, consumer studies, innovation studies and transition studies will be addressed in order to portray the conceptual framework which is used to analyse transition processes at the level of everyday life. We will describe the insights that can be gained from viewing mobility as a form of consumption. Analysing (the history of) automobility as a form of conspicuous consumption provides information about the social-cultural meanings associated with the car and its uses and functions in society. The viewpoint of conspicuous consumption is especially helpful in understanding the acquisition process of commodities. Thereafter the focus will shift to the role of routines to describe the habitual, taken-for-granted character of the vast bulk of activities of day-to-day social life. Drawing on contemporary theories of practices it will be argued that a practice-based approach provides a fruitful perspective to bridge the divide between actor-oriented and system-oriented approaches to sustainable consumption and production. In this chapter we will also describe the key elements

[5] The Dutch context is, amongst others, visible in the mobility culture (such as 'the bicycle culture'), the nation-specific policies, and the Dutch style of consensus-based policy and decision-making.

of the social practices model, the research framework adopted in this thesis to analyse sustainable consumption and production in the mobility domain.

In Chapter 5 the components of the social practices model will be re-conceptualized for the domain of everyday mobility[6]. In this chapter the notion of mobility practices will be more specifically defined and its basic characteristics will be elaborated. Here we will also focus on the everyday mobility strategies and patterns that citizen-consumers employ to realize their projects and plans. Depending on the skills, preferences and specific contexts, citizen-consumers may differ in the access to mobility options. This concept of 'mobility portfolios' is explained as a means to describe the variety in the ways that people participate in mobility practices.

Subsequently, in three empirical chapters, we will show how the theoretical framework can be applied to study sustainable development in the domain of everyday mobility. The main focus of each of the three cases is on situated interactions taking place at the crossroads between modes of access and modes of provisioning in the domain of everyday mobility; that is, the ways that citizen-consumers get access to forms of mobility on the one hand and the ways that mobility is supplied by mobility service providers to them on the other. The first empirical chapter, discussed in Chapter 4, centres on the dynamic relation between consumption and production in the automotive sector by analysing the practice of new car purchasing in the Netherlands. We focus specifically on the question how environmental information is presented and used in this practice. In addition, attention is given to the analysis of different consumer-oriented strategies which aim to influence consumer's car purchasing decisions. The analysis is based upon interviews with key respondents and a focus group research conducted with Toyota car salesmen on the one hand, and Toyota purchasers on the other.

Chapters 6 and 7 build on the conceptualisation of mobility practices described in the second theoretical chapter. Chapter 6 elaborates on the practice of commuting, one of the defined mobility practices. Based on three case studies, various initiatives to orchestrate alterations in commuting practices are described, predominantly involving a shift from single car use towards other modes of transport and telecommuting. One case centres on temporary measures to reduce congestion during road construction works. The other two cases provide examples of company mobility plans where mobility innovations are actively provided to the employees. The primary aim of this chapter is to show how innovation processes in mobility practices may take shape. Furthermore, the three case studies shed light on the barriers and opportunities for mobility management as one of the transition management pathways.

The case studies in Chapter 6 provide a descriptive analysis of access and provision of socio-technical innovations in everyday mobility. In contrast, Chapter 7 is based on a large-scale quantitative survey amongst Dutch citizen-consumers. The primary aim is to determine the role of contextual factors in mobility behaviour by examining (consumer's perception of) the quality and quantity of the available sustainable products and services on offer in combination

[6] Chapter 5, the operationalisation of mobility practices, follows after the first empirical chapter. The reason for this choice of outline is that the first empirical chapter, on new car purchasing, builds on the theoretical framework of Chapter 3, especially with regard to the conspicuous consumption. The second and third empirical chapters follow directly from the conceptualisation of mobility practices described in chapter four and therefore focus on travelling routines and how they change.

with the possible strategies of producers and suppliers in the modes of provision. By using the aforementioned mobility portfolio we also aim to analyse individual variety in the conduct of social practices in the domain of mobility while avoiding the traps of the voluntaristic approaches which dominate the sustainable consumption debate.

In the final and concluding chapter we will return to above formulated research questions. By reflecting on the theoretical and empirical outcomes of this thesis, building blocks for a scientific research agenda on sustainable everyday mobility will be postulated. In addition, recommendations for a citizen-consumer-oriented environmental policy in the domain of everyday mobility will be provided.

Chapter 2. Shifting policy perspectives on sustainable mobility. Towards a demand-oriented mobility policy?

> *For a long time the tenor of public policy was: mobility is a problem. If you have to leave home, use the bicycle, bus or train. Then came the phase in which mobility was accepted. The current cabinet has the point of view that mobility is a must. Mobility – of persons and goods – is an absolute precondition for the functioning of society and economy.*
>
> Public brochure of the Nota Mobiliteit (Dutch Mobility Policy Plan, 2004)

2.1 Introduction

Sustainability is most usually described as a concern for the long-term impact of human activities on the environment, social life, well-being, and prosperity. Current mobility patterns are generally considered to be unsustainable as they place enormous demands on the environment, safety, health and welfare. Furthermore, the capacity of the road networks is exceeded which leads to reduced accessibility, congestion and economic costs.

How the multi-faceted concepts 'mobility' and 'sustainable mobility' are defined and approached in this thesis will be described in Paragraph 2.2. Because the focus is predominantly on the ecological dimensions of sustainable mobility, these ecological dimensions are described in more detail as well. Here, attention is given to the main trends in mobility and its environmental impacts. This paragraph will make clear that structural changes in current transport systems and mobility patterns are necessary in order to attain the sustainability goals in the domain of mobility.

Due to improved knowledge and awareness about the detrimental effects of current systems and patterns of mobility the issue of sustainable mobility has received substantial attention from policy makers, private sector actors and the general public, especially in the last three decades (though it has not always been indicated with the term sustainable mobility). Throughout the years OECD countries have developed a number of pathways, that is a combination of approaches and policies that potentially reduce the negative impacts of mobility growth on society, through which sustainable mobility is pursued. Though most countries have developed and implemented nation-specific transport policy plans, in general the sustainability strategies that have been pursued show some common characteristics. In the Paragraph 2.3 a closer look is taken at these dominant paradigms in mobility policy, especially in relation to sustainable mobility. In this paragraph the transport policy responses of the Dutch, and – to a more limited extent – also of the UK government with respect to sustainability challenges are described. From a historical perspective, starting from the 1960s onwards, sustainable mobility governance and research in these two European countries are explored. This paragraph will show that the relative weight that is attached to the various dimensions of sustainable mobility has differed greatly over time. More importantly it will show how over the years (sustainable) mobility policies, due to long-standing scepticism towards the possibilities of directly targeting mobility practices of citizen-consumers, have become strongly supply- and technology-oriented. Only in the most recent period of time, the demand-side of sustainable mobility practices has gained more emphasis as an indispensable element of the agenda for mobility policy.

In Paragraph 2.4 the limitations of a supply oriented mobility policy are exposed while arguing for the need to further elaborate the emerging orientation on citizen-consumers and their everyday life mobility practices. Because of the crucial relevance for mobility policy, the actual and potential roles of citizen-consumers in transitions towards sustainable passenger mobility are at the core of this thesis.

2.2 The issue of sustainable mobility

2.2.1 Defining sustainable mobility

Before we describe the dominant environmental impacts that are implicated by current systems of mobility, it is useful to first take a closer look at what exactly is meant with the terms 'mobility' and 'sustainable mobility'. In scientific literature and policy documents the words traffic, transport, transportation, mobility and mobilities are often used as synonyms or at least as partially overlapping concepts. Furthermore, the literature on sustainable transport and sustainable mobility is immense and rapidly growing without unanimously accepted definitions and clearly defined end-goals. Whether this is a problem or not depends on the specific approach applied in research and policy, a matter which is addressed shortly hereafter. Without going into too much detail in this section the abovementioned concepts are clarified and a position is taken on the perspective of sustainable mobility transitions.

Transport, brought back to its essence, is generally defined as the physical movement of persons and goods between geographical locations. Traffic is the accumulated movement of transport units that make use of socio-technical transport infrastructures. Traffic therefore is the appearance of transport in the shape of movements of transport modes (Van Wee & Dijst, 2002). Traffic and transport are generally expressed in quantitative units: number of vehicle kilometres per day or the number of kilometres that are travelled by a group of travellers. Over the years the use of mobility (both in its single and plural form) is increasingly adopted, especially on the European continent. This increased usage of the mobility terms mostly displays a shift in how movement is perceived in both science and policy. Studies of transport often have a strong technological orientation (Thomsen *et al.*, 2005; Urry, 2007). They for example focus on the changing nature of infrastructures, and on the quantitative mapping of transport patterns. Furthermore, as the definitions of traffic and transport already indicate, there is a strong focus on the relations between transport technologies (modalities) and geographical locations. The term mobility indicates a gradual incorporation of the relation between the transport systems and societal developments. It also represents a more social perspective on the movement of people and their social networks, thereby indicating that 'being mobile' entails more than just physical movement. The social perspective understands mobility as a social concept (Vogl, 2004) and leads to a different definition of movement. From this perspective, mobility can be defined as an actor's competence to realize projects and plans while being on the move (Bonss & Kesselring, 2004).

The term mobilities (the plural form of mobility) refers to the provocative idea that mobility is not only about movement of people and goods, but also about information and ideas. 'The concept of mobilities encompasses both the large-scale movements of people, objects, capital and information across the world, as well as the more local processes of daily transportation, movement through public space and the travel of material things within everyday life' (Hannam, Sheller & Urry, 2006, p. 1). Furthermore, the separation between geographical mobility and social mobility (change of status within life's social hierarchy) is discarded. The mobilities concept also focuses not only on actual movement of these various elements described in the definition, but also about potential movement (see Kaufmann, 2002) and blocked movement. In this perspective on mobility, understanding social life starts with analysing the relationships between the various forms of mobility. Corporal, social, and virtual mobility are strongly related and can be perceived as different ways in which communication may take shape.

In sum, there is wide range of mobility-related concepts in use, from the narrow defined term 'traffic flow' to the more encompassing concept of 'mobilities'. This thesis is written predominantly from the mobility (singular) perspective as described above. The plural form, which distinguishes between corporeal mobility (persons), object mobility (freight transport), virtual mobility (information), and imaginary mobility (media and ideas) (Thomsen *et al.*, 2005; Urry, 2000) is made selective use of when developing the analytical and conceptual frameworks. In general however, the mobilities perspective seems to stretch the focus too far beyond this thesis' object of investigation: everyday mobility practices of citizen-consumers in the context of sustainable transitions. Nonetheless, the main message of the mobilities perspectives – 'putting the social into travel' (Urry, 2003b) – is taken serious and applied in the broad, also social way in which the concept of 'mobility' is used throughout the thesis.

With mobility-related concepts already showing a wide variety of meanings, the diversity seems only to increase when the issue of sustainable transport or mobility is brought up. Relevant questions in this respect are: what exactly should be made sustainable? When are the different elements contained in mobility or transport systems considered to be sustainable (or not)? Do (growths in) current patterns of mobility need to be preserved or can the societal and economic functions they fulfil be substituted? Against the background of these kinds of questions we shortly discuss how the concept of sustainable mobility is used throughout this thesis.

Sustainability is generally characterized by the pillars People, Planet and Profit (increasingly also referred to as prosperity) representing the social, ecological and economic dimensions of the sustainability concept. Since the publication of Our Common Future by the Brundtland Commission in 1987 the most widely used definition of sustainable development refers to a 'development that meets the needs of the present without compromising the ability of future generations to meet their own needs' (Brundtland Report, 1987). Inspired by Our Common Future, the earlier definitions of sustainable transport closely resembled the concept of sustainable development. One of the first widely accepted definition stems from an OECD conference in 1996. Here sustainable transportation was defined as 'transportation that does not endanger public health or ecosystems and meets mobility needs consistent with (a) the use of renewable resources at below their rates of regeneration; and (b) the use of non-renewable resources at below the rates of development of renewable substitutes' (OECD, 1996). More recently, policy and scientific publications have generally incorporated the social dimensions of mobility. This is reflected in the more frequent use of the term sustainable mobility instead of sustainable transport in most policy documents. One of the more influential definitions in effect today is construed by the World Business Council for Sustainable Development who defines sustainable mobility as 'the ability to meet the needs of society to move freely, gain access, communicate, trade, and establish relationships without sacrificing other essential human or ecological values today or in the future' (WBCSD, 2001, 2004).

Though informative, these definitions do not provide answers to the concrete questions stated above. To answer this type of questions, sustainability indicators are often used to formulate, measure and monitor end-goals for specific elements of sustainable mobility. According to Gudmundsson (2004, p. 4) measuring and monitoring sustainable mobility involves the components of conceptualization (defining what is to be monitored), operationalization (making concepts measurable by selecting parameters and indicators) and utilization (the ways in which the indicators are drawn upon in analysis or policy). However, the detail to which these components are elaborated in policy and research depends on the specific approach to sustainable mobility being applied. In his analysis of sustainable mobility indicators, Gudmundsson describes three approaches that are used in scientific and policy literature on sustainable mobility.

Sustainable mobility can serve as a metaphor where the mobility policy agenda takes into account sustainable development concerns. The metaphorical approach is typically adopted by policy administrations who use indicators, for example growing transport volumes and carbon dioxide emissions, as general guidelines for sustainable mobility.

Sustainable mobility can also be approached literally with explicit references to the meaning of sustainability for mobility. In this approach sustainable mobility entails specific limitations set by the environment and demands of society. This approach is mostly pursued by in academia. The last approach is a blending of the first and second approach in which the literal meaning of sustainability guides the construction of indicators for research and policy assessment of sustainable mobility.

In this thesis sustainable mobility is approached from the latter, a blending of the first and second approach to sustainable mobility. Generally speaking, the main emphasis is on sustainable mobility as a metaphor, encompassing a broadly defined and envisioned end-goal and less so on sustainable mobility or transport in its literal sense. This means that sustainable mobility is mostly used as a conceptual tool to refer to and indicate a continual process of societal change towards sustainability (with a specific focus on the environmental dimension of sustainability as an object of research). To view sustainable mobility in this way, in our view, corresponds with the two main theoretical approaches adopted in this thesis, transition theory and ecological modernisation theory, both of which are theories of social change which look at structural changes of socio-technical systems (Mol, 1995; Mol & Spaargaren, 2000; Kemp & Loorbach, 2006; Loorbach, 2007). Especially the reflexive ecological modernisation approach with an emphasis on social learning, cultural politics and new institutional arrangements (Mol & Spaargaren, 2000) in many ways has a process-oriented approach in common with the basic elements of transition management.

The adoption of a process-oriented approach to sustainable mobility also derives from the characteristics of current mobility problems. Problems associated with mobility can be characterized as multidimensional, complex, and persistent[7] (Avelino *et al.*, 2007). This means that they occur

[7] Environmental problems are considered persistent when current policies are not succeeding in achieving the desired environmental qualities, even in the long-term (see VROM-raad, 2005).

on various scale-levels and involve many different actors with varying perspectives, norms and values (Loorbach, 2007). To indicate the difference in perspectives among stakeholders Table 2.1 shows that different stakeholders (in this case mobility experts, policy makers and users) to a certain extent attribute different priorities for sustainable mobility policy. Policy actors give a higher priority to environmental problems and to neighbourhood quality. Citizens tend to attach more importance to social aspects of mobility such as social exclusions and traffic safety (PODO II, 2004).

Therefore, sustainable mobility objectives and indicators should be flexible and adjustable, as they are bound to change because of the long-time period needed to achieve the envisioned levels of sustainable mobility, and because of the diversity of the vested interested involved. 'The complexity of the system is at odds with the formulation of specific objectives. With flexible evolving objectives one is in a better position to react to changes from inside and outside the system' (Loorbach, 2007, p. 73).

So, an important part of the process in answering the types of questions concerning sustainable mobility stated above involves the formulation of shared future visions, projects and frames with respect to sustainable mobility. These visions do not so much entail or have at their core some concrete solutions to realize a specific sustainable mobility system. Instead they refer to the general qualities such a system must possess (CE, 2008a). The development and use of objective sustainability criteria are not excluded but used primarily as instruments for constructing policies, projects and visions on sustainable mobility. For example it is one of the purposes of this thesis to investigate how – with the help of what kind of concrete criteria and frames – an environmental rationality (Mol, 1995, 2005) can be brought (and increasingly already is brought) into the everyday life mobility practices of citizen-consumers. This aim will be pursued throughout the upcoming chapters. These general quality standards for sustainable mobility are listed in Figure 2.1. This could be seen as the conceptualization step as put forward by Gudmundsson (2004).

While recognizing the interdependencies and forms of overlap that exist between the three dimensions of sustainability, in the next paragraph the changes throughout the years within the 'planet dimension of sustainable mobility' will be explored in more detail.

Table 2.1. Priority attributed to societal problems for sustainable mobility policy by different stakeholders (PODO II, 2004).

	Mobility actors	**Policy actors**	**Users**	**χ^2-test**
Traffic accidents and safety	1.3	1.2	1.3	P=0.44
Congestion	1.5	1.8	1.8	P=0.53
Quality of neighbourhood (land use, noise)	2.2	1.5	1.9	P=0.03
Air quality	2.0	1.6	2.1	P=0.18
Climate change	2.0	1.9	2.2	P=0.69
Social exclusion	2.3	2.3	1.8	P=0.21

1 = absolute priority / 2 = priority / 3 = less priority / 4 = no priority.

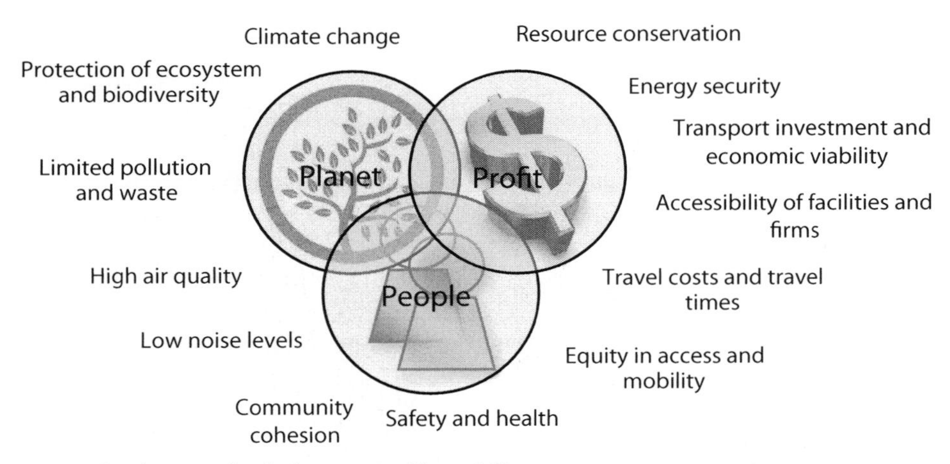

Figure 2.1. Quality standards for sustainable mobility.

2.2.2 Environmental dimensions of sustainable mobility

Passenger mobility contributes significantly to local, regional and global environmental problems. In the Netherlands, the transport sector as a whole is responsible for 40 to 60% of the various emissions that are harmful for human health. Furthermore, about one fifth of all greenhouse gas emissions can be attributed to this sector. The contribution to global warming and climate change has grown substantively over time and will likely continue to grow in the upcoming decades considering current worldwide trends. In this section the growth in transport volumes and the corresponding environmental impact of passenger mobility are portrayed in more detail. The attention focuses specifically on the three problems that provide the greatest challenges for transitions towards sustainable mobility, namely local air quality, climate change and noise (CE, 2008a).

Growth in mobility and future trends

The historical growth in personal mobility, which is expressed in terms of rising transport kilometres, is a worldwide phenomenon. Figure 2.2 clearly shows that on the one hand people increasingly use resource-intensive transport modes to move around and on the other hand that people travel more kilometres in general. The graph also shows that everyday mobility has become almost synonymous with automobility. However, there are important differences between countries, especially the differences between the patterns of transport growths in developed countries on the one hand and developing countries and so called transitionary economies on the other are remarkable.

Most OECD countries have already realized the largest growth in everyday mobility over the past decades. In the European Union as a whole, car use makes up around 78% of the share of

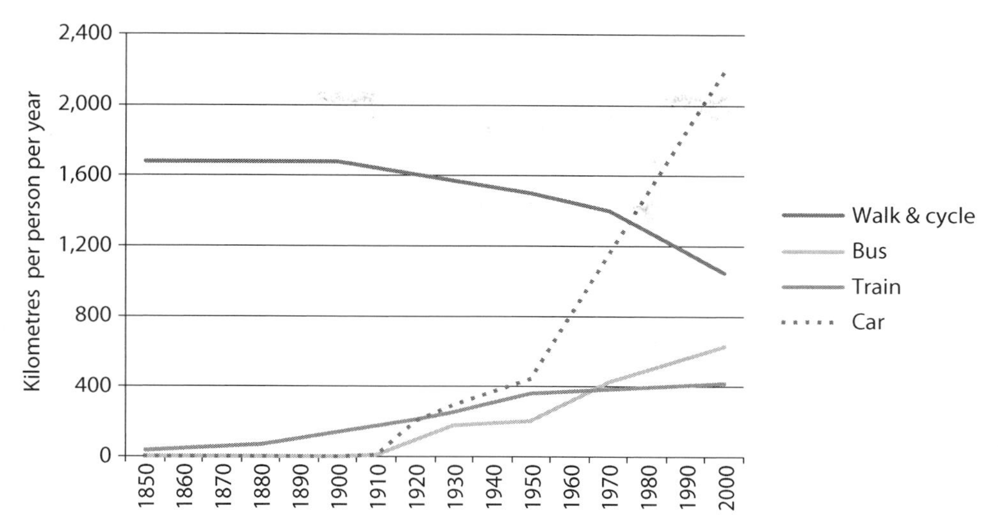

Figure 2.2. Worldwide per capita movement of people, 1850-1990 (adapted from Gilbert & Perl, 2007).

passenger (non-aerial) transport when measured in kilometres (EEA, 2006)[8]. In the EU, member states passenger transport volumes grew by 30% on average between 1990 and 2002 (*ibid.*). The increase in kilometres travelled is most often explained by a combination of demographic changes (population growth and age distribution), socio-demographic developments (economic growth and rising net-spending on consumption), spatial developments (large-scale transport infrastructure construction and the separation between workplace and home), and finally, social-cultural trends (individualisation, increase in double-income families, emancipation, temporal intensification of daily life/culture of haste) (CE, 2008a; Harms, 2003, 2006, 2008). Though economic growth is often seen as the major driver behind transport growth, Germany is one country were a reduction in transport kilometres has been accomplished despite of continued economic growth, a clear sign that a decoupling is possible (EEA, 2006). The decoupling between economic growth and transport growth is related to the more stabilised mobility patterns of many OECD countries.

In the Netherlands passenger kilometres between 1985 and 2009 has increased by approximately 40% as a whole, and increased by 54% for car kilometres specifically (KiM, 2010), see Figure 2.3. The 54% growth in car kilometres can be attributed to three aspects, two of which are the direct result of changing consumer patterns: 24% is the result of increasing travel distances (for example due to longer commuting distances), 20% is the result of higher frequency of travelling (people travel more often), and 10% is the result of population growth.

[8] The modal split is quite different when the number of transport movements instead of transport kilometres is used for comparison. In the Netherlands the modal split in 2005 was: car 48%, bicycle 28%, walk 17%, bus/tram/metro 3%, train 2%, other 2%.

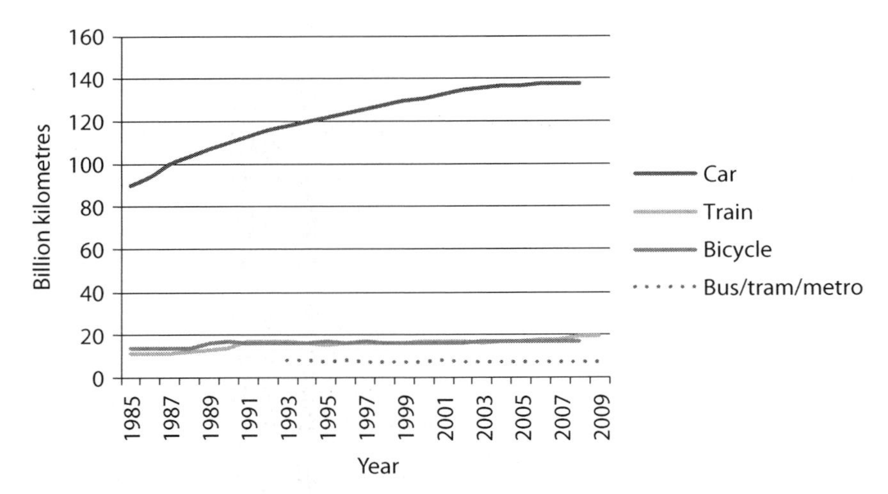

Figure 2.3. Passenger kilometres in the Netherlands, 1985-2009 (adapted from KiM, 2010).

The transport growth in the last 25 years has taken place primarily in the late nineteen eighties and early nineties. From 2000 onwards there is a stabilization in the growth of personal (auto) mobility for daily activities in the Netherlands (Figure 2.3)[9].

Developing countries (especially transitionary economies such as China, Russia, former Eastern European countries) on the other hand are currently witnessing unprecedented transport growths. In China, for example, car sales have risen by 20% per year without any sign of stagnation for the near future. There were over 10 million sales of automobiles in China in 2010, making China the world's largest car market. Predictions for China's car market for 2020 indicate sales numbers that exceed twenty million cars (see www.aftermarketnews.com). The prognoses for passenger transport, especially when taking a long-term and global perspective, show an alarming increase in mobility growth. For the next fifty years there is an expected yearly worldwide growth of 1.7% in passenger transport kilometres (Figure 2.4). In the European Union per capita transport activity per year is projected to grow from around 12,500 km in 2005 to around 18,000 km in 2030 (DGET, 2008). The environmental impact of (the growth in) mobility for the Netherlands is discussed hereafter.

2.2.3 Trends in environmental impacts of mobility

Air quality

Local air quality is a much debated topic, especially in the Netherlands, since the European Space Association used their satellites in 2004 to make air pollution in Europe visible. These maps placed local air quality instantly on the national, and to a lesser extent European, agenda. Large parts of the Netherlands, Belgium and Western Germany were shown to be among the areas with the

[9] These growth rates exclude freight transport and aero-mobility, both of which are areas which continue to show steep growth rates in the Netherlands.

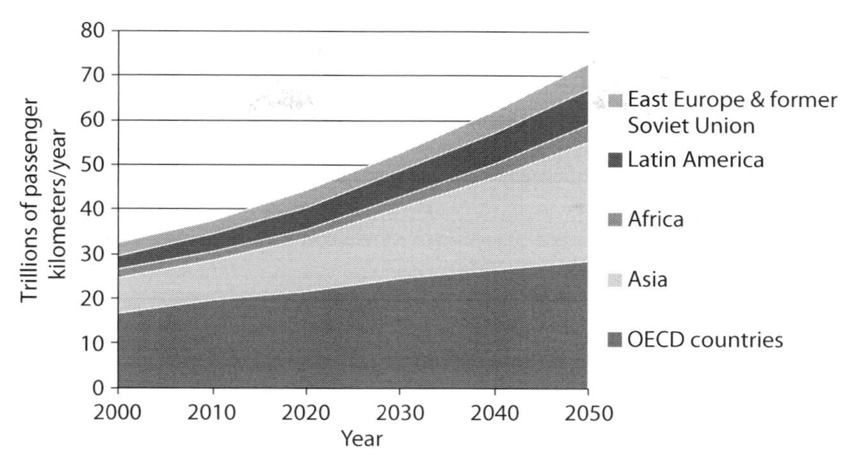

Figure 2.4. Projected worldwide growth in personal transport activity (adapted from WBCSD, 2004).

poorest air quality in Europe. The large media attention focused specifically on the dangers of poor air quality for public health. It was estimated that thousands of people in the Netherlands, especially those living in large urban centres and near intensively used roads, were suffering from a premature death due to high concentrations of harmful emissions. European-wide the number of premature deaths was estimated around 370,000 (EEA, 2006). The transport sector (including freight transport) is responsible for 40% of PM_{10} emissions. Interestingly enough though, air quality has actually improved over the last decades, both in Europe and in the Netherlands (see MNP, 2007; CE, 2008a; CPB, 2006; EEA, 2006). Figure 2.5 shows that from 1990 onwards there has been a significant reduction in the most harmful emissions for local air quality emitted by road transport vehicles: NOx, SO_2, volatile organic compounds (VOS) and fine particulate matter

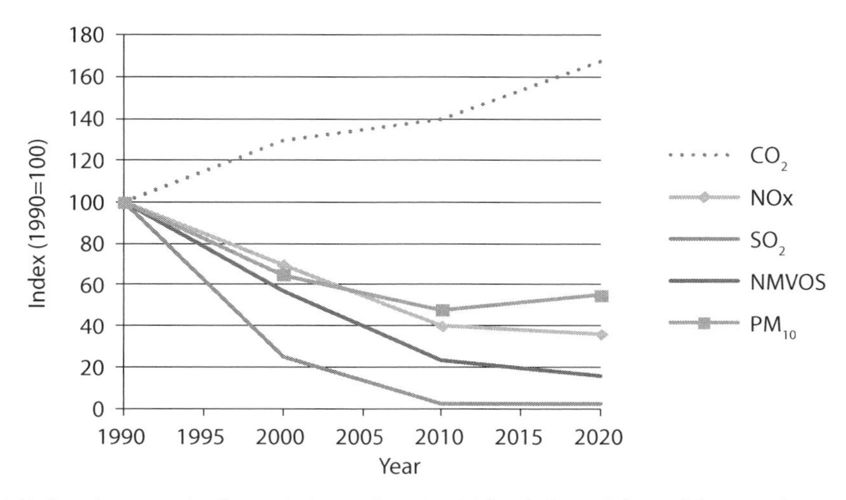

Figure 2.5. Development in the emissions of road vehicles (adapted from CE, 2008a).

(PM_{10}). European Union wide from 1990 to 2003 this reduction was between 30 to 40%. This reduction has been achieved predominantly because of stricter vehicle emission norms (Euronorms) which stimulated technological innovations for automobiles, busses and trucks. The Euro norms led to relatively simple end-of-pipe technologies such as the three-way catalytic converter for petrol cars and the oxidation catalytic converter and diesel particulate filter for diesel cars (CE, 2008a). Reduction in SO_2 emissions have also been accomplished because of a lowering of the sulphur content in fuels (MNP, 2007). Continuation of emission limits on the European level will likely result in a continued reduction of the harmful transport emissions.

Despite these positive developments, air quality continues to be a problem on the short term as PM_{10} and NOx emissions are highly concentrated in specific areas. While the reductions have mostly been achieved on the national levels, in the Netherlands little reduction in fine particulate matter has been measured since 2000 (MNP, 2007; CE, 2008a, EEA, 2006) on the local (city or street) levels. Furthermore, while on a transport vehicle level large reductions have been accomplished; there is a so called rebound effect due to the increase in transport kilometres.

To reduce concentrations in fine particulate matter and NOx, EU-wide limits for PM_{10} and NOx emissions have come into force. A number of cities in the Netherlands currently exceed these limits. To reduce these local concentrations many European cities currently target urban traffic as this is the largest source of pollution that can be influenced by local policy measures.

Climate change

In recent years climate change has become *the* dominant environmental problem. In many public debates the environment has become almost synonymous with climate change. It is by now widely acknowledged that the global climate system is changing and that this is very likely the result of human behaviour. To limit the effects of climate change the European Union strives for a maximum temperature increase of two degrees Celsius. The EU target for the reduction of greenhouse gasses in 2020 is between 20 and 30% (as compared to 1990). However, the growth in transport-related climate change emissions seriously threatens the likelihood of accomplishing these climate change goals. In the European Union transport accounts for 28% of climate changes emissions, half of which can be attributed to passenger cars. Between 1990 and 2003 these emissions increased by 23% in the European Union (EEA, 2006). Figure 2.5 shows that the increase in CO_2-emissions in the Netherlands has been even larger. Surprisingly enough, climate policy for the transport sector is still at its infancy (CE, 2008a). Though progress has been made to increase fuel efficiency of passenger cars, generally these reductions have not been sufficient. A decoupling between transport growth and emissions, which has been accomplished with air polluting emissions, is currently out of reach (*ibid.*). The voluntary agreements made in 1998 between the EU and the car manufacturer associations have not led to the desired target (140 gram CO_2/km) in 2008/9. The postponed EU objective for 2012 (120 gram CO_2/km) was also out of reach (Figure 2.6). Nevertheless, the newly appointed EU target for 2015 (130 gram CO_2/km[10]) is likely to be met in time.

[10] The EU targets for 2012 were applied to the car company level, in contrast the new targets for 2015 are differentiated on the basis of the weight of the vehicles produced in 2015 compared with the average weight of the vehicles the entire industry will produce over the 2011-13 period (see T&E, 2010).

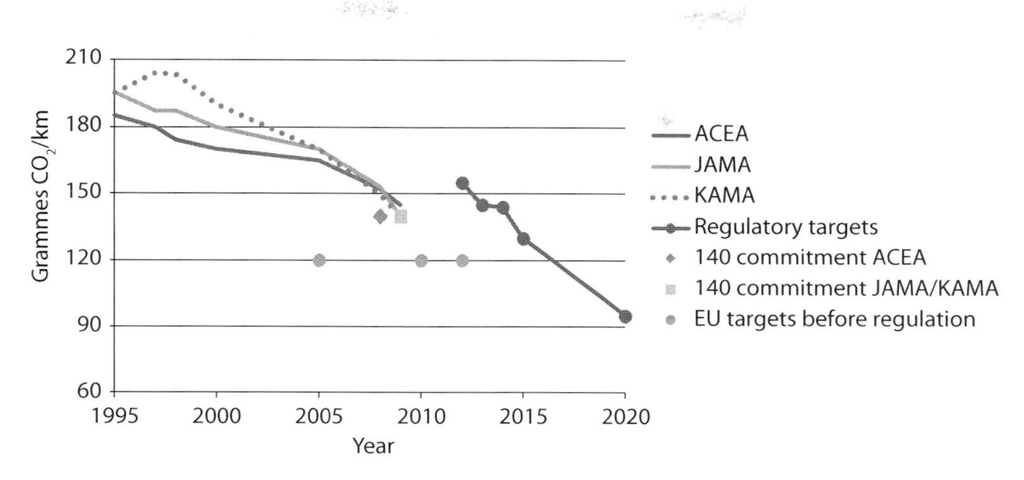

Figure 2.6. Progress over time in the CO_2 commitment of three car manufacturer associations in Europe, including historical and existing voluntary and regulatory targets (adapted from T&E, 2010).

Looking at car sales in the Netherlands specifically, a similar trend is visible. While the sales of energy efficient domestic appliances show substantial increases in the Netherlands (currently 90% of washing machines, dishwashers and refrigerators sold have an energy class of A, A+ or A++) since the introduction of the energy efficiency label in 1995, the sales of energy-efficient passenger cars remained relatively modest until quite recent times. However, the rapid growth of the share of energy efficient cars since 2008 might indicate a trend shift (Figure 2.7). The (changing) role of the environment in the practice of car purchasing will be discussed in more detail in Chapter 4.

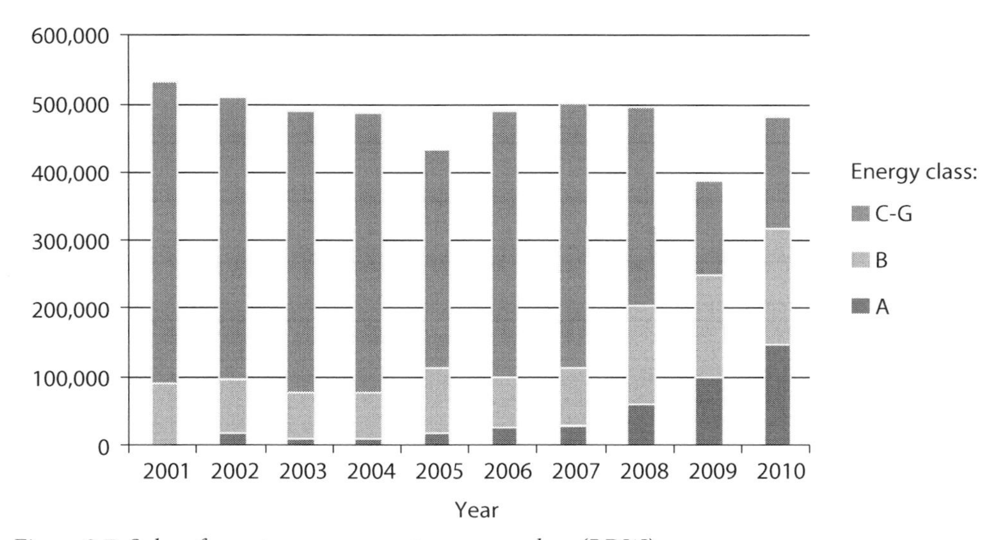

Figure 2.7. Sales of new passenger cars per energy class (RDW).

Transport noise

In Europe 40% of the population is exposed to noise levels that are potentially harmful to human health (T&E, 2008). Most of these noise levels are the result of road, rail and air traffic. In the Netherlands, 29% of the population experiences serious nuisance from one or more sources of traffic noise (CE, 2008a), see Figure 2.8[11].

Overall noise levels are on the rise since the increase in transport volume outweighs the decrease in the noise level per unit of transport. Especially in densely populated areas exposure to noise can be a serious problem. Gradually, the influence of traffic noise on the quality of human and natural life is becoming manifest and acknowledged. Exposure to noise levels can lead to nuisance, sleep deterioration, and on the long-term to a raised chance of cardiovascular diseases, and diminished cognitive functioning (MNP, 2007). It is estimated that in the EU over 231,000 people are affected by an ischemic heart disease due to traffic noise annually. About 20% (almost 50,000) of these people suffer fatal heart attacks (T&E, 2008). Nevertheless, in transport policy the reduction of noise is not given as much attention when compared to climate change and air quality. Surprisingly, currently no European standards exist to protect citizens against unhealthy noise levels (MNP, 2007). The only EU obligation is to measure traffic noise and to develop action plans to reduce exposure to noise. EU member states do have their own national norms and policies. However, these norms are generally not strict enough to sufficiently negate the negative effects of transport noise. Furthermore, transport noise policies are mostly focused on reducing the effects of noise, not on prevention them. Adaptation measures have to deal with the problem of restricted space

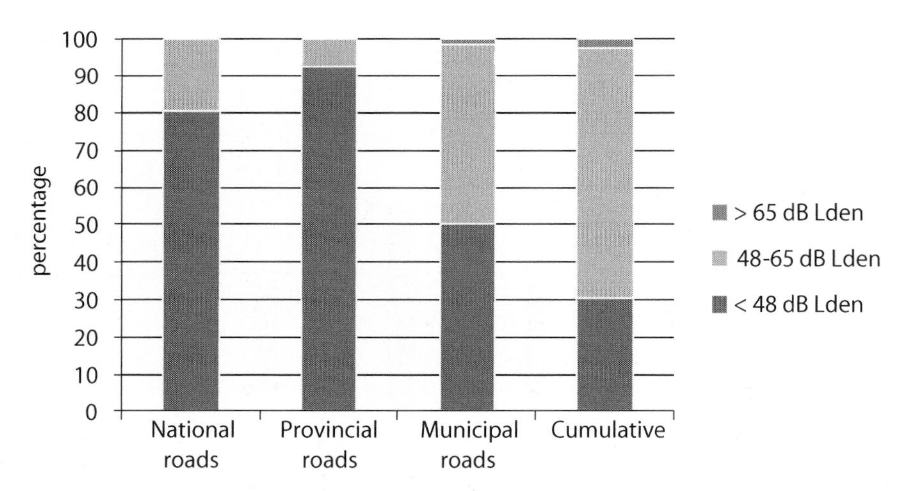

Figure 2.8. Acoustic load for houses due to road transport in the Netherlands in 2005 (MNP, 2007).

[11] Environmental noise above 40-50 dBA is likely to lead to significant annoyance. Outdoor noise levels of 40-60 dBA may disturb sleep. Noise levels between 65-70 dBA may be risk factors for school performance and ischemic heart disease. Traffic noise of 70 dB(A) may cause hearing impairment (T&E, 2008).

and the high costs of installing noise barriers and insulating houses. More silent vehicles and tyres are already available on the market (with equal characteristics and often equal purchase price). However, prevention at the source, though likely more cost-effective than adaption measures, so far has not taken off because of stagnating and ineffective noise policies (CE, 2008a; T&E, 2008).

2.2.4 To shift or not to shift?

In this section the relation between the various transport modes and their environmental impact will shortly be addressed. Contemporary research and policy making are directed to understanding and influencing choices for specific transport modalities from an environmental point of view. One of the dominant paradigms for reducing the environmental impact of mobility is rooted in the shift from automotive transport to public transport and non-motorized forms of transport. Key element of this paradigm is the assumption that public transport modes are more environmentally friendly when compared to automobiles. Without doubt non-motorized forms of transport (cycling and walking) are more environmentally friendly than motorized forms of transport. However, it is very hard to make generalizations about the most environmentally friendly forms of motorized transportation. In an attempt to calculate the environmental effects of different modalities, CE conducted a thorough life cycle analysis in 2003 and 2008 (CE, 2003, 2008b). The main message is that the cleanest modality does not exist (CE, 2008b). The number of passengers and the specific technology often play a more important role than the specific transport mode. Per capita, a car occupied by four persons is also four times as efficient as a car occupied by one person. Considering the average occupation rate of passenger cars, the private car generally has higher CO_2-emissions than public transport (*ibid.*). However, occupation rates are very different for distinct mobility practices. During the daily commute a passenger car on average shows an occupation rate of 1.1. During leisure time related mobility practices car occupation rates are 2.0 or higher in 54% of the movements (Harms, 2006). If the occupation rate is higher than two persons per car, public transport is not necessarily the better option from an environmental point of view. Furthermore, a passenger car which is fuelled by compressed natural gas and is in compliance with the new Euro 5 standards is four to ten times as clean as a passenger car which is just in compliance with Euro 1 standards. Electricity driven rail transport is relatively clean and fuel efficient. However, diesel trains, though fuel efficient, show high levels of air polluting emissions. Similarly, the average coach – if not furnished with specific clean technologies – represents a rather polluting form of passenger transport. Within the build environment, transportation by bus in general tends to be more detrimental to local air quality when compared to transportation by private car.

2.2.5 Sustainable mobility: resumé

In this paragraph various perspectives on sustainable mobility have been addressed. On the one hand the transport trends described indicate that structural changes in current mobility systems are necessary. On the other hand, it is hard to develop and monitor strict indicators for sustainable mobility: not only is knowledge about sustainable mobility indicators constantly changing, the weight attached to each of these indicators varies as well (e.g. per stakeholder, per country and per period of time). Finally, solutions for one mobility related problem can have adverse effect on other dimensions of sustainable mobility. These characteristics provide some serious dilemmas for sustainable mobility policy, and, as we will see in the next paragraph, have led to fluctuations in the storylines about how sustainable mobility should be perceived and dealt with in real life.

2.3 Storylines with respect to the governance of sustainable mobility

In this paragraph a short overview is presented about the ways in which European mobility policies have addressed the environmental impacts of mobility over the years. More specifically, in this overview mobility policies in the Netherlands and the UK are described and analysed, paying particular attention to the storylines behind mobility policies in both countries[12]. Interestingly, throughout the years significant changes in the political storylines behind (sustainable) mobility have taken place in both countries. Over time, specific problems seem to come and go, resulting in a highly dynamic agenda for mobility policy. Similar problems are often framed differently and targeted with different expectations and suggested solutions. To analyse these changes in policy discourses, this paragraph makes extensive use of the works of Vigar (2002, 2001, 2000) and Peters (1998, 2003) who conducted discourse analyses of UK mobility policy and Dutch mobility policy respectively. According to Hajer (1995, 1993) a discourse is defined as an ensemble of ideas, concepts, and categories through which meaning is given to phenomena. Thus, discourses frame certain problems. In line with Hayer's approach, both Vigar and Peters have deployed a policy discourse analysis as a tool for interpreting policy changes. When presenting the discourse analysis we aim to make three points. First, mobility policy is dominated by issues of accessibility and congestion while an integrated policy on the planet-side is still in its developing stage. Second, sustainable mobility policy is primarily supply-oriented and lacks a convincing consumer orientation. Third, this lack of a consumer-orientation is rooted to a considerable extent in the disappointing experiences with and the lack of success of the demand-oriented policies which for a short period of time acted as the central focus of mobility policy.

2.3.1 Discourses in UK mobility policy

With the use of transport planning literature, Vigar (2002, 2001, 2000) provides an historical overview of the UK transport planning policy from the 1950s to 2001. In his analysis he assesses the extent to which UK transport planning could be conceptualized as having a dominant paradigm, or a hegemonic discourse maintained by an institutionalized discourse coalition (2002, p. 42). Furthermore Vigar explores whether or not a paradigmatic shift has taken place in transport policy and practice from the 'predict and provide' towards the so-called 'new realism' paradigm (see Table 2.2).

[12] Though it would be a simplification to suggest that these two countries can be seen as representative of all the diversity in Western European transport policies, they do form good examples. Furthermore, the two countries show interesting similarities and differences when compared by a short discourse analysis.

Table 2.2. Policy goals for mobility within two paradigms (adapted from Vigar, 2002, p. 191).

The 'predict and provide' approach	**The 'new realism' approach**
Network congestion is a problem in and by itself	Congestion is a problem because of the external effects it brings along for the environment, economic activity and quality of life
Travel demand is an expression of social and market dynamics and cannot be prevented without serious and unknowable consequences	Travel demand can be influenced by public policy, e.g. with price signals, user prioritization, and spatial organization
The appropriate policy response is to change the network capacity	The appropriate response is to develop travel demand management, stabilize mobility levels, and provide a modal shift. Network expansion plays only a supporting role and only under certain conditions
Policy issues concern the rate, priority order and scale of changes with respect to network capacities	Policy issues include broader economic, social and environmental impacts

1950-1987: the hegemony of 'predict and provide'?

Vigar argues that for most of the twentieth century nations dealt with the continued increases in demand for mobility by building more roads, the so-called 'predict and provide' policy. This policy was characterized by a strong roads program largely disconnected from considerations related with other transport modes and from other spatial developments. In the predict and provide approach the existing demand for mobility was extrapolated into the future and then attempts were made to match the supply of infrastructure to that potential demand. The same type of models were used to project further declines in passenger numbers for public transport companies, thereby justifying cuts in rail and bus networks (Vigar, 2002, pp. 1-2). However, the extent to which the predict and provide approach dominated transport policy differs per time period and also for intra- and inter-urban forms of traffic.

During the 1960s, 1970s and the 1980s UK central government policies on inter-urban transport were clear and consistent. Movement by cars was promoted by massive investments in road infrastructures, while investment in public transport was seen as a duplication of the road network and as uneconomic (i.e. too expensive to support apart from a core network of inter-city routes) (Vigar, 2001, p. 280). Complex transport models, which predicted dramatic increases in mobility were used to support the extensive road construction policies. The main explanation for this approach was that central government, considering the public opinion to mobility restrictions, viewed demand-management tools as inappropriate. With regard to intra-urban transport the picture, however, is more complicated. During the 1950s and early 1960s, the building of urban roads was, comparable to interurban transport, seen as the principal solution to problems of movement within towns and cities (Vigar, 2001, p. 274). However, in the late 1960s and in the

1970s a change in central government policy occurred in relation to problems of intra-urban movement. Road construction in line with traffic forecasts were seen as unacceptable for the environment and suggestions were made that traffic should be controlled instead of provided for. Furthermore, the Transport Policy White Papers in 1966 and 1977 focused on restricting car use in (potentially) congested urban areas and on promoting public transport. Maintenance of local rail services were regarded essential for social reasons as well (Vigar, 2001, 2002). Nevertheless, even though local authorities were encouraged to invest in public transport, due to financial deficits many suburban rail lines were closed, thus undermining any public transport emphasis. Also in the 1970s municipal authorities showed little interest in restraining car use as congestion levels were not reaching the intensity as suggested by transport models in the 1960s[13]. The support for public transport ended with the Thatcher government. Throughout the 1980s the national government stimulated local authorities to designate money to road construction while simultaneously targeting municipal transport operators, resulting in the privatization of many public transport providers (Vigar, 2002, pp. 48-52).

1987-2001: towards a paradigmatic change in transport policy?

Despite the dominance of the predict and provide approach as the outcome of transport policy until the end of the 1980s, the previous sections showed that even in the 1960s and the 1970s a competing storyline emerged which argued for a more balanced and people-centred approach (see Vigar, 2002). However, it was not until the 1990s that this approach, since then labelled 'new realism', came to be a modus operandi for UK transports policy. The year 1987 can be marked as the moment of change, since in this year criticism about the existing road-construction policy reached new heights. The ensuing 'new realism' within transport planning consisted of growing concerns among transport professionals that road construction could no longer solve the problems of increased mobility. During the 1990s this concern resulted into a major policy shift whereby environmental considerations played a key role in providing the legitimation and line of argumentation underpinning the shift (Vigar, 2002, p. 84). The new realism represented an alternative solution to mobility problems since it focused on a set of measures which are provided as a package instead of working with individual policy measures and targets. The approach has been described in terms of 'a demand management discourse' (Vigar, 2000, p. 28) and it involved the combined measures of traffic calming, road pricing and improving the provision of non-automotive transport providers including public transport, walking and cycling (Vigar, 2002, p. 67).

Interestingly, throughout the 1987-2001 period the conception of mobility and the environment changed in various ways. In the 1989 Road Programme environmental effects were seen as unavoidable by-products of inevitable growth in mobility, which could best be solved via technological fixes. In the 1990s more emphasis was placed on the balance between the three pillars of sustainability. More specifically, local air quality and global ecological conditions received greater emphasis. However, although the failures of the 'predict and provide' approach were already noticed in the 1980s, actual reductions in spending money on road construction did not occur until 1994! Only after 1996 the UK Department of Transport recognized that, in order to lower

[13] This was partly due to overestimates of economic growth and population, and traffic management policies.

environmental impact of transport, the rate of transport growth itself needed to be influenced (see Vigar, 2002, p. 85) as well.

In sum, the recognition of ecological limits in the context of UK transport policy contributed to the 'realism' that capacity for traffic growth should not be (automatically) provided for. At the very same time technical studies verified the fact that capacities matching existing patterns of traffic growth could not be provided for (Vigar, 2002, p. 180). These combined facts provided the political space needed for the development of new story lines such as the 'user pays' principle and the 'demand side management' approach into more prominent policy alternatives. Increasingly, these new storylines contained in the new realism discourse became incorporated within actual UK transport policy. These storylines were consolidated in the 1998 Transport White Paper. Vigar concludes that by the end of the period 1987-2001 the 'predict and provide' paradigm had largely been replaced. Furthermore, the full range of ecological concerns arising from continued growth in mobility had become firmly embedded in policy discussions, a noticeable shift compared with the previous decade. However, strong elements of the 'predict and provide' approach continue to linger in the minds and practices of transport policy actors (Vigar, 2002, p. 185, 2000, p. 30). According to Vigar the new story lines associated with ecological conditions and demand side management did not became fully institutionalized as transport policy largely remained under the dominant influence of congestion issues (Vigar, 2002, p. 184). Nevertheless, Vigar does conclude that the 'predict and provide' paradigm has run its course, with the new paradigm still lacking coherence. Finally, he concludes that one of the key issues remaining is the deeply rooted cultural resistance within transport policy against attempts to change travel behaviour. At the end of this chapter I will discuss in some more detail to the roots of this cultural resistance within transport and mobility policies.

2.3.2 Discourses in Dutch mobility policy

In this section the discourses in Dutch mobility policy are examined, mostly through the work of Peter Peters. In his analysis Peters uses Hajer's discourse framework to analyse the changes in the storylines and discourses in Dutch mobility policy until 2001. Peters (1998, 2003) starts from the premise that Dutch mobility policy from the 1970s onwards can be characterized by an innovation deficit which is represented by the inertia of the mobility debate, a stagnation in thought and a lack of effective solutions (Peters, 2003, p. 42). In his historical overview of Dutch mobility policy Peters tries to find out why Dutch mobility policy continuously gets stuck in the same pattern of thought.

When compared to the UK, until the 1970s the growth of mobility was not seen as a complex and persistent problem in the Netherlands. On the contrary, individual mobility was seen as the foundation of the commonly shared dream of material progress in Dutch society. As with the 'predict and provide' paradigm described by Vigar, in the successive national road programs the length of the new to build roads was simple derived from the growth prognoses of automobile transport. So, until the 1970s mobility-related problems of car use were defined as capacity problems (Peters, 1998, p. 41). Influenced by the Limits to Growth Report by the Club of Rome and the world oil crisis, during the first half of the 1970s the desirability of unlimited growth in car use was put into question. From that moment onwards, transport policy objectives changed quite rapidly. It was deemed necessary to reduce the need for movement and the position of public transport and slow transport modes was to be improved (Peters, 2003, p. 44). Comparable with the UK transport policy in the 1970s, the following years were characterized by a certain dualism. On the one hand the road network was extended, the Dutch car park and car mileage increased, and the share of public transport decreased. On the other hand the Dutch government did succeed in limiting some of the negative consequences of automobile growth by limiting transport emissions and the number of transport casualties, and also by putting a stop to the hegemony of the car in inner-urban centres (Peters, 1998, p. 43).

Between 1988-1990 the 'Tweede Structuurschema Verkeer en Vervoer' (SVV2) was developed. This strategic policy plan can be regarded as the most ambitious mobility policy document in the history of the Netherlands. In the 1980s environmental problems were intensively researched, leading to alarming and highly influential reports on the state of the Dutch environment, most notably concerning the influence of acid rain (RIVM, 1988). The government by now was convinced that mobility-related environmental problems posed a serious threat to society (Nationaal Platform Verkeer en Vervoer, 1991). Peters describes how between 1988 and 1996 the policy discourse on mobility changed significantly due to this shift in the societal debate (Peters, 1998, p. 58). The newly emerged storylines of the 'car as polluter' and 'the dying of forests' indicated a strong shift in the ways in which mobility problems were perceived. From now on mobility-related problems were predominantly defined in terms of emissions (*ibid.*). In the political discourse the system of automobility came to be described as a threat to liveability (Peters, 2002, p. 56). Interestingly, congestion had almost disappeared from the national political agenda. The pursuit of sustainable development formed the point of departure of the SVV2 (SER, 1994). There was a strong sense of urgency about the need to restrict the projected growth in mobility. This resulted in the ambitious goal of the SVV2 to halve the projected 70% increase in mobility growth. Table 2.3 shows the five strategic steps which simultaneously should improve liveability and accessibility. The high level of ambition was deemed realistic by the policy makers because they assumed that there existed broad public support for imposing strict policy measures. Next to the technical measures to reduce emissions levels (such as the introduction and financial support of the catalytic converter), price mechanisms in the form of road pricing and excise duties should make car use less attractive and improve the relative competitive advantage of public transport (Peters, 1998, p. 61). Improvements in public transport were suggested on the basis of the assumption that car drivers could and should be offered a fully equivalent alternative to the use of private cars.

Behavioural change of travellers was targeted explicitly via mass communication and via obligatory company transport plans.[14] The ambitious goal of the SVV2 with regard to company transport plans was that every private corporation and governmental agency with over fifty employees should formulate a transport or mobility plan. This mobility plan aimed to map out the commuting activities taking place within the organisation and to provide the data for determining which modal-type of transport would be most suitable for an individual employee when it comes to reducing transport kilometres, time and costs (SVV2, 1990). However, between the high ambitions of the mobility reduction goals of the SVV2 and the actual implementation of these

[14] In reality the obligatory transport policy plans often resulted only in voluntary carpool programs for employees.

Table 2.3. Overview of the proposed measures in the SVV2 to be achieved by 1995 (adapted from SER, 1994).

Step 1 **Emission reduction at the source**	**Step 2** **Reduction of growth in automobility**	**Step 3** **Non-automotive modal split**	**Step 4** **Selective accessibility**	**Step 5** **Strengthening elements of policy**
• catalytic converter • lighter cars • safer road network • 30 km zones • touristic bicycle zones	• ABC location policy • telework • road charges • increase in the variable costs of transport • stricter parking policy • company transport plans • rush hour tax	• improvement of regional public transport • integrated passenger information • carpooling • safe bicycle roads • group oriented transport	• facilities for specific target groups • carpool lanes • traffic control	• conscious choice of transport use • company transport plans • maximum of parking norms • regional transport plans

mobility policies into concrete policy measures there was a world of difference. The ambivalence of the Dutch cabinet resulted from the large public resistance to the generic mobility reduction goals (Peters, 2002, p. 46). In response the Ministry of Transport indicated that one of the main issues remaining was to create large-scale public support for the policy plans.

By the mid-1990s environmental issues had disappeared from the main political agenda again. The apocalyptic tone of voice in the SVV2 was gradually replaced by concern over the increase in traffic jams and car driving was increasingly seen as essential to modern daily life (*ibid.*). The storylines of the 'car as polluter' and 'car driving as an environmental crime' moved to the background of mobility policy. The moral appeal from the government had evaporated by 1996 with the launch of the new intermediary mobility policy plan entitled 'Samen werken aan bereikbaarheid' (Working together on accessibility). This new policy plan marked a drastic policy shift towards the political acceptance of the growth in mobility, a storyline which would determine transport policy for the next decade. Not only the problem definition shifted towards congestion, also the ambition level was significantly reduced. The cabinet recognised that the basic structure of the mobility patterns of travellers could not be changed overnight (Peters, 1998, p. 51). Therefore, the use of mass communication as a policy tool to influence mobility behaviour, typical of the 1990s period, came to an end. Instead of awareness raising as a policy tool (based on social-psychological theories on behavioural change), mobility behaviour was now predominantly understood and targeted from an economic perspective. In 2001 the SVV2 was officially replaced by the National Traffic and Transport Plan (Nationaal Verkeers- en Vervoersplan, NVVP), entitled 'from A to Better'. The main message of the NVVP consisted of a further acceptance of autonomous mobility growth. Politicians and scientific advisors generally agreed that ten years of attempts to reduce the growth in car transport turned out to be a complete failure (Peters, 2003, p. 49). The acceptance of mobility growth was accompanied with a further professionalization of personal mobility. Car driving was accepted; however, the negative consequences for accessibility, safety and the environment had to be paid for by that same car user. The introduction of market principles to individual mobility choices (by way of variable road charges) is indicative of the economic perspective as the new steering paradigm for mobility behaviour. The scheme for charging road use was also made possible by new developments in information and communication technology. According to Peters a discourse coalition had emerged between the storylines of 'employment', 'economic growth', 'accessibility', 'infrastructure', 'market function' and 'technology' (Peters, 2003, p. 58).

The discourse analysis of Peters ends in 2001, nevertheless it is safe to say that the main storylines in Dutch transport policy were strongly present until at least 2006. In 2004 a new traffic and transport policy plan was presented, the much debated 'Nota Mobiliteit' (Memorandum on Mobility). Its subtitle 'towards a reliable and predictable accessibility' signifies a further strengthening of the position of transport growth as a prerequisite for a well-functioning economic structure. The topic of sustainable mobility is only mentioned in the last chapter of the policy plan. Interestingly, this is also the sole chapter in which transitions and transition management are mentioned as governance tools. Furthermore, the ambition level with regard to environmental aspects remains low: the negative effects of mobility growth should be mitigated. The strategy consisted of just complying with international standards while avoiding additional obligations and regulations at the national level. The economic storyline of 'the user pays principle' was further institutionalised in 2005 after a stakeholder-platform reached a monumental agreement on a new method of paying for (auto) mobility. The key point of the policy advice was that payments for mobility services delivered by the state should no longer be based on car ownership but on actual car use instead (paying per kilometre driven). The advice was adopted by the transport Minister and the new system was aimed to become operational for freight transport in 2011 and for passenger transport in 2016.

Mobility policy from 2006 onwards has increased in complexity and taken some remarkable twists and turns. As prescribed by the Memorandum on Mobility (Nota Mobiliteit), transport policies were to be based on three pillars:

1. building: creating and adapting infrastructure to increase road network capacity;
2. road pricing: implementation of road pricing according to the 'user pays principle';
3. utilisation: optimizing the use of available road capacity.

Especially the first pillar clearly shows similarities with the predict and provide paradigm. Congestion is framed as the dominant mobility problem and the increase of road capacity as its preferred solution. In 2009 road construction received a further impetus with the policy of the 'Spoedaanpak', a 'priority or emergency treatment' of thirty projects in order to increase road capacity within a few years. This approach was made possible by restricting and shortening some of the legal trajectories for getting permissions.

While the first pillar is based on increasing capacity, the two other pillars aim to optimise the use of the existing infrastructure, either through economic stimuli (road pricing) or by traffic management (utilisation). Even though preparations were in an advanced stage, road pricing as yet never reached the phase of implementation. In a policy dossier which can best be described as 'history repeating itself', the support for road pricing was suddenly withdrawn by the Dutch cabinet in 2010. Lack of end-user support was one of the main arguments used in the abortion of the whole endeavour. In the eyes of the general public – so it was argued – road pricing had increasingly become an undesirable and over-expensive 'digital panopticon'. With the implementation-failure of road pricing policy, and with the accelerated realisation of many road works from 2011 onwards, utilisation has become the prime pillar for Dutch mobility policy in the contemporary era.

The end-user of mobility services

Paradoxically, although the emphasis on predict and provide has been strengthened in recent times, Dutch mobility policy since 2005 has not become entirely supply-oriented. While the increase of road capacity has been the dominant policy focus, simultaneously a user- and demand-oriented approach became more influential as well.

Initially this new orientation was most prominent in the strong focus of the Dutch Road Administration (Rijkswaterstaat) on the road user, for example to reduce nuisance during road construction works. Under the heading of 'public-oriented network management' the demands of the road user was given a central place within the organisation[15]. However, this perspective of the road user as a client of the road network manager has little to do with travel demands and mobility choices of travellers.

This broader perspective on the mobility choices was taken up after the policy advice of the Social Economic Board in 2006 on mobility management (SER, 2006). The Social Economic board described mobility management as the whole of activities of governmental bodies and social partners which – wherever possible and desirable – is focused on the stimulation of conscious behavioural choices of employees (SER, 2006). The choices in this context refer to the question if and when the trip is made and with what kind of transport modality it is organised. Interestingly, the responsibility for a more sustainable mobility system is assigned to all parties involved in the process: governmental bodies, employers, transport companies, and employees.

As a result of the SER advice in November 2007 the Taskforce Mobility Management was initiated. The TFMM aimed at developing an encompassing set or package of concrete measures in order to reduce the amount of car kilometres and the accompanying excretion of environmentally harmful emissions during rush hours by 5%. This initiative has recently been supplanted by the programme Better Utilisation which has an even stronger focus on the traveller and its behavioural choices. While this programme reveals are renewed demand-oriented mobility policy, the aim of the Better Utilisation explicitly is to reduce congestion in the most densely populated urban areas; environmental improvement is merely a potential positive side-effect.

[15] The change towards a public-oriented network management was initiated after strong criticism on the inward focus and infrastructure orientation of the Dutch Road Administration. The new orientation was accomplished after a large-scale reorganization which included steering on the measuring and monitoring of performance indicators of user satisfaction.

Tekstbox 2.1. Transition management and mobility: the Platform Sustainable Mobility and Transumo.

Aside from the policy orientations reflected in the national transport policy plan and the environmental policy plan, two multi-stakeholder public-private platforms were initiated to act as an action and coordination institution for the realisation of the sustainable mobility goals. Although both platforms focused on transitions to sustainable mobility, their problem definition and approaches were different. The Platform for Sustainable Mobility, launched in 2005, was part of the Dutch Energy Transition, a government interdepartmental initiative backed by the Ministries of Economic Affairs, Transport, and Environment. The Platform supported transitions towards sustainable mobility, that is 'to generate an impulse towards accelerated and more substantial market penetration of technologies and fuels that faster than projected today improve the outlook for considerable gains in air quality, health conditions, noise contours, and in the reduction of greenhouse gas emissions, as related to traffic and road transport' (Action Plan for the Platform Sustainable Mobility, 2005). However, the Platform explicitly focuses on 'cleaner cars, cleaner fuels, cleaner usage' and aims for the following reduction targets: a factor 2 reduction of greenhouse gas emissions for new vehicles in 2015, a factor 3 reduction of greenhouse gas emissions for the complete Dutch car park in 2035. To reach this ambition four transitions pathways were formulated: (1) hybridization of the car park; (2) application of bio fuels; (3) driving on hydrogen fuels; (4) intelligent transport systems. The Platform therefore clearly has a strong technological emphasis in trying to improve the energy efficiency of the Dutch car park.

The second multi-stakeholder public-private platform called Transumo (an acronym for transition to sustainable mobility) aimed to provide a significant contribution to the necessary knowledge to enable the development of a sustainable mobility transition. Very interesting is the Transumo mission: 'to accelerate/encourage the transition to sustainable mobility. This will be achieved by initiating, and establishing for the long term, a transition process that leads to the replacement of the current, supply driven, mono-disciplinary technology and knowledge infrastructure, with a demand driven, multidisciplinary and trans-disciplinary, participative knowledge infrastructure' (www.transumo.nl). The storylines behind sustainable mobility in this platform consist mostly of 'prevention of economic losses', 'accessibility', and 'construction of innovative infrastructures'. The planet-side of mobility is only slightly addressed (CE, 2008a). Furthermore, the strong infrastructural focus is in contrast to the aim of demand-orientated mobility. Currently, both platforms to stimulate sustainable mobility transitions have ended.

2.3.3 Synthesis

In the preceding sections the shifting storylines and discourses in UK and Dutch transport policy have been discussed. The overview showed that the importance attached to the various dimensions of sustainable mobility differed greatly over time. So the choice for a certain problem definition implies that certain aspects will and other aspects will not be on the societal and political agenda. Therefore, what it needs for a transport or mobility system to be labelled 'sustainable' shows significant variation over time, as is the case with ideas about the possibilities and limitations of influencing actual mobility behaviours.

Figure 2.9 shows a summary of the patterns in mobility policy in the Netherlands and the UK (see also Bouwman, 2004). This graph generally confirms the discourse analyses of Vigar and Peters with respect to the periods of demand- and supply-oriented mobility policy. When the two countries are compared on the basis of their mobility policy we can see that until the 1970s the problem definitions are basically similar. In both countries there was a strong emphasis on road construction, not surprising considering the enormous growth in mobility and the lack of knowledge about the negative consequences of mobility growth. In both countries from the late 1980s onwards the detrimental effects of mobility on the environment became more prominent. In both countries policy plans were developed at the national level which aimed for radical policy shifts towards mobility reduction and a modal shift to public transportation and slow transport modes. However, the countries diverge with respect to the actual implementation of their transport policies. Vigar assessed that the 'predict and provide' paradigm within UK transport policy only came to be challenged seriously after 1994. The newly developed storylines of the Labour government did not sink in, mainly because it was not implemented at the local level (also referred to as the implementation deficit). This was to a large extent based on resistance by the local population to traffic calming measures.

The storylines behind the new realism paradigm takes the form of a move away from a technical approach characterised by an emphasis on supply-side initiatives, 'predict and provide', to a more demand-management orientation, 'predict and prevent' (Vigar, 2000). The principal elements of the new approach are: less reliance on road construction, the adoption of a package of solutions to given problems, and in general terms a preference for the active management of travel demand rather than just providing the capacity for it. Vigar suggested that by 2001 the idea of 'roads as a problem' had permeated the mind-sets and practices of transport practitioners. The

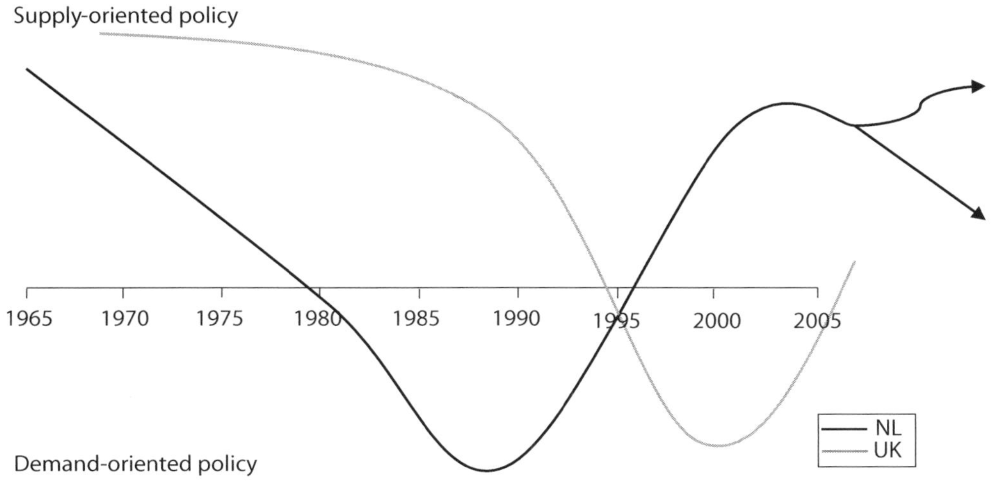

Figure 2.9. Developments in mobility policy in Western Europe, 1965-2010 (adapted from Bouwman, 2004).

penetration of these policy notions into the mind-set and practices of the general public and in local communities, however, seemed to be much more questionable (Vigar, 2002, p. 193).

With regard to the Dutch situation in the late 1980s and early 1990s, demand-oriented mobility policy was implemented by way of promoting carpooling, the designation of carpool lanes, and the introduction of company transport plans. A change in the mobility behaviour of travellers was also attempted by mass media campaigns promoting a reduction in car use and a modal shift towards public transport. Characteristic of the mobility debate in this time-period was the strong moral appeal to the individual car user to restrict car use in order to preserve the environment. In this approach, based on social-psychological theories of behavioural change, a strong emphasis is put on the motives, values and beliefs of individual humans.

However, the SVV2 lacked a strong economic program to implement the firm objectives and was based to a significant extent on voluntary behavioural change from travellers. Furthermore, the disappointing results of the mass communication programs in the 1990s – intended to raise awareness about the negative consequences of car driving – for many professionals confirmed the strategic importance of a technology oriented mobility policy (see also Peters, 2003, p. 47). Boelie Elzen summarized the shift to technology provision in the following quote: 'a shift has taken place towards almost exclusively relying on technical solutions to tackle mobility problems' (Boelie Elzen, 2006, p. 332). The success of the catalytic converter as an end-of-pipe technology contributed to this policy shift significantly. In the mobility policy plans following up to the SVV2, demand-oriented policies (directly) targeting the mobility behaviours of travellers were explicitly dropped. It was deemed more effective to directly focus on environmental impacts instead of trying to influence modal choices (Nota Mobiliteit, Deel III, 2004, p. 151). Policy makers in this period had become increasingly sceptical about the possibilities of deliberately influencing traveller's behaviour.

In the last five years multiple discourses have come into existence, resulting in a complex, multi-layered mobility policy. The dominant layer concerns the continuation of the predict and provide paradigm with an emphasis on road construction works to increase road network capacity. Accessibility as a prerequisite for economic development serves as the main storylines in this respect. However, at a different layer simultaneously more emphasis is given to the end-user and his or her travel demands. Environmental factors in the current discourses play on a very limited role.

To a large extent the supply-oriented approach was also present in the two platforms focusing on transitions to sustainable mobility (Platform Sustainable Mobility and Transumo). For example, the Platform sustainable mobility clearly focused on facilitating the adoption and diffusion of technological innovation within the Dutch car park. Modal shift as a shift in behaviour was not a pathway which was explicitly targeted. Furthermore environmental problems were defined as emission problems, most notably greenhouse gas emissions, to be solved via technological means. Citizen-consumers were targeted only in their role of purchasers of automotive innovations, not in their practices of movement or in their everyday life as citizen-consumers. Within Transumo, the planet-side was very poorly developed as sustainable mobility was predominantly approached as economical and accessibility problem. With a few exceptions, the infrastructural and technological emphasis was even more present within Transumo, despite its explicit mission of a move towards demand-oriented mobility systems.

The analysis leaves us with the following conclusions. Firstly, in spite of the enormous policy efforts invested in mitigating the negative consequences of mobility, the effects remain far behind the formulated policy objectives. While during the transition to a car based mobility system images of autonomy, flexibility and speed were dominant, in the current era mobility and mobility policy are first and for all about recognizing and confronting mobility problems as being structural, long term problems. The domain of daily passenger mobility clearly is faced with the existence of a set of persistent and complex problems (see also Geerlings & Peters, 2002; Avelino *et al.*, 2007). Secondly, mobility policies show shifting emphases and problem definitions over time, while also the suggested directions for solving the problems vary accordingly. The policy measures implemented in the domain of passenger mobility reflect the ways in which the role of mobility in society is perceived. The adopted policy approaches also seem to reflect the changing perception of the role of citizen-consumers in everyday mobility practices. The disappointing experiences with the failed attempts to change consumer behaviour in the context of a demand driven mobility policy in the late 1980s and early 1990s resulted in a drastic shift in the dominant policy approaches in the Netherlands. At the start of the 21st century, a strong supply orientation in mobility policy characterized Dutch mobility policy, showing little or no interest in attempts to change consumer behaviour in the field of mobility. The possibilities to effectively influence everyday mobility practices of travellers along the suggested policy lines (modal shift, polluter pays, etc.) were judged to be very limited. It is only in recent years that the attention for demand-oriented approaches re-emerged again. In the next paragraph it is explained why a focus on to the role of citizen-consumers and social practices in transitions to sustainable passenger mobility, despite of the hesitations expressed by policy-makers, deserves special attention.

2.4 Innovation and mobility, the role of citizen-consumers in sustainability transitions

The previous paragraph showed how contemporary mobility policies tend to be strongly supply-oriented in character, thereby lacking a clear cut and convincing orientation on citizen-consumers and their everyday life mobility behaviours. Even worse, in many occasions citizen-consumers are considered as barriers to the successful realisation of sustainable transitions. Citizen-consumers, as the argument goes, are not willing to accept and/or able to implement structural changes in their everyday life mobility patterns. In this paragraph I will provide some arguments why an orientation on the perceptions, behaviours and experiences of citizen-consumer can be regarded as of crucial importance to sustainable mobility transitions. Such a focus on citizen-consumers is not pre-given and obvious, since both researchers and policy makers have been keen in formulating and emphasizing legitimate arguments for deliberately moving away from citizen-consumer oriented environmental policies. Drawing upon the previous sections and reflecting upon more general debates on the role of citizen-consumers in sustainable consumption policies, I start by describing some of the reasons for not bothering citizen-consumers in the context of a behavioural approach to sustainable mobility transitions:

1. There is a gap between citizen and consumer, and between saying and doing. By far the most cited argument is that the behaviour of consumers is often not in line with the ideas and viewpoints which that same person has a citizen. As citizens and voters we would like, and even demand, a high environmental quality. However, as consumers we predominantly

tend to shop with our wallet. Therefore there clearly is an inconsistency in the everyday behaviour of citizen-consumers[16]. Research has also indicated that the environmental impact of consumption patterns is mostly dependant on the household income levels and does not show a (direct) relationship with environmental awareness, value patterns, problem perception and motivation (Vringer, 2005; Vringer *et al.*, 2007; MNP, 2007b). Thus, as the argument goes, consumers are clearly unwilling to change their existing behaviour and are too opportunistic to start consuming in a more sustainable manner. To indicate this viewpoint: Rein Willems (at the time President-Director of Shell, and Chair of the Dutch Energy Transition) during the launch of the transition networks in 2005 requested: 'I hope that the knowledge networks within the transition network will explicitly unleash their scientific expertise on the 'schizophrenic consumer'[17]. This consumer invariably mentions that behavioural change is tremendously important. However, with the same breath he states that this should begin with his neighbour, to unfortunately halt at his own door because he is an exceptional case' (2005, author's translation). In the same year, the Director Climate Change and Industry of the Dutch Ministry of Environment expressed that: '(1) consumers hold on to mobility and car preferences, (2) consumers are unwilling to pay for fuel-efficient cars and technology'.

2. The focus on behavioural changes is not effective. As the previous paragraphs indicated, technological innovations, such as the catalytic converter, have been responsible for large reductions in environmental impact. In addition, policies directly targeting behavioural changes so far have generally shown limited effects and are therefore considered, in comparison to technological innovations, a less feasible and effective option to realize the substantial reductions needed.

3. There should always be 'freedom of choice'. A more principal argument is that a government has no right to impose policies which directly inhibit or exclude certain behavioural options or types of (environmentally relevant) practices as it erodes the fundamental freedom and sovereignty of the consumer. For example, Van Soest (2007) rebels against policies that prohibits terrace heaters, regular light bulbs, and SUVs in urban centres, also arguing there is no legal basis to do so.

4. Citizens demand national government to take action. When asked who is mainly responsible for solving current structural environmental problems, citizens tend to collectively point towards the national government and to the larger corporations (see for example Insnet, 2007). Furthermore, citizens would prefer policy measures which do not direct interfere with their everyday lives. E.g. with regard to climate change policy, the largest public acceptance is for

[16] Spaargaren (2003) distinguishes two types of inconsistencies: between attitudes and behaviour, and between citizen roles and consumer roles. After portraying these inconsistencies he describes how a social practices approach could address these issues.

[17] Hopefully this thesis indeed is considered as providing new knowledge about the role of the 'schizophrenic consumer' in sustainable transitions.

reduction targets in electricity production and energy savings amongst producers which result only in moderate increase in household energy prices (MNP, 2007b).

5. Information about pro-environmental behaviour and technology. A final argument used is that environmental problems are too complex for consumers to understand. Because environmental problems and policies are subject to permanent changes, consumers cannot be informed in a reliable way about the best environmental choices. It takes a lot of time and effort for people to get accustomed to a certain environmentally friendly social practice, while the best available environmental technologies and options are subject to quick changes. Furthermore, ordinary people do not understand the feedback mechanisms and (negative) relationships between the various environmental problems and problem solutions[18].

Considering these five arguments, at first sight there seem to be good reasons for not developing a consumer-oriented sustainable mobility policy. However, it can also be argued that the arguments are invalid or out-dated and connected to (mobility policies) that are in urgent need of reform. Numerous scientific and policy studies in one way or the other have stressed the importance of a citizen-consumer orientation in sustainable transitions (see for example Hoogma & Schot, 2001; Jackson, 2006; MNP 2006, 2007a; Putman & Nijhuis, 2006; SER, 2003; Spaargaren, 1997; Spaargaren & Van Vliet, 2000; VROM-Raad, 2005). Citizen-consumer involvement in environmental and mobility policies can be argued to be relevant and even necessary for again five (interrelated) reasons:

1. The environmental impacts of contemporary consumption behaviours are too significant to be ignored. A general but very convincing argument for taking the changing practices of citizen-consumers into consideration is the fact that consumption patterns form the largest contribution to global environmental impacts (World Watch Institute, 2004; MNP, 2006, 2007a). Sustainable consumption and production policies traditionally focus on the supply side (SER, 2003). According to the Netherlands Environmental Assessment Agency however, technological improvements in the production processes have been insufficient to compensate for the increasing environmental pressures originating from an increase in consumption growth (MNP, 2006, 2007). According to the VROM-raad (the primary advisory council for the Ministry of Environment) it is exactly this aspect which makes contemporary environmental problems so persistent (VROM-raad, 2005). Persistent environmental problems are thus inherently consumption related.

2. Understanding the interaction between innovations and everyday life practices is necessary for the development and diffusion of sustainable innovations. As potential buyers citizen-consumers play an obvious role of importance in the adoption or rejection of sustainable innovations. However, providers of sustainable innovations also have to take the use of a product or service in the everyday life into account. Firstly, the diffusion of sustainable

[18] The information deficit and the complexity of environmental impacts also play a significant role in Chapter 5 where these aspects are related to the purchase of a new car.

innovations may fail because they do not fit into the lifestyles and everyday routines of modern citizen-consumers. Secondly, the use phase is relevant because it can greatly influence the effectiveness of an environmental innovation. Well-known are the rebound effects that can take place when environmental advantages disappear because of increase in use with increased eco-efficiency. Also, environmental advantages can be dependent on the correct use of sustainable innovations[19].

3. Connecting to the life world of citizen consumers is a prerequisite for developing societal support and public acceptance of unavoidable hard-to-digest sustainable policy measures. Societal support and public acceptance for stringent measures, is a well-known and well-researched topic in the domain of mobility, especially in relation to various road pricing schemes. While transport economists have campaigned for decades to implement different forms of road pricing, tradable emissions permits etcetera, these measures have only been marginally implemented, usually due to a lack of public and political acceptability. Research into public acceptability (Schade & Schlag, 2003; PODOII, 2004; Whittles, 2003) shows how various citizen-consumer related aspects (e.g. equality and fairness, privacy issue, ways of charging, trust, problem definition) have an enormous effect on public acceptance.

4. Understanding the behaviours of citizen-consumers is essential for dealing effectively with the new relations between private sector, public policy, and citizen-consumers. A large part of the (Dutch) population feels adequately informed about the environment. However, environmental policy is seen as abstract and to a significant extent incomprehensible which has resulted into a gap in the communication between the environmental professional and citizen-consumers (Vrom-raad, 2005, p. 52). The traditional approach based upon a strong technocratic steering by the national government, with an environmental policy dominated by environmental professionals in which the citizens are only addressed in their roles of builders of public support, does not seem the best way for dealing with the persistent and complex environmental problems facing contemporary society (*ibid.*, p. 9). Adding to the complexity is the globalisation of environmental problems and politics and the consequences for a citizen-consumer orientation in environmental policies (see Spaargaren & Martens, 2004). Globalisation of environmental politics can offer opportunities, but also demands a further strengthening of the relationship between 'the environment', environmental policies and everyday life of citizen-consumers.

5. Citizen-consumers are crucial for developing attractive long-term sustainability visions in an interactive way. In addition to the previous point, Ingold has argued that the current discourse of global environmental change which depicts the environment as a set of issues, global in scope and physical in origin, is a configuration that remains detached and abstract from everyday

[19] In the Netherlands around 300,000 houses have been built with a so-called 'balance ventilation system' which automatically regulates the supply and outlet of air. These systems have been built in highly insulated houses to reduce the need for manual ventilation (by opening windows) and thus contributing to the energy efficiency of newly build houses. The practical use of these systems can differ from the prescribed (theoretical) use in many ways, thereby influencing not only the energy efficiency but also the health conditions of the indoor climate (see Soldaat, 2007).

life (Ingold, 1993, in Macnaghten, 2003). Various authors (Macnaghten, 2003; Spaargaren & Martens, 2004; Steg & Gifford, 2005; VROM-raad, 2005) all stress the point that there should be more attention to the social-cultural context in which sustainable innovation takes place. Existing environmental policy goals, formulated in abstract and 'eco-technocratic' terms, should be better adjusted to the everyday life-worlds of citizen-consumers. Citizen-consumers do not think in terms of environmental impact but in terms of needs and functions (VROM-raad, 2005). To strengthen the involvement and commitment of citizen-consumers in sustainable transitions, instead of a goal in itself, environmental policy should therefore be formulated as a means to accomplish inspiring future visions. 'The environment' should be translated into quality of life indicators, or in terms of the VROM-raad: the art of the good life.

With these arguments I have tried to stipulate that a citizen-consumer orientation is not only a fruitful but also a necessary part of (policies targeting) transitions towards sustainable mobility. This requires a different policy approach as conventional policy frameworks are not compatible with a citizen-consumer oriented approach. However, this implies that citizen-consumers are seen in a broader perspective than purely as adopters or rejecters of sustainable innovations, or to solely see them as users or non-users, for that matter. Also the dominant distinction of citizens acting as voters, and consumers acting as economists is only part of the whole story (see Spaargaren, 2003).

In Chapters 3 and 5 the theoretical underpinnings of how a citizen-consumer oriented policy focused upon transitions towards sustainable mobility could look like are addressed. There, a social practices approach towards transitions to sustainable mobility is introduced.

Chapter 3. Analysing mobility as an everyday life consumption routine

All human beings are knowledgeable agents.

Routine is the predominant form of day-to-day social activity.

Structures are always both enabling and constraining.

Anthony Giddens (The Constitution of Society, 1984)

3.1 Introduction

The preceding chapter was concluded by discussing the various arguments why a transition towards sustainable mobility can only be realized when citizen-consumers are actively involved in this process. Only when a connection can be made with the cultural concerns of citizen-consumers, and when the inclusion of citizen-consumers is no longer approached as an obstacle to change but rather as a source for it, transitions to sustainable development can be realized. However, this conclusion, important as it is, presents us with a rather difficult starting position when situated against the backdrop of some early variants of transition theory. Whether transitions are defined as a shift from one socio-technical system to another (Geels, 2005a,b) or as a continuous process of structural change in a societal sub-system (Rotmans *et al.*, 2001), it is clear the definitions indicate that transitions are inherently systemic by nature. Indeed, a precondition for transitions is that innovations take place on a societal level, the so-called system innovations (Rotmans, 2003). So, while there is a quickly expanding literature in transition studies 'in this new research field transition processes are studied from a variety of system-perspectives: socio-technical systems, innovation systems and complex adaptive systems' (Loorbach, 2007, p. 17).

Analysing societal change as long-term co-evolutionary processes of complex systems at first sight seems to leave little room for the actions of (organised) individual citizen-consumers. How does a proposed transition to a hydrogen society relate to the everyday actions of individual car drivers? And, *vice versa*, how do information and awareness campaigns, such as the introduction of a fuel efficiency label for new cars, relate to long-term co-evolutionary processes of system change? In short, there is a gap to be filled to connect (policies aiming to influence) individual everyday action of citizen-consumers with technological transitions and system innovations. Similarly, it is evident that there is a strong focus on technological change in analysing and managing transition processes and patterns in the current literature.

Not surprisingly these original conceptualizations of transitions have been criticized for their lack of agency, consumer-orientation and user-involvement (Mont & Emtairah, 2006; Shove & Walker, 2007; Smith *et al.*, 2005; Spaargaren, 2005, 2006; Van Vliet, 2002). Geels (2005a) agrees that the multi-level perspective has almost solely been used to understand the emergence and diffusion of novelties. While few would disagree that technological change plays a crucial role in any sustainable development policy, an almost exclusive focus on technological innovations has its drawbacks. As one of the critical reviewers of transition management as a new mode of governance, Shove & Walker (2007) point out that in the early formulations of transition theory there is almost no reference to the ways of everyday living that are implied in the technological futures. In reviewing strategic niche management, Hegger stresses the point that niche managers always redesign system and life-world, both at the same time. Technology therefore should be deprived of its privileged position (Hegger, 2007, p. 29). The basic idea behind these criticisms is that consumers are not external to system innovations; on the contrary, they are central to it.

Putting technological change as the point of entry for discussing sustainable mobility pathways often results in a bias (Hoogma *et al.*, 2002)[20]. For example, the distinction between incremental innovation leading to system improvement and radical innovation leading to system innovation depends partly on the perspective chosen. What is radical from a technological point of view is not necessarily so from a citizen-consumer point of view. However, the point we want to stress is not that user preferences, consumer practices, and structures of meaning are more important than technological and infrastructural regimes, quite the reverse, they are equally important and should be analysed in relation to each other. Because the technological trajectory of systems has much more prominence in transition studies we put greater emphasis on these citizen-consumer related aspects. Such an approach to the relation between agency and transitions is in line with more recent formulation of transition theory as put forward by Grin *et al.* (2010) in particular.

To understand the role of citizen-consumers in transition trajectories it is important to delve deeper into the literature on consumption. The previous chapter already hinted at the fact that the structural nature of mobility-related problems to a large extent is strongly related to the fundamental nature of consumer society and consumer behaviour in the domain of mobility. Passenger mobility is inherently related to consumer cultures, the symbolic and affective motives for car usage, the functional aspects of trips, consumer perceptions and appreciation of different modalities, and so forth. As the thesis title suggests, mobility is a form of consumption and there are important benefits when analysing mobility as such. Therefore, to get more insight in these dynamics, in Paragraph 3.2 the nature of consumer society and the dynamics of consumption behaviour are explored in more detail and its relevance to the domain of mobility is elaborated[21]. Next, I will shift the focus to the role of routine in structuring the daily lives of citizen-consumers. In line with Giddens' theory of structuration it is argued that consumption behaviour should primarily be understood as a set of social practices structured in time and space. Especially contemporary theories of practice provide important impetus for understanding the connecting between everyday practices and socio-technical systems and technologies (Spaargaren, 2011). In Paragraph 3.4 we will build upon the approach developed in the CONTRAST research program, which is a specific form of practice theory based on the social practices model (Spaargaren *et al.*, 2007). The aim of this project is to analyse and understand transition processes at the level of

[20] These authors rightfully state that other choices, for example by focusing on the emergence of new green consumer values or new management practices within industry, have their own specific bias and often lead to a neglect of technological pathways.

[21] In line with the philosophy of the CONTRAST research program I deliberately do not take sustainable consumption behaviour as the point of departure for this chapter. Sustainable behaviour as such does not exist and is not a separate category in the everyday behaviour of citizen-consumers (Spaargaren *et al.*, 2002). It is the other way around; nearly all human behaviour is in principle environmentally related.

everyday practices. This approach aims to bridge the divide between actor-oriented approaches and system-oriented approaches to sustainable consumption. By relating these insights from consumption and practice theory with perspectives on system innovations and transitions, we intend to make real and elaborate the expression that 'users matter' (Geels, 2007).

The perspectives described in this chapter form the theoretical building blocks of the thesis. They are discussed separately and at a general level in this chapter. In chapter five the building blocks will be nit together and further specified for the analysis of mobilities.

3.2 What is this 'thing' called consumption?

When advocating a more citizen-consumer oriented approach to sustainable transitions, one of the prime claims made by the CONTRAST project is that not taking a consumer perspective implies ignoring one of the most fundamental dynamics of current society. Contemporary society consists of consumer cultures in which the demand for material goods plays a very important role. Sustainable transitions, so we argue, run into the dynamics of consumer society all the times for example when increases in eco-efficiency are negated or neutralized by increases in the level of consumption.

However, what exactly do we mean when talking about consumer society and the dynamics of consumer cultures? What is this 'thing' called consumption? And how do general consumption issues relate to sustainable consumption and the consumption of mobilities in particular? In this paragraph a short review of the consumption literature is offered in order to shed some light on these questions. Considering the vast amount of literature we tend to restrict the review to the various approaches which have been developed in recent years. Consumption, and to a lesser extent sustainable consumption, has been researched from economic, psychological, sociological, and anthropological perspectives. Each of these perspectives places emphasis on a specific element of consumption behaviour and will be explored for their potential relevance to the study of mobilities. Throughout the paragraph we want to show that the increase in consumption patterns can be understood as resulting from the dynamic relationship between consumption, production and consumer culture.

3.2.1 Producer-led versus consumer-led approaches to consumption

Increasing consumer demand can be seen as an important feature of consumer culture (Lury, 1996). While the increase in consumer demand has been well-documented, the explanations of this growth vary tremendously. Generally, explanations tend to be centred on one question: are consumption and consumer culture determined by producers or is production merely the result of the combined action of individual consumers? Or as De la Bruheze *et al.* (2004, p. 4) phrase the question: 'do producers control consumers, or do consumers dominate producers?'. Producer-led explanations build upon economic and social theories which see mass consumption predominantly as the result of mass production. Neo-Marxist approaches focus on the systemic internal dynamics of capitalism which, in a one-way relationship, shape consumer needs. According to Adorno (1974), through practices of packaging, advertising and promotion the relation between products and human wants and needs are manipulated (in Lury, 1996, p. 42)[22]. Most of the producer-led approaches tend to be negatively oriented towards capitalist modes of production and provisioning. Critiques on the continual bombarding of consumers with brand advertisements by captains of industries remain influential (Klein, 2000). However, also in studies on the history of technology consumers are mostly attributed a passive role. While engineers, planners and producers tend to be portrayed as the designers and makers of modern society, the role of the consumer is limited to that of the purchaser of novelties (Schot & De la Bruheze, 2003). Also, some approaches in technology studies which focus on user behaviour and technology development can be attributed this label due to the limited attention for the active role of users. For example, some approaches in the sociology of technology (like actor-network theory and theories on scripting) have been criticized because design is represented as a one-way process in which the power to shape technological development lies in the hand of design experts (Oudshoorn & Pinch, 2003). In script approaches the activities of the designer are guided by various user representations which become materialized in the design of the technology (Geels, 2007). While the notion of scripting gives explicit agency to the user, this agency is restricted solely to the interaction of the isolated individual with the specific technology (Oudshoorn & Pinch, 2003)[23]. In short, producer-led explanations see the modes of design, production, distribution and marketization as the dominant factors in shaping consumer culture and behaviour. These systemic approaches have been criticized for the limited or passive role attributed to consumers. Apart from being seduced, most producer-led approaches provide no clear explanations why consumers actually consume as they do.

Situated on the other side of spectrum, consumer-led approaches show more consideration of the active role of consumers in processes of consumption, though the interpretations of (the motives behind) consumption behaviour show tremendous variation. Psychological studies on consumer behaviour traditionally seek explanations of consumer behaviour in the fundamental nature of human needs and values. These studies build upon a typology of needs and values (e.g. Maslow, 1954) as the guiding principles of consumer behaviour. These needs and values range from physiological needs (hunger, thirst, etc.) to self-actualisation (realisation of one's potentials). Individuals consume energy, materials, products and services because they are driven by basic needs and wants (Vlek *et al.,* 1999). In this approach consumption satisfies the various needs and wants that individuals have and which constitute their quality of life. Deeply rooted in psychological theories (mostly cognitive and motivational theories), and taking needs as the starting point, the goal of the marketing domain is to understand and influence consumer buying behaviour. Traditionally, analyses of purchasing decision processes follow a five-stage model known as the EKB-model: (1) need recognition; (2) information search; (3) pre-purchase evaluation of alternatives; (4) purchase decision; and (5) post-purchase alternative evaluation (Engel & Blackwell *et al.*, 1995). These types of models focus on the individual decision making process and the role of information in these processes behaviour. However, the most influential

[22] For neo-Marxist approaches on the impact of increasing production processes on global ecosystems, see the Treadmill of Production perspective (e.g. Schnaiberg *et al.*, 2002).

[23] For a more detailed account on approaches in user-technology interaction, especially those that do more justice to the co-construction of user behaviour and technology, see De la Bruheze *et al.* (2004), Geels (2007), Oudshoorn & Pinch (2003), Verbeek & Slob (2006).

models of consumer behaviour are expectancy-value theories upon which most of the sustainable consumption policies are based. In these theories choices are made on the basis of the expected outcome from a choice and the value that is attached to those outcomes (Jackon, 2005)[24]. These models are strongly voluntaristic in the sense that sustainable behaviour is expected to be the result of pro-environmental attitudes of the individual consumer. The assumption behind policy strategies such as information provision, awareness raising and internalizing external costs is that actors will act according to their beliefs and are able to adapt their behaviour using the information provided (Nye & Hargreaves, 2008).

While economists and psychologists tend to study consumer behaviour in isolation from other consumers, sociologists and anthropologists perceive consumption as being socially grounded (Mont & Plepys, 2003). The first contribution of the sociological view on consumption is that goods also play a central role in the social distinction between different groups and in strengthening the identity of particular groups (Bourdieu, 1984). They argue that there is more to consumption than the functional utility of products and services. Thorstein Veblen (1899) was one of the first to point out that consumption is more than just an economic activity or a satisfaction of internally driven human needs. Looking at consumption as a method of displaying one's social status, Veblen initiated a whole range of studies on conspicuous consumption and the relation between consumption and identity formation. Veblen connected conspicuous consumption with the rise of the leisure class. The nouveaux-riches publicly demonstrate their status by using consumer goods during their leisure activities (Lury, 1996). So, consumption is motivated primarily by the social status that goods provide. Therefore, consumption choices are not merely the result of individual choices, but they are shaped by the social context in which people live. The point of stratification of classes and lifestyles has been elaborated further by Bourdieu in his book Distinction (1984). Strongly focused on class struggles and power relations, Bourdieu showed that social differences are at the same time expressed by, as well as constructed through consumption patterns. Put in other words, our (publicly displayed) lifestyles are both expressed by and confirmed through our consumption patterns. According to Bourdieu in consuming we exercise and display our tastes and styles (Slater, 1997). An important addition to the work of Veblen is that Bourdieu considers social status and social stratification in broader dimensions than just economic capital. For Bourdieu consumption is as much about the cultural dimensions of display. He is well-known for the classification of social groups by the composition and the level of their cultural and economic capital. Cultural capital can be seen as the accumulated knowledge and competence required to make distinctions and value judgements (Lury, 1996). It is about knowing and using the cultural codes of conduct (e.g. distinction between art and non-art, ways of behaving in different cultural settings, etc.). So, to Bourdieu there is not only a difference of tastes, which is expressed through consumption of goods and services, but also a normative and hierarchical structuration related to these tastes. Taste, therefore is not a matter of intrinsic value, but of classifications grounded in class struggles (Slater, 1997). One of the most important terms in the work of Bourdieu is the concept of 'habitus', which is defined as a system of (pre-)dispositions which organises the individual's capacity to act (Lury, 1996). The habitus, shaped in childhood by family and the social

[24] For an extensive overview of social psychological models of consumer behaviour and behavioural change, see Jackson (2005).

class position, can be seen as the 'classificatory grid by which the individual cognitively maps the social world and orients action within it' (Slater, 1997, p. 162). The way we consume – the type of newspaper you read, the car you drive, the neighbourhood you live in – indicates and make visible to others our social position. With Bourdieu culture is the battleground of class struggles and competition with conflicts over tastes (*ibid.*). For example, which car brand is the 'legitimate' car (for you) to drive in and who (co)decides about this legitimacy? Who decides about the cultural meanings and functionalities which become attached to a specific car as a technological artefact?

Emulation is a social phenomenon, closely connected to the thoughts of Veblen and Bourdieu, which offers an interesting explanation of increasing consumer demand from a sociological perspective. The above makes clear that goods and services have the ability to mark status especially when they are part of the lifestyle of a high status group (Slater, 1997). Consequently, the desire for goods and services is derived from the desire to emulate the consumption styles of these higher status groups. As a result of emulation the relative value of certain consumption behaviour decreases as more individuals imitate the behaviour of the higher status group. Indeed, if certain consumer goods or services become too widespread they are no longer discriminating and lose their function as discriminating for social status. Following this line of argumentation, the dynamics of emulation ensure there is no saturation level of consumption and this will lead to the continuous expansion of consumption levels.

This theory of imitation and emulation has been heavily criticized, amongst others by Colin Campbell (1989). According to Campbell the theory is empirically incorrect as (fashionable) innovations are not by any means always introduced by the social elite. Instead of inborn needs and wants (bio-psychologically), actively created needs and wants (through advertising), or passively created needs and wants (imitation and emulation), Campbell seeks to locate the explanation of modern consumerism in hedonism and the romantic self. Modern hedonism, according to Campbell, is marked by a preoccupation with 'pleasure', envisaged as a potential quality of all experience. Thus, Campbell's argument is that modern consumerist hedonism is not about the satisfaction of needs, but about the pursuit of experience (Slater, 1997, p. 96). This 'illusory' form of hedonism manifests itself in day-dreaming and fantasizing. According to Campbell hedonism explains why an individual's interest is primarily on the meaning and images of novel products. However, imaginative hedonism leads to an inevitable gap between the perfected pleasures of the dream and the imperfect joys of reality (Campbell, 1989, p. 95). This continuous cycle of pleasurable fantasies, expectations and inevitable disappointment leads to the insatiable character of modern consumerism (Lury, 1996). Importantly, Campbell sees the character of consumption as a voluntaristic, self-directed and creative process in which cultural ideals are necessarily implicated (Campbell, 1989, p. 203). It is interesting to see the close connection between the ideas of Campbell and product advertisements which indeed are far more focused on fantasies and desires than rational arguments. Advertisements are literally filled with (unrealistic) promises, building upon various personal longings to become another or a better self. Written twenty years ago, Campbell's ideas also come back when contemplating the recent growth of the experience economy which, according to Jeremy Rifkin, is transforming the very nature of business and everyday life: 'today's consumers don't ask themselves as often, 'what do I want to have that I don't have already; they are asking instead, 'what do I want to experience that I have not experienced yet?' (Rifkin, 2001, p. 145).

While the sociological, psychological and economic perspectives described above predominantly focus on the functional aspect of goods and services, either as fulfilling a person's wants and needs or as establishing and confirming a person's social position (with Campbell as an obvious exception), the anthropological perspective focuses specifically on the social meaning of goods. Douglas & Isherwood (1979) suggest that the utility of all goods are framed by the cultural context. Even the ordinary objects of daily life have a cultural meaning attached to them. The view that goods make and maintain social relationships is a much richer idea than mere individual competitiveness. According to Douglas & Isherwood, goods present a more or less coherent set of meanings. These meanings are read by those who know the code, and who know how to scan it for the information contained in it. Thus, goods are part of everyday life information systems, and consumption can be conceived of as a series of rituals[25]. Goods are carriers of social meaning through which social identities are construed. Another interesting anthropological perspective is provided by Appadurai (1986), who shifts the focus of attention from the social lives of individuals to the social lives of objects. Appadurai differentiates two kinds of trajectories: (1) the life history of a particular object leading to the identification of a specific cultural biography of the object; and (2) the social history of a particular class of object in the light of large-scale dynamic transformations (Lury, 1996, p. 19). The important contribution is that the meanings related to objects are not static, but change as objects circulate through various networks and cycles of production, provisioning and consumption. Furthermore, through this historical process objects may acquire a certain weight and authority which they can exert in our lives, thereby influencing our beliefs and directing our attention and actions (*ibid.*).

Finally, studies on consumer culture investigate the role of consumption in (post-)modern societies as a way of emphasizing the self-identity of the individual. Lury (1996) states that processes of stylization, or aestheticisation, are among the most distinctive traits of consumer culture. They refer to the design, making and use of goods as part of the self-conscious creation of a lifestyle (*ibid.*, p. 77). Lury describes how consumer society provides new sources of identity and has contributed to a 'politics of identity' (*ibid.*, p. 227). The strong role of consumption in self-identity is often related to the emancipation from the fixities and hierarchical domination of traditional society leading towards a politics of choice (Giddens, 1991, p. 214). While Bourdieu makes a clear distinction between popular (with which he means low) culture and high culture, in the postmodern era the cultural elitism of modernism has become less and less meaningful. It is argued that the strong categorical differences have disappeared and easy reference points from which to judge cultural practices are no longer present (Storey, 1999). The secure order of hierarchical values and positions is being replaced by an overwhelming variety and fluidity of values, roles, authorities, and symbolisms out of which an individual has to construct and maintain a personal and social identity (Slater, 1997, p. 83). Modernity has led to a pluralisation of life worlds and lifestyles and the resulting threat of identity crises.

[25] One example of the system of symbolic exchange given by Douglas & Isherwood (1979) is the difference between professional and personal services. Professional services are paid with money and are classed as commerce. In personal services there is a moral judgment about gifts: it is alright to give flowers, but not to give the same amount of cash with a note attached 'get yourself some flowers'. Another example is the Christmas dinner: the turkey is not eaten en masse in December because of its good taste, but because of its meaning related to the Christian heritage and family closeness.

According to Giddens (1991) lifestyle choices have become increasingly important for the reflexive constitution of self-identity and daily activity. The shift from simple to reflexive modernity (Beck, Giddens & Lash, 1994; Giddens, 1990) indicates that the concept of identity is replaced by multiple and mobile identities and consumption is one of the most significant ways we perform our sense of self (Storey, 1999, p. 136). In this sense the relation between consumer culture and modernity is discussed by some authors also with respect to its negative consequences. As all acts of consumption are decisions not only about how to act but as well about who to be (Warde, 1994, in Slater, 1997), we have no choice but to choose (Giddens, 1991) and this might result in (new) anxieties from the side of consumers.

In this paragraph the various perspectives on consumption behaviour and the development of material culture have been shortly introduced. These different perspectives are summarised by Ben Fine in terms of the following quote: 'for economists, consumption is used to produce utility; for sociologists, it is a means of stratification; for anthropologists a matter of ritual and symbol; for psychologists a means to satisfy or express physiological and emotional needs; and for business, it is a way of making money' (Fine, 1997, in Mont & Plepys, 2003, p. 3). Our overview also showed and discussed some interesting developments in the way consumers and their behaviour are perceived and analysed. While for a long time consumers have been analysed as actors at the downstream end of producer controlled capitalist production chains or as victims of their own bio-physiological and emotional needs, recent sociological and cultural studies tend to emphasize the agency of consumers: 'the presentation of the consumer as a dupe of capitalist production has been supplanted by a model of an active, creative, self-reflexive agent' (Gronow & Warde, 2001, p. 2)[26]. This 'cultural turn' in consumption theory in the 1990s explicitly deals with the various ways in which consumers consciously and reflexively construct meanings as they result from the dialogue between material products and images of the self in various social settings. In the next section we will elaborate on the relevance of the cultural turn in perspectives on consumption for the analysis of car mobility in particular.

3.2.2 The relevance of consumption theories for car mobility

In this section a connection is made between the psychological, sociological and anthropological perspectives of consumption and the role of the car as the most crucial and omnipresent mobility object in contemporary societies. This means that the focus will be on the historically and social-culturally defined symbolic meanings associated with the automobile as an object of acquisition and use. So, what images of the self and society have become attached to the automobile over time? Many authors have pointed out that the automobile is much more than a functional object to be understood with the help of just rational arguments. If the car would involve just rational and functional uses, it would be difficult to understand why policy measures aiming to reduce or restrict car use tend to elicit such heavy emotions and contestations among citizen-consumers.

[26] As we will see in Paragraph 3.3, this conceptualisation of consumers has become criticized by various approaches in sociology of consumption and sociology of technology for overstating the level of agency attributed to individuals.

While most researchers working on the issue of sustainable mobility seek for ways to get people out of their car, psychologist, anthropologist and sociologists have highlighted some elements which help explain why people actually tend to stick to their cars and car uses. Diekstra & Kroon (2003) have meticulously described the car as an object of desire and happiness. They delve deep into motivational psychology to explain car ownership and driving behaviour. According to these authors 'rarely…has technology provided a more successful satisfier of basic human needs and motives than the car and it is unlikely that the feat will ever be repeated' (*ibid.*, p. 17). The ability of the car to reinforce already existing qualities in human beings is seen as an important factor for explaining the global diffusion of the car. The importance attached to individual freedom of movement, which the automobile clearly offers, is explained from the hunter/gatherer perspective: the car increases our 'auto'-regulative functioning which in nomadic societies was of crucial importance (*ibid.*). Furthermore, the car increases our territory tremendously, and with the speed and power of the car the individual driver has a strong weapon at his or her disposal. The technical design of the car, with its engine power and protective capabilities sheltering us from the outside world, provides us with feelings of omnipotence (Sachs, 1983). Road rage, therefore, is seen as a conflict over territorial disputes, while the car-driver is the current knight in shining armour (an archetype based on the desire to be admired and be heroic) (Diekstra & Kroon, 2003).

The longing for the car can also be explained by using the line of argumentation put forward by Colin Campbell. Since consumption is argued to result from a desire for novelties, car owners take delight in their cars because they enjoy possessing the latest technology. This gives them the satisfaction of having the most sophisticated technological devices available (Sachs, 1983, p. 352). However, in line with Campbell's theory, the pursuit of novelties is not a stable process: 'taking into consideration this continuous process of obsolescence and renewal it should be noted that the relationship between cultural meaning and its technical object is by no means stable. On the contrary, the symbolic power of a particular car is degraded in the course of time, since new products appear on the market which promise to meet the consumer's expectations even more perfectly' (Sachs, 1983, p. 356). Next to being an object of desire, also the sensation of high speed is said to be a source of excitement due to the stimulation of the central nervous system which provides a narcotic effect. Diekstra & Kroon conclude their analysis with the statement that, compared to the car 'public transport clearly is at such a great and insurmountable psychological disadvantage that it can never hope to close the gap on its own' (Diekstra & Kroon, 2003, p. 17).

Motivational and trait psychology are widely used in marketing and communication strategies. As almost any car salesmen will be quick to tell you, purchasing a car is dominated more by emotions than by rational decision-making. There has to be a personal connection or fit between the car and (the lifestyle of) its potential buyer. Therefore, car manufacturers are very keen on developing new and individually distinctive car models with a profile that fits the specific personality of the consumer. The concept of lifestyles has been used since the 1960s for market segmentation within a research approach called 'psychographics'. Cluster analysis of personality values (e.g. Rokeach values) result in lifestyle typologies with presumably similar behaviour and consumption decisions (Zemp, 2002). So, value based segmentation is used to position car brands and models (Figure 3.1).

While the psychological perspective is interesting and contains arguments which seem valid (such as the positive evaluation of the sensation of speed and the perception of safety provided by

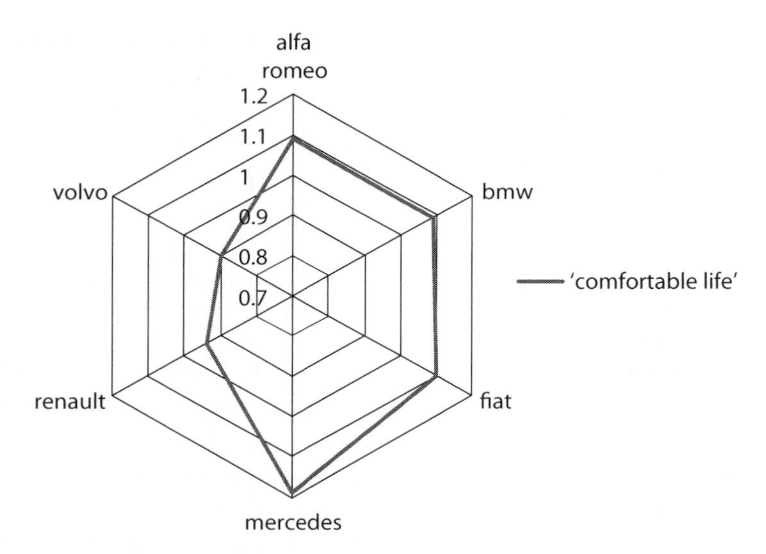

Figure 3.1. Position of car brands by one of the Rokeach values (adapted from Schipper et al., 1998).

the car[27]), to attribute the success of the car completely to its ability to satisfy basic human needs and motives does not seem very convincing in the end. Thinking along the lines as suggested by Appadurai, we can try to get a further grip on the symbolic meanings which have become attached to the automobile by following the social and cultural life of the car as material objects The question then becomes how the automobile has acquired, throughout its social life, the cultural meanings and symbolisms that are attached to this technological object. Wolfgang Sachs (1983, 1984) explains that why people desire cars and how they want to use them is largely structured by the cultural meaning attributed to the automobile. Therefore, the changes in consumer preferences over time are no longer considered as rooted in individual needs, but analysed instead as resulting from changes in the order of cultural meanings. These cultural meanings of the car are intricately linked with initial functions and users during the introduction of the automobile. The first automobiles were developed in an era in which racing was tremendously popular. Around 1900 bicycle, automobile, powerboat and air racing were very popular and widely reported about. The internal combustion engine car was an adventure machine, not only because of the high speeds involved, but also because of the technical skills required to repair the automobile (Mom *et al.*, 2002; Schot *et al.*, 2002). The initial use of the automobile forced the car manufacturers to build vehicles suitable for racing by building increasingly powerful engines and shock-absorbing car parts (*ibid.*). Interestingly, this technical trajectory, set in motion by the initial users, is still visible after more than one hundred years of car based mobility. Car magazines and television programs (e.g. Top Gear) still test and review cars as if they will be used for everyday racing instead of everyday commuting. The qualities of the car are primarily assessed with respect to aspects such as maximum

[27] For example, one of the explanations for the seemingly contagious spread of SUVs is the self-proclaimed feeling of safety that these large cars provide to the predominantly female drivers.

speed, torque, and the lap times on the race circuit. Redshaw (2007) concludes that the racing attributes being articulated help constitute aggressive, competitive styles of driving and that car advertisements illustrate the emotional themes and the expectations of faster mobility. While Diekstra & Kroon (2003) dwell into motivational psychology to explain car characteristics and car driving habits, analysing the socio-cultural meaning of the car can also be shown to relate to the intrinsic connection with racing as manifest in historical analyses of the car.

Next to the association with speed and adventure, the car has been widely discussed for its connection with display. Cars fulfil a 'function' as material objects used for conspicuous consumption by the rich upper classes. Especially during the first years of the developing car culture, promenading with an automobile in public places was a favourite pass-time activity of the wealthy class (Mom *et al.*, 2002; Timmer, 1998). In line with Bourdieu, Sachs (1983, 1984) describes how the car served as privileged status symbol for its potential to command public attention when used. Importantly, while at first the automobile caused no immediate revolution in mobility itself, it did so in the dominant order of symbols of prestige (Sachs, 1984, p. 11). An automotive lifestyle was identified with social superiority. Cars came to represent the possibility of moving around without any (apparent) constraint. They brought about the joy of shaping your life with your own decisions and mobility patterns. Furthermore, Sachs describes that processes of emulation took place in the diffusion of the automobile: 'ownership spreads along the lines of social stratification; the automobile has become a tangible symbol of upward mobility ... Unfortunately, however, since the distribution is highly dynamic, things don't stay as they are: inflation is built in and the symbolic value of differences in ownership decreases as soon as the majority has caught up' (Sachs, 1983, pp. 351-354). The role of the car as a status symbol obviously differs in various contexts. In countries with a developing car culture, such as some Asian and formerly Eastern European countries, the role of status tends to be much more prominent when compared to countries where cars are part and parcel of everyday mobility[28].

Cultural studies have focused on how the automobile has become an integral part of our socio-cultural lives and environments (Miller, 2001). After the pioneering phase, the use function of the car becomes more important in the construction of cultural meanings and in the development of an encompassing (auto) mobility culture. As soon as the automobile becomes more affordable its cultural meanings are attached to the utilitarian function next to the symbolism of adventure, speed and social distinction (Mom *et al.*, 2002). The car was not only a useful and efficient transportation vehicle for business purposes (doctors, lawyers), but widely used as well by the new middle-classes for leisure and tourism travel. Cars acquired utilitarian functions and became affordable, but did not lose their initial images. On the contrary, the utilitarian and adventurous character of the car blended into the emerging car culture (*ibid.*). The resulting 'car culture' is visible in many aspects of contemporary social life. Drive-in restaurant and motels have become cultural icons of a mobile lifestyle which is also a typical American way of life (Peters, 2003, 2004). Travel myths (e.g. Route 66) and images of freedom have come to relate to car driving (*ibid.*). These images clearly had an impact on conceptions of social identity and images of the self. Especially

[28] Shove (1998) illustrates that in the Turkish context (28 cars per 1000 inhabitants) probably any car will be a status symbol, while in the Danish context (380 cars per 1000 inhabitants) it is the specific make and type which counts.

for the American adolescents of the 1960s freedom and adventure as connected to cars mixed perfectly with the possibility of travelling independently and maintaining social contacts and activities away from home. Finally, the deep embedding of cars in the culture of a society can also be read from the different meanings associated with cars in different countries (Timmer, 1998)[29]. Cars as material goods perform a series of functions in the culture of a society which sometimes relate to mobility only in an indirect way and which take the form of specific rituals (Douglas & Isherwood). An illustrative example is the meaning attached to attaining your driver's license: 'passing a driving test is a far more important social ritual today than exercising the vote for the first time!' (Wickham & Lohan, 1999, p. 3). Obtaining one's driver's license is regarded as a 'rite of passage', as the transformation from childhood to adulthood.

Resumé

Analysing (the history of) automobility from the perspective of different consumption theories provides information about the social-cultural meanings associated with the car and its uses and functions in society. Thinking about the values and images attached to the car in terms of individually experienced emotions and associations is widespread. Also the conception of cars as technological devices fulfilling basic human motives and needs tends to be prevalent. As an alternative perspective we put forward the suggestion to follow the social and cultural history of the car in a way as suggested by Appadurai (1986).This helps us explain why and how the symbolic meanings attached to current automobiles have been construed. By taking the changing structures of car related meanings and images into account, it becomes possible to deliver a satisfactory analysis of the diffusion of the car and its contemporary cultural functions. Images of the car as a material object related to freedom, adventure and especially speed, are reproduced over and over again through practices of car design and advertising, in car magazines, and through social talks[30]. As a result, Jensen (1999) was able to show that no less than 80% of Dutch car drivers feel that the car symbolizes freedom and independence. They even share this feeling with 50 to 60% of the cyclist and public transport users[31].

---

[29] See 'Car cultures' edited by Daniel Miller (2001) for various accounts of car consumption in different cultural contexts.

[30] In contrast to the car, cycling, which also originated from the function of racing, has largely lost its initial images of racing and adventure in the everyday design and use of bicycles. While the image of independence and flexibility has remained, symbolic and affective motives for the bicycle seem to be less important when compared to the automobile.

[31] Studies have also indicated that even public transport users value the car higher than public transport in most occasions. For example, in a recent study only one out of ten regular public transport users considered this modality type as the most pleasant (KiM, 2007). Though it might seem likely that this appraisal is based upon the differences in functional qualities of the modalities, Steg (2005) argues that the symbolic and affective functions play a more important role in explaining the level of car use than instrumental motives.

Meanings associated with automobiles however are not fixed but subject to changes[32]. Heffner *et al.* (2006) have argued that car attributes and meanings are neither (solely) individually construed, nor stable over time. They develop and change over time while meanings depend on the social and cultural context of the consumer, and on important actors influencing this cultural context (e.g. fashion and advertising systems, journalists, opinion leaders). It is impossible to talk about cars like the Toyota Prius or the Hummer H2 without taking the associated societal and personally attributed meaning structures into account. In line with Giddens, Heffner *et al.* (2006, p. 32) describe that consumers 'play an active role in determining what products mean, and in sharing those meanings with others. Thus, the process of using products to define our identities can lead to alterations in the cultural significance, or socially-shared meanings, of those products'. The relation between symbolic meanings of consumer goods and the environment has also been investigated by these authors. Their research has shown that at least some of the owners of hybrid electric vehicles (HEV) have purchased their vehicles because the car has a green image, is perceived as socially responsible and represents environmental stewardship (Heffner *et al.*, 2005). Interestingly, this research showed that the green image of the consumer good HEV has different connotations for various lifestyle groups of consumers[33]. The relation between symbolic meanings and environmentally friendly vehicles is further explored in Chapter 5.

3.3 From conspicuous consumption to social practices

The cultural turn in consumption theories supplanted the view of the consumer as a spineless victim of capitalist producers and advertising industries by the image of a consumer as being an active, self-reflective and knowledgeable agent. Especially the late-modernist approaches to consumption (e.g. Giddens, 1991; Lury, 1996) indicate that consumption choices have become the most important aspect in the reflexive constitution of self-identity. However, the heavy emphasis on symbolic meanings, desires and communicative values of consumer goods has recently been put to question: 'the sociology of consumption has concentrated unduly on the more spectacular and visual aspects of contemporary consumer behaviour, thereby constructing an unbalanced and partial account' (Gronow & Warde, 2001, pp. 3-4). This might to a large extent be related to the fact that the sociology of consumption has tended to concentrate on moments of acquisition, rather than on the ways of doing things in everyday life. Shove & Pantzar (2005, p. 44) indicate that 'while acquisition and ownership are undoubtedly important in signalling all manner of statuses and identities, it is also clear that many products are quite directly implicated in the conduct and reproduction of daily life'.

[32] For example, while SUVs for years in a row have been the fastest growing car market, especially in the United States, the rise in fuel prices and the increase in environmental concerns related to climate change have completely altered the images of SUVs. The SUVs original image, consisting of a curious symbiosis between off road adventure and the culmination of a luxurious lifestyle embodied by Hollywood's rich and famous, has become more and more infused with images of waste and uneconomic driving.

[33] Heffner *et al.* (2005) separated household into three different groups with varying meanings attached to the green image: the purchases of *natural greens* were motivated primarily by the vehicle's environmental benefits, the purchases of *money greens* were motivated mainly by the potential for cost savings, and the purchases of *light greens* were motivated by a combination of both.

These authors use the concept of 'ordinary consumption' (Gronow & Warde, 2001) and 'inconspicuous consumption' (Shove & Warde, 1998) to signify the role of the material (products, technologies) in the constitution of normal everyday life. In Table 3.1 a summary is given of the elements associated with ordinary consumption which, according to Gronow & Warde (2001), have been neglected and the elements associated with conspicuous consumption which have been overemphasized in the sociology of consumption. Ordinary consumption differs from extraordinary consumption in its focus on products and behaviour which are not highly visible or seen as anything special. More importantly, the focus on the role of routines in structuring the everyday activities of citizen-consumers has provided important opportunities to cross the divide between voluntaristic and systemic approaches to consumer behaviour, the central problem in this chapter. This point will be explored in the following sections.

3.3.1 The role of routines in consumption

The clear point made in the literature on ordinary consumption is that the major part of what is labelled 'resource consumption' by the environmental sciences is but the result of mundane, everyday life routines. This focus on everyday life routines is a response to the perspective on behaviour as a rational decision-making process which has been criticized both by psychologists and by sociologists, though from various points of departure[34]. Social psychologists have pointed towards the important role that must be attached to habits in processes of individual decision-making and behavioural change interventions (Aarts *et al.*, 1998; Verplanken, 2006; Verplanken & Wood, 2006). Verplanken (2006) describes that habits develop when behaviour is repeated in a stable context. Habits are associated with automated behaviour, fewer thoughts and fewer emotions as the behaviour is performed. Indeed, the function of habits is to decrease the complexity of human decision-making. If contextual circumstances are similar it makes sense to follow existing

Table 3.1. Focus in sociology of consumption according to Gronow & Warde (2001).

Too much focus on	**Too little focus on**
Extraordinary	Ordinary
Conspicuous	Inconspicuous
Conscious rational choice	Routine, repetitive behaviour
Purchase	Practical contexts of appropriation and use
Commodified	Other forms of exchange
Personal identity	Collective identification

[34] It is interesting to see that while both social psychologist and sociologist focus on the role of habits and routines in relation to (consumption) practices, these two bodies of literature exist independently without noteworthy cross-referencing.

cognitive constructs based on positive previous experiences thereby reducing the cognitive load. Habits exist because they make life easier and more predictable.

What then are the implications of habit formations for (sustainable) mobility behaviour? The influence of habit formation on behavioural change policies can be substantial. Based on their previous experiences consumers with strong habits develop expectations for certain outcomes in a specific decision context (Verplanken & Wood, 2006). This often leads to a tunnel vision where consumers expect prior experiences to repeat and therefore become blinder to minor contextual changes. Furthermore, this tunnel vision reveals itself in a more limited search for (information on) alternative behaviour. Finally, the information search is often biased in favour of the existing habit (*ibid.*). Research by Aarts *et al.* (1998) on habits and information use in travel mode choices indicates that when mobility behaviour becomes habitual, less information and reasoning is needed to perform the action. So, drivers with strong habits use significantly less information to make a travel choice. Therefore, the effectiveness of behavioural interventions such as information campaigns is severely limited by everyday life habits if the contextual condition and social structures maintaining the habits remain unaddressed (Verplanken & Woods, 2006). One could say that routinized behaviour leads to a blind spot for alternative modes of doing things.

Policy-making which takes car use habits into account focuses on various ways of breaking habits. Thøgersen & Møller (2008) report about an experiment which aimed to break car use habits by introducing a free one-month travel card for 1000 car commuters. The idea behind the experiment was to 'make drivers skip the habitual mode of decision-making to let them make choices more frequently based on reasoning and intentions' (*ibid.*, p. 332). The expectation was that a significant number of car drivers would continue using public transport after the promotion period had come to an end because unjustified negative expectations and a lack of knowledge about public transport use no longer played an important role. While the free one-month pass did increase public transport use during the promotion offer, which means it was able to effectively 'break' car use habits, when the normal public transport fare had to be paid these car drivers resumed to their old habits (Thøgersen and Møller, 2008). While this type of research and policy is interesting it remains focused on the individual decision processes of consumers without taking the context sufficiently into account. This experiment also shows that the assumption behind this approach, that habits are illogical and can be reformed by bringing consumers from practical consciousness into discursive consciousness, seems to be too simplistic. In Chapter 6 we will elaborate on this perspective when we describe three cases of mobility management in which the aim was to reconfigure commuting practices.

Now, we will shift the attention to the role of collective routines as proposed in Giddens structuration theory. While the social psychological perspective focuses on individual routines, Giddens and other theorists working in the tradition of practice theory (Bourdieu, 1984; Schatzki, 1996) also emphasize the function of routines on levels stretching beyond the individual.

The role of routines in behaviour has also received greater attention in some branches of the sociology of consumption. According to Ilmonen (2001) the newly developed interest in consumption routines, in contrast to reflexive agents making conscious consumption choices, results from the overriding emphasis on the importance of agency in the sociology of consumption. In social sciences Anthony Giddens was one of the first to ponder upon the role of (behavioural) routines in society. Routines play a central role in the theory of structuration as presented in the

Constitution of Society (Giddens, 1984) in particular. In his theory, Giddens brings together two (opposing) theoretical branches of sociology. He builds a bridge between functionalism (including systems theory) and structuralism on the one hand, and interpretative sociology and actor-centric (or voluntaristic) approaches on the other. In the Constitution of Society, Giddens considers routines as one of the fundamental elements of day-to-day social activity (Giddens, 1984, p. xxiii, p. 19): 'routine is integral both to the continuity of the personality of the agent, as he or she moves along the paths of daily activities, and to the institutions of society, which are such only through their continued reproduction' (*ibid.*, p. 60). The importance of routines for Giddens is rooted in the fact that they provide a form of safety, and a feeling of power which is vital for maintaining ontological security[35].

To explain differences in social activities Giddens makes a distinction between actions and routines on the one hand and discursive and practical consciousness on the other hand. Working on tacit knowledge, routines consist of repetitive day-to-day behaviour which is carried out primarily in the mode of 'practical consciousnesses'. Practical consciousness is to be distinguished from discursive consciousness. Working on discursive knowledge, actions consist of activities which are more intentionally chosen and build upon discursive consciousness (*ibid.*, pp. 5-14). Practical consciousness is defined as 'what actors know (believe) about social conditions, including especially the conditions of their own action, but cannot express discursively' (*ibid.*, p. 375). Giddens notes that the distinction between discursive and practical consciousness is not a rigid one. Indeed, knowledge of actors for Giddens refers both to tacit and to discursively available knowledge. The routinized activities of day-to-day life are 'made to happen' by human agents who are involved in a constant process of reflexive monitoring of their actions. People possess an on-going theoretical understanding of the grounds or reasons 'behind' their day-to-day activities without this knowledge being necessarily discursive in nature. Nevertheless, when asked to, most actors will usually be able to explain what they do and why they do so (Giddens, 1984, p. 6). This means that people are quite able to indicate the reasons behind their routinized activities, however, under normal circumstances these reasons are not discursively reflected upon.

There are some fundamental differences in the use of habits and routines between the social-psychological tradition and the sociological tradition as represented by Giddens. Whereas the social-psychological perspective on daily consumption routines fits more into the actor-centric approaches, with structures as inhibiting, or at the least stable, contexts, Giddens emphasizes that structures are always both constraining and enabling (*ibid.*, p. 25), nor are they external to individuals. In contrast to functionalist and structuralist approaches, Giddens conceives of structures not in terms of stable patterns of social relations but as properties of social systems. Structures are both the medium and the outcome of the social practices they recursively organise (*ibid.*, p. 25). This means that social structures, by enabling, constraining and streamlining social action, both make routinized social practices possible and are at the same time reproduced by

[35] Giddens describes how the destruction of existing routines, without allowing the build-up of new routines in Second World War concentration camps had a devastating effect on concentration camp prisoners. The deliberate and continuous attack upon the ordinary routines of life removed all the predictable frameworks and produced an immense degree of anxiety among the prisoners. The disruption from normality and the unpredictability turned some of these prisoners into 'walking corpses', no longer able to behave as human agents (Giddens, 1984, pp. 60-64).

these practices, a process labelled 'duality of structure in interaction' by Giddens[36]. The duality of structure indicates that any form of agency presupposes a form of structure (instead of being opposed to it) and vice versa. Routinized social practices can be considered as social systems with a specific pattern in time-space. When human agents perform everyday social practices they draw upon the virtual existing sets of rules and resources which make possible the routines. Routinized social practices occur in co-presence with other actors and they display a specific pattern of time-place. In that sense routinized social practices differ from social-psychological habits: where a habit is largely considered to be an individually repeated behaviour in a more or less stable context in which the individual relies upon a specific mental construct, a social practice is inherently not an individually but a collectively shared routine. As Giddens himself indicates (1979, p. 219) a social practice is close to the concept of Bourdieu's habitus in the sense that practices are habits shared (and recognized) by a group or community of actors.

This shift from individually to collectively shared routines is one of the fundamental characteristics of contemporary practice theory which is currently being developed (Shove, 2003a,b,c; Spaargaren, 2003; Warde, 2005). So, while an individual agent participates in routinized social practices, the practice is not purely the result of the activities of the individual at that moment in time. The structures which enable the performance of that specific practice are themselves the outcome of historically performed practices. It are these historically constructed rules, resources and knowledge which the individual agent builds upon when enacting the practice. The routinized character therefore applies both to the individual who carries out the social practice, in the sense that he or she has sufficient knowledge and skills to routinely perform the practice, and likewise the routinized character applies to a specific social practice itself in the sense that shared practices can be regarded as institutions reproduced by groups of actors in a similar way.

In discussions on consumer behaviour and the role of social-technical contexts, the two different scale-levels of routines and routinization are frequently intermingled. To clarify, a social practice such as cooking at home, to take an example from the domain of food consumption, is a social practice which is widely institutionalized in the Western world and involves more or less standardized activities and knowledge. Though there are countless variations of how cooking as a practice should be exactly executed it is clear that this practice involves certain objects and requires a specific knowledge of doings which take place in a specific setting in time and space. The cooking practice by way of routinization provides the continuity of social reproduction, in this case with respect to the ways in which we handle in designated rooms (kitchens) at specific moments in time (lunchtime, dinnertime). According to Giddens, routinization is thus not linked to whether or not one specific individual has formerly engaged with ways of cooking, but routinization is the 'habitual, taken-for-granted character of the vast bulk of activities of day-to-day social life' (Giddens, 1984, p. 376). Thus, the social practice is routinized and can mostly be seen as the taken for granted way of doing things. With Giddens routines bear close resemblance to conventions that exist and which are drawn upon as mutual knowledge by actors engaging

[36] The reproduction of social structures by human agents is most of the times not intentional, but is the result of unintended consequences of action. When speaking a language a person uses collectively shared linguistic rules. While the person's goal is limited to communication, the unintended consequence of speaking is that the linguistic rules are reproduced.

in the social practice. This is a different way of speaking about routines when compared to the social-psychological tradition of habits and routines, meaning whether or not an individual can make use of his personal past experiences and therefore can rely on the 'automatic nature of behavioural habits' (Verplanken, 2006, p. 653). An individual who very rarely prepares a family meal cannot rely upon his or her personal behavioural routine but nevertheless recognizes the existence of the social practice as a collectively shared routine: in that sense the norm is the routine (while the person's activities are not)[37]. Likewise it is possible to formulate individually routinized activities which are not collectively shared and recognizable. This is for example the case with consumers of niche-innovations (driving an electric vehicle, car-sharing) or consumers involved in novel lifestyles (living in a car-free neighbourhood) which require new mobility routines. While these consumers perform individual routine behaviour, in the sense that the activity is habitually undertaken, the activities themselves cannot considered to be collectively shared routines.

In sum, while both levels of routines (individual routines and collectively shared routines) are used in consumption theory, as is the case with the associated terminology of de- and reroutinization, they represent significantly different ideas of thought. The intermingled use and meaning of the word routine is also visible in the Table 3.1, which showed the eclectically grouped issues separating the sociology of ordinary consumption from conspicuous consumption, wherein the former is associated with routine and repetitive behaviour and the latter with conscious and rational choice[38].

This detailed account on the relevance of routines and routinization is important as it brings us to one of the core elements and discussion surrounding theories of practice[39]. In the next section we will elaborate on the state of the art in contemporary theories of practice which build upon the social practices as formulated by Giddens and Bourdieu, but are significantly different in their emphasis. In recent understandings of practice 'the actor is more viewed as a carrier of routines, than as an independent individual and this has importance for the understanding of how to make individuals change their routine' (Gram-Hanssen, 2006, p. 3). Another difference is the status of objects and technologies in theories of social practice, an aspect which has not been sufficiently developed in earlier theories of practice (Reckwitz, 2002a). Finally, an aspect which will be addressed is the question how practices develop and change. This section seemed to imply that routines are more or less stable. While routines are indeed subject to inertia and resistance to change, history shows that consumption routines change continuously. As Gronow & Warde (2001) rightfully state, most of the research on consumption routines is focused on established

[37] In line with Giddens one could say that this person nevertheless has some basic understanding of what this social practice should entail, however he or she simply has not the knowledge and skills necessary to be able to perform the practice.

[38] Giddens himself does not always make a clear distinction between the two levels of routines. In the introduction of the Constitution of Society he writes: 'The routine (whatever is done habitually) is a basic element of day-to-day activity' (Giddens, 1984, p. xxiii). Thereby Giddens contributes to the confusion whether or not routines are considered to be properties of the individual, of social practices or of both, without giving a clear indication of how the two are related to each other.

[39] The term 'social practice' is by some scholars replaced with the sole word 'practice'; as practices are inherently social by nature, the use of the term 'social practice' is by these scholars considered to be a tautology (see Reckwitz, 2002b).

activities and objects which have been in use for a long time, while there is less attention to processes of routinization of novelties. From the perspective of transitions in consumption, the key point is exactly how processes of routinization of novelties can take place. These three points will be addressed in Section 3.3.2.

3.3.2 Contemporary theories of practice

As the heterogeneous bundle of theories of practice betrays, there is not one common understanding of what practice theory is about. One could say that what binds theories of practice is that they place practices at the centre of the understanding of the social (Gram-Hanssen, 2007, p. 2). So, in line with Giddens, all theories of practice consider practices as the key unit of analysis to understanding social life, in contrast to individualistic approaches which place emphasis on individual actions. While Giddens and Bourdieu are much appreciated for their conceptualizations of social practices, contemporary scholars on theories of practice also bring other aspects to the forefront. One of the most important critiques with respect to the early formulations of practice theory is the lack of attention for the material dimension: 'these are thoroughly social theories in the sense that material artefacts, infrastructures and products feature barely at all' (Shove & Pantzar, 2005, p. 44). Influenced by Latour's work, the definition of practices by Reckwitz shows the integration of the material dimension in theories of practice: 'a practice is a routinized way in which bodies are moved, objects are handled, subjects are treated, things are described and the world is understood ... A practice is social, as it is a 'type' of behaving and understanding that appears at different locales and at different points of time and is carried out by different body/minds' (Reckwitz, 2002b, p. 250). What is important in this theory of practice is that a practice not only consists of a routinized type of behaviour, but that objects and materials (referred to as 'things' by Reckwitz) are important elements of a practice. Without the object, or without the necessary capabilities and know-how of how to handle the object, a practice cannot be undertaken. Or as Gram-Hanssen (2007, p. 7) has phrased it: 'things act as resources which can both enable and constrain practices'. A practice to Reckwitz thus consists of the interconnection of several different elements whose existence cannot be reduced to any one of the single elements. In line with the previous, Reckwitz sees individuals as carriers of a practice. Individuals can carry a wide variety of practices, without there being a need of coordination between these practices: 'the social world is first and foremost populated by diverse social practices which are carried by agents' (Reckwitz, 2002b, p. 256). However, as the definition by Reckwitz indicates, agents are not the only carriers of practice, the same applies for the objects as well. In this variant of practice theory, neither the individual nor the object should be analysed independently, only when the elements are investigated as co-existing elements of a practice can we understand their functioning. Only when individual human agents have learned to know and use an object and have the capabilities of usage can the agent become a practitioner.

What then is the relation between contemporary approaches of practices and consumption? With the definition of Reckwitz at hand it is easy to see that nearly all practices involve consumption of products and services in one way or the other. That is why the abovementioned writers on 'ordinary' consumption see the purchase and use of goods not as something which can simply be reduced to supply and demand, but as an elementary component which is necessary to perform a certain practice. Or put in line with Warde (2005) and Shove (2003a), it is the standard and convention of a certain practice which generates the level of consumption, rather than individual desires. So, goods and services are not important for their own sake, but for the practices they make possible (Shove, Watson & Ingram, 2005, p. 7). To many contemporary authors on theories of practices, consumption is nearly always a necessary part of practices but not a practice in itself. That is, the appropriation of products and services occurs within practices[40]. Indeed, in order to be able to perform a practice, a practitioner is required to appropriate objects and services. According to Warde, a competent practitioner has acquired the necessary services and objects, possesses the right competences and know-how to handle the tools and devotes sufficient commitment to the conduct of the practice (Warde, 2005, p. 145).

While by now it has become reasonably clear what practices are, little has been said about trajectories in practices. On the one hand practices have a certain robustness; in line with Giddens' theory of structuration, processes of routinization keep practices stable to a certain extent. By conducting the practice (appropriating and using the objects needed to conduct the practice, conforming to the conventions and rules of the practice, and by actively performing the practice) the outcome is that the practice is reproduced in time-space. However, these practices are also subject to constant change and simultaneously contain seeds of change (Warde, 2005). While Schatzki and Reckwitz predominantly theorize about the stability in practices, Elizabeth Shove on the other hand has conceptualized trajectories of change in practices. Central question in her work is the question how practices change, which trajectories can be discerned and what consequences these trajectories have for forms of consumption. Bringing together sociology of consumption and sociology of technology, Shove (2003a,b) starts from the idea that people do not consume resources, but pay for services like heating, lighting and cooling in order to sustain conditions and experiences of comfort, cleanliness and convenience. While these conventions now seem to be non-negotiable, these are by no means stable over time. Focusing on the collective redefinition of conventions in household practices, Shove's (2003a,b) central thesis is that what people take to be normal is completely malleable. Not only are these conventions to large extent social constructs, also the technologies deployed in the practices to acquire certain levels of comfort, convenience and cleanliness may vary (depending on the practitioner's own way of doing things and on the local context). By analysing different pathways in the co-evolution of technology and practice Shove describes how taken for granted conventions of comfort and cleanliness change and why they are becoming increasingly resource-intensive.

Because Shove can be considered as a key author on dynamics in practices and trajectories of practices we want to describe her approach in more detail. In Comfort, Cleanliness and Convenience, Shove (2003a, pp. 49-64) starts with an overview of the various forms of co-

[40] Whether the appropriation of products and services is a practice or not is open to debate. Even Warde, who does not consider consumption to be a practice in itself, sees 'shopping activity', indeed as a practice. This distinction could be important as in Chapter 4 the purchase of a new car is conceptualised indeed as a practice in its own right. However, more important than the question if the purchase of a car is indeed a practice or not, is the acknowledgement that the purchase of a car is also part of wider processes involving mobility practices (the function and meaning associated with the purchased car is also reflected in the practices of commuting, business travel, leisure travel, etc.). Secondly, it is important, when analysing the purchase of a new car as a practice, to indeed analyse it as a practice with specific routinized ways of understandings, know-how and structures of meaning.

evolution between three elements: (1) habits, practices and expectations of users and consumers, (2) symbolic and material qualities of socio-technical objects and (3) socio-technical systems and collective conventions (Figure 3.2). Shove indicates that accounts of co-evolution can be discussed by the processes of difference (de/revaluation) and coherence (de- and restabilization). Difference, focusing predominantly on the co-evolution between objects and practices, explains changes in terms of a cycle of evaluations. People distinguish themselves by acquiring new products and services. As these new objects spread through society they become normal, thereby changing collective meanings. This process clearly corresponds with the dynamics of conspicuous consumption described earlier in this chapter. The second process, coherence, describes co-evolution between systems and objects and how this indirectly influences practices. This process of change emphasizes consistency, or path-dependency, in the sense that dominant designs have more or less fixed socio-technical trajectories. The process of coherence also can be used to explain the co-evolution between objects and practices, not from the viewpoint of acquisition, but from the viewpoint of user-technology interactions. Technological innovations can script the practices of the users, or, alternatively, users may 'resist' scripting and by processes of domestication reinvent the use of an object.

Though the co-evolution of practices, objects and systems provides meaningful insight into the ways expectations and technologies change, Shove argues that it does not provide insight in the mechanisms of system level changes: 'more is needed if we are to understand how entire cultures and conventions of comfort are redefined and reproduced' (Shove, 2003a, p. 64). What makes the analysis interesting is that, instead of following the trajectory of socio-technical regimes (such as the transition from horse-based carriages to automobiles as described by Geels, 2005a), Shove investigates trajectories of conventions. Conventions of comfort, for example, are represented by the one-direction-only ratchet representing an escalation in comfort-related practices and technologies. With a case on indoor climate management Shove describes how the air-conditioning industry, by prescribing global standards of narrowly defined comfort conditions, is hard-wiring an air-conditioned way of life into people's everyday practices. Homes and offices are increasingly designed with global standards of comfort in mind (thus enabling global 'nine-to-five' working hours and the end of coping mechanisms such as the siesta).

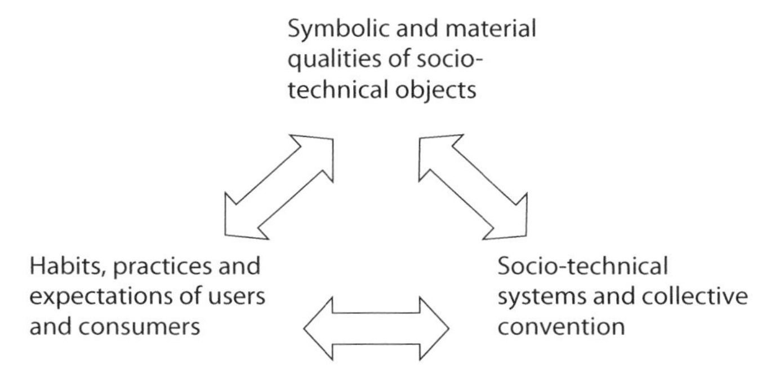

Figure 3.2. Three dimensions of co-evolutions (Shove, 2003a, p. 48).

More relevant for mobility practices, under the heading of convenience Shove also examines the relation between practices, time and technology. Convenience is related to devices and services which help people to shift or save time such as microwaves, freezers, washing machines, and automobiles (Shove, 2003b). Thus, these devices promise on the one hand to alleviate the constant pressure of time, and on the other hand provide the ability to construct and determine personal time schedules. The difference between public transport, whereby people need to adhere to collectively determined time schedules and routes, and private transport, whereby people travel when they choose and along personally determined routes, is a clear example of convenience in the domain of mobility. According to Shove these convenience devices have escalatory consequences for the fragmentation and coordination of time, leading to a 'collective societal drift towards a do-it-yourself mode of coordination' (Shove, 2003b, p. 414). As collective modes of coordination decline and normal practices are constantly redefined, the demand for convenience devices increases further leading to a corkscrew-like pathway from which there is no clear way out.

The lessons from Shove's work for sustainable consumption policies and research are clear and numerous. The dominant paradigm sees consumers as autonomous decision-makers, motivated by economic and psychological factors. An outcome of this paradigm is reflected in the questions stated in Tim Jackson's review of consumer behaviour in the report Motivating Sustainable Consumption: 'What factors shape and constrain our choices and actions? Why (and when) do people behave in pro-environmental or pro-social ways? And how can we encourage, motivate and facilitate more sustainable attitudes, behaviours and lifestyles' (Jackson, 2005, p. iii). The provocative thought within Shove's work is that policy-makers should shift their attention from the green possibilities by which existing expectations and conventions can be met (for instance by purchasing energy-efficient appliances), to what those expectations and conventions actually are and how they have come to require greater consumption of environmentally relevant resources. Shove makes the convincing argument that from an environmental point of view the eco-efficiency of one technology matters less than the pathway of the service which the technology sustains. Even more, policy tools designed to influence consumer choice are not only likely to be ineffective, they may even prove to be counterproductive as they legitimize, and take for granted, existing non-sustainable conventions. So, instead of focusing on green consumer choice the attention should be on the actual processes through which habits are constructed and on the relation between convention, technology and practice (Shove, 2003a, p. 199).

The focus on the relation between structures of meaning, everyday practices, artefacts and convention brings us to the heart of the matter. Also, the question behind the analysis, (how might whole constellations of practices, which make up daily life, be reconfigured on a large scale?) is of crucial importance when thinking about transitions in everyday consumption behaviour. Nevertheless, the work of Shove also contains some serious limitations. First, the analyses of comfort and convenience, where processes of standardization and escalation of convention and technology work as self-sustaining attractors, portray little agency to the role of consumers and users in everyday practices. Practitioners are clearly at the losing end of the spectrum, even though Shove acknowledges that people may have their own way of doing things. While practitioners may have their own context-specific ways of appropriating technologies and conventions, their actions seem to exert little influence on the direction of change at the regime level. Human agents are made to fit within dominant socio-technical systems and processes of escalatory resource-consumption. Second, the role of government, governance and policy in the structuration of practices, objects and systems is hardly discussed in the work of Shove. It is also unclear how, if at all, policies may actually influence the self-propelling trajectories of societal conventions. Though intricate in analysis, the described trajectories of complex constellations of practices provide us with few indications about which policy measures should actually be taken and by whom they should be taken up. Third, where relevant forms of policy interventions and appropriate measures of sustainable consumption are being discussed, this is done in a way that many questions remain. Shove suggests that policy makers should focus more on the temporal organization of society, for example by introducing technical systems which help restructure the use of time or by enforcing social measures such as fixed working hours and holidays which help reinstitute collective coordination (Shove, 2003a, p. 185). Furthermore, social and cultural diversity should be facilitated to stimulate multiple meanings in conventions and expectations. Without it actually being framed as such, and without the strong normative component which is usually attached to anti-consumerism, the policy interventions suggested (such as advocating the slow movement and reinstalling the siesta; Shove, 2003a, 2003c) seem to come close to suggestions put forward by the advocates of downshifting consumption. The important difference is that Shove does not imply an ideological shift away from consumerism, but aims to create diversity in interpretations of normality and collective understandings of how practices should be constituted.

In sum, by connecting practice theory with theories on system innovation and transitions, Shove has provided a world of knowledge which help us understand why and how practices are on the move and how new routines and practices emerge and old practices fade away. Shove has done much to conceptualize the internal dynamics of practices and the co-evolutionary relation between practices and socio-technical structures. What remains a bit unclear is the ways in which the impact of individual actions on collective practices, conventions and socio-technical systems should be conceptualized. How do individual differences in habits and routines relate to the specific way of using objects in practices and in the formation of new practices? As the internal structure of practices can be defined as the active integration of objects, meanings and skills (Reckwitz, 2002b; Shove & Pantzar, 2005), it is clear that practitioners show differences in levels of understandings, in access to objects and in levels of competences of handling the objects. As Warde indicates: 'social practices do not present uniform planes upon which agents participate in identical ways but are instead internally differentiated on many dimensions. Considered simply, from the point of view of the individual person, the performance of driving will depend on past experience, technical knowledge, learning, opportunities, available resources, previous encouragement by others' (Warde, 2005, p. 138). What needs to be conceptually refined are the (differences in) individual actions on the one hand and the structuration of collectively shared routines which takes shape in the form of social practices, on the other hand.

Firstly, this is related to the fact that practices are always context-specific in the sense that socio-cultural differences can play an important role in the way practices are conducted. That the global socio-technical regime of the automobile has a strong contextual component is easily understood when one, for instance, compares the car systems of Ghana and the Netherlands (see also Miller on car cultures, 2001). While in both countries people travel predominantly by car, there is a world of differences, not only in road rules, customs, meanings and infrastructures, but also in attitudes to technology and competences in handling technology (*ibid.*). On a national level the influence of social-cultural factors is revealed in the difference between the conduct of mobility practices between natives and immigrants in the Netherlands. The native Dutch population use bicycles as a transport mode two to three times as much as the Dutch immigrant population (constituted predominantly by Moroccan and Turkish Muslims), while the opposite is true for walking and public transport (Harms, 2008). In the Moroccan community the image, status and social acceptance of the bicycle is low when compared to the native Dutch who have an established and renowned 'bicycle culture' (Verbeek, 2007). Notwithstanding the important policy relevance, the point we want to make here is that the lack of a bicycle culture in the Moroccan community influences the competences and the access to materials needed to conduct mobility practices by bicycle. So, even though the Netherlands has one of the most elaborate and well-maintained bicycle infrastructure, many Moroccans lack the skills to use a bicycle, lack access to a bicycle and attribute negative meanings to its use (especially for women it is considered inappropriate to ride a bike)[41], all of which are elementary components to conduct a practice. I will return to this aspect in Chapters 5 to 7 when we discuss differences in mobility portfolios and motility, people's potential to be mobile.

Secondly, while contemporary theories of practice consider the social world primarily as a constellation of practices which are carried by practitioners, this does not limit the description and characterization of the consumption behaviour of an individual (Warde, 2005, p. 144). Without this consideration the development of a consumer-oriented policy, targeting specific groups, is hard to imagine. As practices are internally differentiated, with individuals differing in levels of understandings, skills and attributions of meaning, there are multiple ways of doing things in practices. There is a plurality of lifestyles in modern societies which is reflected in the way which practices human agents participate in, how practitioners connect the elements of practices together, and the degree of their involvement in the practice. Therefore, to understand variations in practices the integration of the individual action with collectively shared routines is important. Furthermore, while Shove & Pantzar (2005) rightfully conclude that new practices consist either of new constellations of existing elements or of new elements which combine with those that already exist, this does not explain who will likely be the creators or the carriers of the new practices.

The social practices approach, developed by Spaargaren (2003, 1997), is a specific type of practice theory which aims to take the abovementioned considerations into account when analysing practices. While this approach is certainly not opposed to contemporary theories of practice (as developed by Shove and Warde), the emphasis shifts from escalating standards of conventions and expectations to the active involvement of practitioners in the trajectories of change implied in the ecological modernisation of consumption and systems of provision. This means that the normative element, the active involvement of consumers and producers in sustainable transitions, in the analysis of practices becomes much more prominent. The possibilities of influencing social practices are explicitly present in this practice-approach. In the next paragraph I will elaborate on the social practices approach and Contrast-research program which is based on this perspective.

[41] In an attempt to overcome some of these barriers, recently a number of large Dutch municipalities have installed cycling lessons specifically designed for immigrant groups.

3.4 Contrast: a theoretical perspective on the analysis of transition-processes within consumption domains

Behind many of the theories discussed in this chapter runs one fundamental question: is the (changing) social world, the result of supra-individual systems and structures, or is the social world the result of competent and intentionally operating individuals? Theoretical perspectives having a voluntaristic, subjective and actor-oriented interpretation of social phenomena often stand in contrast to theoretical perspectives with a deterministic, objectivistic and systemic interpretation (Munsters *et al.*, 1993). To indicate, what is labelled as external social context by most contemporary psychologists is the sole object of investigation by theorists working in the field of systems approaches such as Large Technical Systems (LTS), and complex systems theory. This dualism is also represented in the debate between consumer-led and producer-led explanations of consumption growth and the debate between technological and behavioural solutions in remedying the environmental effects of consumption growth. De la Bruheze *et al.* (2004, p. 21) conclude that 'the mutual shaping of the processes of consumption and production still does not get sufficient exposure. Production and consumption are treated as isolated identities'.

While it is too simplistic to label studies of system innovations and transitions as purely systemic by nature, there is a recognized neglect for agency in transitions at least in the early formulations of the theory. In response to this criticism Geels & Schot (2007) note that agency is always present in the multi-level perspective though they agree that agency has not been strongly stylized in the case-studies and figures. Because the multi-level perspective is a global model which maps the entire long-term transition process it is difficult to incorporate them in aggregate patterns (*ibid.*). In this paragraph we want to stipulate how the analysis of human agency can be incorporated in transition studies without lapsing into the voluntaristic approaches which continue to dominate the sustainable consumption debate (Spaargaren, 2003). As mentioned in the introduction, the aim of the Contrast-project is to analyse and understand transition processes at the level of everyday practices. Importantly, this project is therefore not outside the tradition of transition studies, but aims to contribute to transition studies by conceptualizing the (possible) roles of citizen-consumers in transitions. In that respect the Contrast-project can be seen as an attempt to bridge the divide between actor-oriented approaches and system-oriented approaches with the aim of attaining policy-relevant knowledge to stimulate the active involvement of citizen-consumers in transition processes to sustainable development. If we want to attribute citizen-consumers their due place (that is on an equal level with producers, government, science and NGOs) as a possible change-agent in transition processes we need to think about how this can be conceptualized and facilitated. To be able to function as change-agents, citizen-consumers (with the aid of information, frames, resources and a sufficient levels of sustainable products and services) need to be invited and challenged with attractive visions and ideas about sustainability, while they are actively supported by governmental agencies and business to get involved in transition processes (Spaargaren *et al.*, 2007). This paragraph gives a summarized outline of the 'contrast-philosophy' and how it can be implemented as an analytical tool to study sustainable consumption transitions[42].

[42] Paragraph 3.4.2 is abstracted from Spaargaren *et al.* (2007a).

3.4.1 The ecological modernisation of consumption

While the social practices model closely resembles Giddens' structuration theory and attributes practices as the key unit of analysis, to understand the social practices model one has to know that its origin also stems from ecological modernisation theory. Ecological modernisation theory is an environmental social theory which analyses the transformation process of modern societies towards more ecologically sound societal forms (Mol, 1995). Central to ecological modernisation theory are processes of disembedding and re-embedding of ecological criteria and rationalities with regard to socio-economic criteria and dimensions of social systems. Environmental destruction is seen to result from a design fault of modern societies in which the pursuit of economic rationalism brought nature and the environment together within a relatively independent economic sphere. To remedy this design fault there is a need to institutionalize and anchor ecological criteria in the organization of production and consumption processes. This institutionalization can only be accomplished if an independent ecological rationality emerges which runs parallel too, and is disentangled from, economic rationality. The emancipation of an ecological rationality and an ecological sphere means that environmental factors have an independent and equal place, next to social and economic factors (Mol., 1995; Spaargaren, 1997). As the ecological rationale becomes analytically equal to the economic sphere, processes of production and consumption will be evaluated both from an ecological and an economic point of view. This means that independent environmental criteria are increasingly incorporated in production and consumption, a process labelled re-embedding of ecology in economic institutions. Processes of disembedding and re-embedding are comparable to the social considerations which have become widely accepted from the beginning of the twentieth century in Western societies (Mol, 1995). The incorporation of social conditions in production and consumption led to 'constrains' on the economy such as working conditions, minimum wages, working hours, and later on maternity leave. One of the central hypotheses of ecological modernisation theory is that, from the 1980s onwards, the design, performance and evaluation of processes of production and consumption are already increasingly based on ecological criteria, next to economic and social criteria (Mol, 1995, p. 58). The ecological restructuring of technological systems takes places via two mechanisms. First, the concept of 'ecologisation of the economy' implies that, enabled by the monitoring of material and energy flows, the development and diffusion of innovations, and a partial de-industrialisation of maladjusted systems, a more ecological rational input and output of the economic sphere should emerge (Mol, 1995). Second, the concept of 'economization of ecology' implies that, by placing monetary values on environmental flows, economic mechanisms are directed at protecting the environment[43].

The social practices model has been developed primarily to conceptualize the ecological modernisation of consumption, although the interdependencies with the transformation of

[43] See Chapter 5 for the implications of the concept of 'economization of the ecology' in the case of new car-purchasing

production are given recognition[44]. The development of an independent ecological rationality since the 1970s is not limited to that of the spheres of design and production but is also increasingly present in the everyday life consumption routines, norms and domestic practices of consumers. This ecological rationality is visible in the so-called first-generation environmental domestic practices such as waste prevention, waste separation, energy-efficient renovation, and because of the importance that is attached to national and global environmental issues (for this last point, see also Visser *et al.*, 2007). Because of its connection with ecological modernisation theory, the practice-approach developed by Spaargaren assumes that every citizen to a greater or lesser extent starts to take into account environmental aspects involved in the reproduction of everyday life routines. Similar to the social-psychological perspectives on consumption behaviour, the social practices model emphasizes the role of citizen-consumers as ecologically more or less committed actors which become increasingly involved in the process of ecological modernisation of production-consumption chains. Spaargaren argues that processes of environmental transformation in everyday life practices are already going on at least in OECD countries of the world (Spaargaren

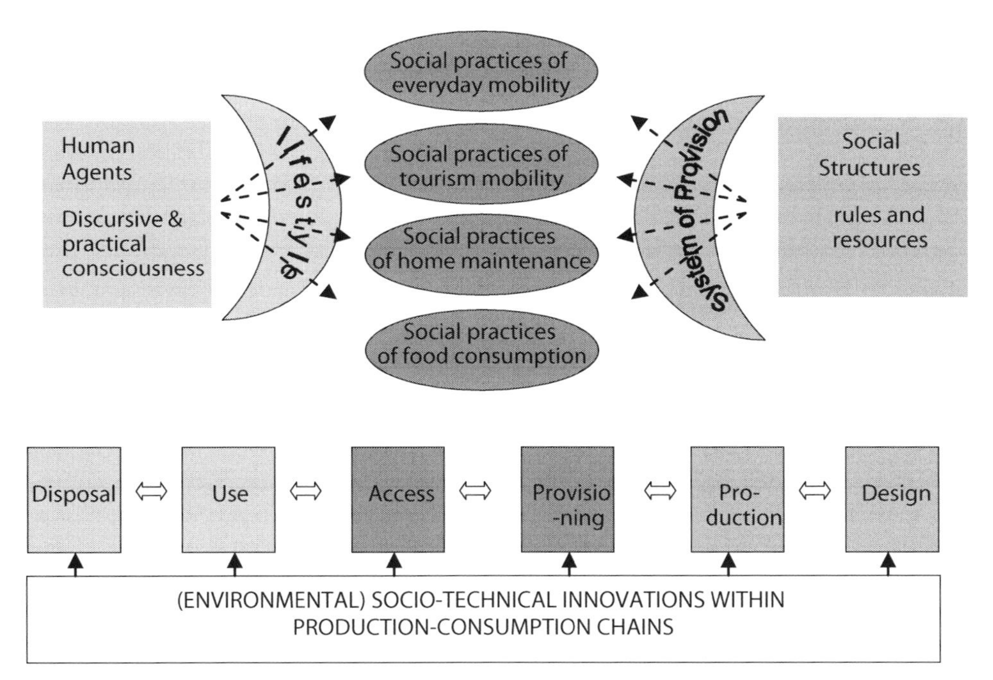

Figure 3.3. Outline of the social practices research model for studying the ecological modernisation of consumption.

[44] Similar to many other approaches in theories on environmental change, the earlier versions of ecological modernisation theory have been criticized because of its narrow focus on technology and overall producer-oriented approach to ecological restructuring. The social practices model, as a product of the second-generation of ecological modernisation theories (Spaargaren, 2006), incorporates the role of consumers in transforming production-consumption chains.

et al., 2002). However, the model differs from social-psychological approaches in the sense that environmental transformations of practices are analysed as resulting not only from ecologically motivated decisions of individuals but equally so forms of changes in which ecological criteria and rationalities become anchored and institutionalized in everyday life practices. These forms of institutionalization depend on practice-specific factors and dynamics which go beyond the individual since they belong to practices. Thus, in line with practice theories, the basic premise of the social practices approach is to analyse consumption behaviour as collectively shared practices which take place in specific time-space settings (Figure 3.3).

The social practices model combines the notion of human agents as knowledgeable, capable actors with an equal emphasis on the influence of the social and technological context on human behaviour (Spaargaren & Martens, 2005). To analyse and specify the agency in processes of transitions the sociological concept of lifestyle is of particular importance. As defined by Giddens (1991), the lifestyle of an individual is defined as the set of social practices which an individual is engaged in, together with the storytelling that goes along with it (Spaargaren, 2003). Individual agents (are increasingly enforced to) also develop storylines with respect to the environmental dimension of their lifestyles. In doing so, they rationalize and make legitimate the 'environmental and climate relevant' choices connected to their daily consumption routines (Spaargaren & Martens, 2005). To analyse and specify the role of 'context' with respect to practices, the concept of 'system of provision' is derived from Fine & Leopold (1993), who stress that different (groups of) commodities (as the objects being used in practices) are distinctly structured by the chains which unite a particular pattern of production with a particular pattern of consumption (Fine & Leopold, 1993, p. 4). This means that each consumption domain (mobility, housing, food consumption) has its own historically developed and distinctive system of provision that determines the level and composition of the objects of consumption and the meanings attached which come along with them (*ibid.*, p. 33).

The system of provision is comparable to the domain-specific socio-technical regime under which domain specific sets of practices are conducted. However, in line with theory of structuration, practitioners involved in practices are not 'outside' the systems of provision. By drawing upon the rules, objects, meanings and infrastructures connected to the systems of provision they are involved in the reproduction of the systems themselves (Spaargaren & Martens, 2005). Furthermore, the system of provision perspective is important because it makes clear that the possibilities for sustainable consumption to a large extent depend on the amount and kind of socio-technical innovations available in the domain specific systems of provision.

3.4.2 Towards a research agenda

To conclude this chapter, we will describe the general dynamics and variables which are researched in the Contrast-research program. The social practices approach – bringing together elements from different streams of thinking as discussed throughout this chapter – aims to provide a model to study the process of greening (ecological modernisation-theory) of social practices (practice theory) of consumption (consumption theory) The model emphasizes in particular the dynamics and factors which are at play on so called 'consumption junctions' as the connection points between consumer rationalities and producer-rationalities. The greening of practices is argued to refer to the introduction into practices of sustainable objects, technologies, ideas and procedures which have to become picked up by the practitioners involved after being made available by actors/experts operating in systems of provision. The term 'consumption junction' is derived from Schwartz-Cowan (1987) who introduced this concept to indicate the specific place and time at which the consumer makes a choice between competing technologies/ideas as being put forward. While Schwartz-Cowan used the concept to understand deliberately made consumer choices, in the Contrast-project the consumption junction is interpreted in terms of representing a particular (interface-like) practice. Consumption junctions are defined as all those places where provider-logics (companies, NGO's, government agencies, designers, etc.) meet the lifeworld-logics of citizen-consumers as the (potential) end-users of the new products and services being introduced: 'marketeers and product developers meet consumers who are not just there to reveal preferences but who are performing their daily routines and from that perspective also look for the products and services they normally use' (Spaargaren, 2006, p. 7).

In the research project five variables have been distinguished that are considered to be most relevant for the analysis of the dynamics at the consumption junction (Spaargaren *et al.*, 2007a):
1. general structural characteristics (structuring principles) of the consumption domain;
2. the stage of the transition process within a consumption domain;
3. quality and quantity of the available sustainable products and services in the social practices;
4. characteristics of the producers and suppliers: system and mode of provision;
5. the characteristics of citizen-consumers: lifestyles and modes of use.

1. General structural characteristics (structuring principles) of the consumption domain as represented by its systems of provision. The general structural characteristics resemble dynamics at the 'landscape level' as distinguished in the multi-level model. In terms of Giddens structuration theory (1984) these structuring principles represent the most abstract level of the structures (rules and resources) implied in the reproduction of social practices within the consumption domain. For these structural characteristics applies that they are involved in the highest level of institutionalization of (series or nexus of) practices. Important characteristics which are relevant for this research are the Western lifestyles and the role of material consumption in consumer cultures, the global use of the car and its related material infrastructures, individualisation, secularisation etc. Though these structural characteristics organise the most deeply-rooted systems and practices of our society, they are by no means fixed, nor is it so that policies aimed at transitions of social practices on the long term cannot lead to fundamental changes of these structural characteristics.

2. The stage of the transition process within the consumption domain. Over the last fifteen years the number of environmental socio-technical innovations which have become available to Western citizen-consumers has increased tremendously. However, in line with the system of provision perspective, there is an uneven development in the availability of these innovations. The level of ecological modernisation differs therefore considerably per

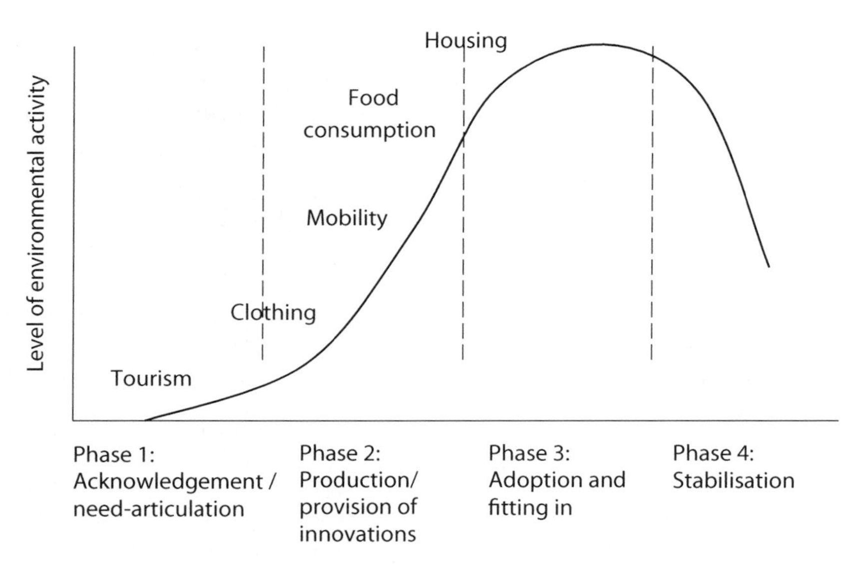

Figure 3.4. Consumption domains characterized by the phase in the process of greening (Martens & Spaargaren, 2006, p. 202).

consumption domain[45] (Figure 3.4). Moreover, the level of environmental activity generally indicates how institutionalized the environmental criteria and considerations are within a specific domain. Apart from the availability of innovations, there is an important role for the already existing sets of environmental products, services and ideas within the domain and the related social practices. For example, in the domain of home maintenance the first-generation environmental innovations have been widely integrated in Western societies. This means that most practitioners, to a greater or lesser account, are used to taking environmental aspects into consideration in domestic practices. It also means that there is more information available to monitor environmental impacts and that a higher amount of innovations have been made available. Thus, one could say that this criterion is an indication of the receptiveness of social practices in a specific domain for environmental change organised through socio-technical innovations.

3. Quality and quantity of the available sustainable products and services in the social practices. The quality of innovations refers to the question whether or not the novelties being made available can become easily integrated by citizen-consumers into the existing social practices/ routines. The quantity of environmental innovations is closely related to the abovementioned transition phase and to the (pro- or low) activity of the suppliers in specific consumption domains. For example, the quantity of sustainable clothes provided in retail outlets is still very modest when compared to the widespread availability of sustainable food products. The

[45] One can compare this with the uneven development of transition processes in different societal sectors as described by Rotmans (2007).

quality of the innovation refers to the question how the novel product compares to the usually employed materials in the practice and to the related conventions of the practice. For example, especially the earlier versions of energy-efficient automobiles had characteristics (e.g. a higher purchase price, lower comfort and convenience levels, and unexciting image) which formed important barriers for potential buyers which resulted in a very low adoption rate.

4. Characteristics of producers and suppliers: system and mode of provision. These characteristics refer to the specific (sustainability) strategies conducted by relevant actors in the system of provision. These strategies can relate to the promotional activities of the providers: information, pricing, constructing storylines (Spaargaren & Van Koppen, 2009). Providers on the one hand differ in the amount of environmental activities they have undertaken (represented by the greening of production processes and by the green product and services they have on offer) and on the other hand have different strategies in the communication with consumers about these activities. Some pro-active providers have (often in collaboration with other actors such as environmental NGOs and/or retailers) explicitly made consumer-oriented environmental information and specific sustainability story-lines available for consumer to facilitate processes of appropriation at the consumption junction. Other providers, for a number of reasons, keep this information internal even when they have environmentally alternatives on offer. In purchase practices the role of the salesmen is important as they can either promote or downplay environmental characteristics of a product or service. Next to advertising and in-store presentation of environmental products the internal training of the staff members is an important element of the system of provision at the consumption junction.

5. Characteristics of the citizen-consumer: lifestyles and modes of use. The characteristics of the consumer are important to understand the diversity within a social practice (why do some consumers purchase and use sustainable innovations in a specific practice while others reject these innovations?), and to understand how on the individual level different social practices are integrated (why does the same individual perform sustainable behaviour in practice A and unsustainable behaviour in practice B?). We already discussed that the concept of the lifestyle refers to the sum of all the social practices in which the individual participates plus the 'storytelling' that goes along with it. In addition, the lifestyle is both individually and collectively constructed: on the one hand the lifestyle is an individual aspect because each person has its own unique ideas, beliefs, competences and identity, on the other hand the lifestyle is a collective aspect because social practices are always collectively shared routines which have a common storyline and belong to a specific lifestyle segment. The lifestyle therefore consists partly of a general lifestyle element or dimension (this is the part analysed in motivational and trait psychology, and studies in market segmentation, for example in typologies of car purchasers) and a practice-specific lifestyle element or dimension which is directly related to a social practice in one of the consumption domains (Spaargaren & Oosterveer, 2010). More specific for sustainable consumption, examples of general characteristics of the lifestyle can be related to a high or low environmental concern, or a preference for a specific strategy on sustainable development, the so-called world visions (see MNP 2007b; Spaargaren *et al.*, 2007a). The practice-specific part of the lifestyle refers to specific values and beliefs related

to a practice, for example, consumer preferences for cars, symbolic meanings associated and valued by a consumer. Furthermore, in contrast to most conceptualisations of the lifestyle (which focus primarily on the 'storytelling'), the definition of the lifestyle by Giddens implies that it is not limited to attitudes and values but also includes both general and practice-specific knowledge and skills. Whether or not an innovation becomes integrated in a specific social practice therefore not only depends on the characteristics of the innovation itself, but also on the historically acquired experience, knowledge and skills, as well as the framing by and evaluation of the individual practitioner. For example, we already indicated how riding a bicycle for some people is an everyday experience which is conducted without much reflexive consciousness, while for the inexperienced rider it forms a complex ensemble of skills, knowledge about traffic rules (both formal and informal), attributed values, and technology (see also Peters, 1999). So, in order to be able to become a practitioner one needs to have the necessary skills and materials required for the social practice. The separation of the lifestyle into a general and a practice-specific part implies that individual practitioners may have a widely varying distribution and integration of green norms, knowledge, skills and priorities over the sets of social practices in which they are involved (in chapter seven both parts of the lifestyle are investigated for the domain of everyday mobility via a large consumer survey). This explains why a fictitious individual on the one hand purchases exclusively sustainable food products while on the other hand uses the car instead of the bicycle to do the grocery shopping.

With this last paragraph I have mapped, though densely summarized, the research perspective as it can be outlined on the basis of the social practices approach. Research based on the social practices model is focused on three core dynamics within production-consumption chains: dynamics at the demand side of environmental products and services (modes of use), dynamics at the supply side of environmental products and services (modes of provision), and dynamics of the social practices in which both come together (the consumption junction) (Spaargaren *et al.*, 2007a).

3.5 Resumé

In this chapter we have stipulated the various building blocks that together make up the theoretical perspective of this study as part of a research program which focuses on the active integration of citizen-consumers in transitions to sustainable development. However, the attentive reader might point out that we have come to a rather awkward conclusion: we started this chapter by claiming that analysing mobility from the viewpoint of consumption provides important benefits because mobility-related problems to a large extent are intrinsically related to consumer culture and consumer behaviour, while we end this chapter by claiming that consumer behaviour is best understood when it is analysed in the (practice-specific) context of a system of provision which unites a particular pattern of production with a particular pattern of consumption. Furthermore, in discussing theories of consumption we started out with theories of conspicuous consumption, emphasizing the various ways in which consumers consciously and reflexively construct meaning from the dialogue between products and images of the self in various social settings, while we end this chapter with an account of practice theories which stress the role of ordinary consumption and routinization in the structuration of everyday life. For understanding this seemingly disparate account it is important to know that on the one hand all consumption is cultural, and simultaneously on the other hand meanings (of objects and activities) are best understood and analysed at the level of social practices. To indicate, Don Slater gives an interesting analysis of the relation between the function and meaning of things in consumer culture. Slater emphasizes that culture does not influence consumption (in the sense that it is an addition to consumption) but that culture constitutes the needs, objects and practices that make up consumption (Slater, 1997, p. 133). He describes the attempt by Baran & Sweezy (1968 in Slater, 1997) to demystify consumer culture by removing cultural meanings from objects, in this specific case the automobile. They compare contemporary automobile production with the cost of producing a 'basic' (read: functional) car, one that is not conceived as an object requiring meaning, but one that is rationally designed in terms of how best to carry out its intended function (*ibid.*, p. 136). Slater criticizes this attempt by arguing that all objects are culturally meaningful and rightfully claims that functions themselves are culturally defined. The distinction between the utilitarian and signification function of practice is a meaningless and non-existing difference. The central problem, according to Slater, in consumer culture is to maintain the position that all consumption is cultural (to avoid fixed needs) without culture to become divorced from social practice (in which culture becomes an abstract phenomenon). Contemporary practice theory can be seen as a good attempt to make this insight work and to work with this insight. To understand the dynamics of social practices it is important to consider that meanings and conventions become normal and taken-for-granted by citizen-consumers when used in the reproduction of social practices in a recurrent way (Spaargaren, 2006). As Shove has shown, there is a strong relation between the reconfiguration of meaning and collective conventions, and the transformation of social practices. When Slater argues that all consumption is cultural because the meanings involved are necessarily shared meanings (Slater, 1997, p. 132), this makes perfect sense from a practice-point of view. Whether or not a person's specific action is in concurrence with the existing norms of the social practice, however, is a different since empirical matter. The same applies for the question whether or not or to what extent citizen-consumers, in line with an emerging ecological rationality in consumer culture, purchase and use products and services to construct parts of their emerging green identity and storylines, as Spaargaren (2006, p. 23) claims.

For now it suffices to say that mobility is analysed as a form of consumption, but with its own domain-specific characteristics and systems of provision. In chapter five the social practices model, as defined in Paragraph 3.4, will be conceptualised for the domain of everyday mobility in more detail. In the next chapter we will first portray one of the three empirical case studies by discussing the ecological modernisation of the practice of new car purchasing.

Chapter 4. Consumer-oriented strategies in the practice of new car purchasing

> *The key is to figure out a way that meets consumer needs instead of just something that would be a 'win' for society.*
>
> Ken Stewart, brand manager of advanced-technology vehicles, GM[46]

4.1 Introduction

One of the transition pathways towards sustainable mobility involves the greening of consumer cars (Kemp & Rotmans, 2004; Nykvist & Whitmarsh, 2007; Urry, 2004). In recent years the attention directed to cars and fuels with low environmental impact has increased rapidly, pointing out that sustainable automobility has become a key topic in governance debates and in business concerns. Even environmental NGOs, traditionally strongly opposed to any form of automobility, have started to promote green cars as a sustainable and feasible option to deal with environmental problems related to mobility. Furthermore, while the fuel consumption of the Dutch car-fleet had shown very little improvement until 2008, the recent acceleration in the sales of energy efficient vehicles, such as hybrid electric vehicles and fuel efficient conventional petrol and diesel cars, seems to point towards a considerable change taking place in the automotive market and consumer car preferences.

To understand the change processes taking place in the greening of consumer cars it is both interesting and necessary to take an in-depth look at the dynamic relation between production and consumption in the automotive sector. This relationship is explored in this chapter by investigating how the social practice of new car-purchasing is construed. Given the domination of the system of automobility, and considering the massive interest in environmental socio-technical innovations in the automotive industry, the limited amount of scientific research on the role of car purchasing in sustainable mobility transitions is surprising. As holds true for most consumption-production chains, the vast majority of policy and research attention has gone to the greening of automotive production processes and the development of sustainable cars and fuels. These business practices can be seen as being part of the broader phenomenon of ecological modernisation of industries in the last quarter of the twentieth century (Mol, 1995; Orsato & Clegg, 2005). Relatively little attention with regard to green cars has been given to consumer aspects such as consumption routines, lifestyles, and social-cultural values. Furthermore, research on environmental dimensions within car-purchasing decisions tends to focus on individual attitudes while the purchase of products and services entails much more than the individual decision of a consumer; consumption takes place within a specific setting which structures and influences these decisions.

[46] Quoted in American Demographics, January 2001 by Dale Russ.

Without doubt, the practice of car-purchasing forms an essential junction between the provision of environmental socio-technical innovations and the access to these innovations by consumers. As described at the end of chapter three, the social practices model emphasises to analyse precisely these connection points between consumers and producers by investigating the various consumption junctions, the specific time and place at which consumers make a choice between competing technologies and are confronted with strategies of providers.

Next to the attention given to the practice of new car purchasing, the empirical analysis focuses specifically on the question how environmental information is provided and used during this purchase practice. In daily life, consumers are more and more 'confronted' with the environmental dimensions of their consumption choices. Furthermore, research on environmental policy points to a change in policy instruments being used, emphasising that economic instruments, voluntary agreements and the provision of environmental information, most commonly through labels, are becoming more important (Van den Burg, 2006). These changes point to new forms of governance that seek to enrol the consumer in the environmental reform of production and consumption chains. As labelling makes more information available to consumers at the crucial places where the consumption end meets the production end of the chain, consumers can play a greater role in co-governing environmental performance through their consumption.

The provision of environmental information has therefore received much attention from policy makers which have developed informational strategies in order to influence consumer's car purchase decisions. Most well-known is the European Union's Labelling Directive (1999/94/EC) which obliges car manufacturers to provide information about the energy efficiency of a car. The focal point in this chapter is on these different environmental information tools and the supporting taxation schemes that were implemented in the Netherlands from 2001 onwards as new consumer-oriented strategies. More recently also non-state actors such as NGOs and car manufacturers have started outlining the environmental performance of cars. This means that the provision of environmental information stems from many different sources and is available in many different forms and formats[47]. Car advertisements, conversations in the showroom, and energy labels on websites and in showrooms all are forms in which information about the environmental effects of cars are communicated. This trend poses interesting questions about the significance of environmental information in the greening of consumer cars. Is the increase of environmental monitoring and monetarisation an indication of a strengthening environmental rationality in the automotive production-consumption chains? In what way do these consumer-oriented strategies (positively) influence the pathway towards the greening of consumer cars; and, under which conditions are these informational strategies successful? To answer these questions, in this chapter the role of different informational and fiscal strategies in the ecological modernisation of the practice of new car purchasing is investigated.

[47] Environmental information is defined by the EU Directive 2003/4/EC as written, visual, aural, electronic and other material forms of information: (1) on the state of the environment; (2) the factors, emissions and withdrawals influencing the state of the environment; (3) environmental measures and policies, (4) reports on the implementation, cost-benefit and other economic analysis; and (5) the state of human health and safety (see Mol, 2005).

In the next paragraph, first, we portray the social practice of car purchasing and the relevant trends influencing this practice. Four different (ideal-typical) routes will be described to picture how consumers may acquire a new or a second-hand car. Then, in Paragraph 4.3, the different formats of environmental information provision at the varying consumption junctions will be described: (1) environmental information tools; (2) face to face information; (3) advertising and in-store presentation. Here we pay specific attention to the implementation and development of the environmental information tools such as European fuel efficiency label. Next, the attention shifts to environmental monetarisation; the different environmental taxations and subsidies which are connected to the environmental performance of cars are described.

To assess the use and impact of consumer-oriented strategies in new car purchasing the empirical analysis takes place on two different levels of scale. First, based upon a focus group research with car salesmen and car purchasers we examine how different actors assess the environmental information tools and taxation schemes that aim to influence consumer car purchase decisions. By zooming in on the showroom we examine how the two worlds of providers on the one hand and consumers on the other hand meet at this consumption junction. This research, conducted in 2006 en 2007 in collaboration with Toyota Netherlands, provides detailed information about the dynamics in these contact points between car consumers and car producers. Though the situation in the automotive sector has changed considerably in the last few years, the analysis provides an interesting insight in the question if and how consumers put available environmental information in automotive production-consumption chains to use. In addition, it provides an understanding of how the provision of environmental information has changed since 2006.

Second, the in-depth analysis at the consumption junction of car purchasing is complemented by an overall analysis of the Dutch automotive market. The impact of the environmental information tools and taxation schemes is analysed predominantly by the sales of hybrid cars and developments in market shares of A-labelled and B-labelled cars from 2001 to 2010. In the conclusions and discussions, we revert back to the original research questions posted in this introduction.

4.2 The practice of car purchasing

4.2.1 Acquiring a car: a typology of the system of provision

The Netherlands has a total car fleet of seven and a half million cars. Yearly, about 2.3 million cars change from one owner to the next (Table 4.1, 4.2 and 4.3). Ideal-typically four different routes can be distinguished to show how consumers acquire a new or used car (Figure 4.1). Though these four routes could be extended with more types, the purpose of depicting these routes is that they will be used to indicate that each of these routes is structured by a specific system of provision which influences the environmental information provision.

The tables and figures show that, not surprisingly, the far majority of passenger cars is sold on the second-hand car market. Roughly 80% of the consumers who purchase a car acquire a used car. Most of the used cars are sold via independent car companies (3,400 registered) and official

Table 4.1. The Dutch car fleet in 2007 (RDC).

Owners:	**Absolute**	**Relative**
Private	6,351,865	83.60%
Lease company	556,334	7.30%
Car company	343,076	4.50%
SME	232,126	3.10%
Fleet owner	72,133	0.90%
Rental	36,823	0.50%
Total	7,592,357	100%

Table 4.2. New car sales in the Netherlands in 2007 (RDC).

Sold from official car dealers to:	**Absolute**	**Relative**
Private	215,142	42.70%
Lease company	160,571	31.80%
Car company	46,743	9.30%
SME	37,071	7.40%
Rental	30,187	6.00%
Fleet owner	11,855	2.40%
Total	501,569	100%

Table 4.3. Second hand car sales in the Netherlands in 2007 (Bovag).

Sold from:	**Absolute**	**Relative**
Private owners	841,960	44.41%
Independent car company ('non-dealer')	693,643	36.59%
Official car dealer	360,246	19.00%
Total	1,895,849	100%
In stock at car companies and official dealers	352,944	

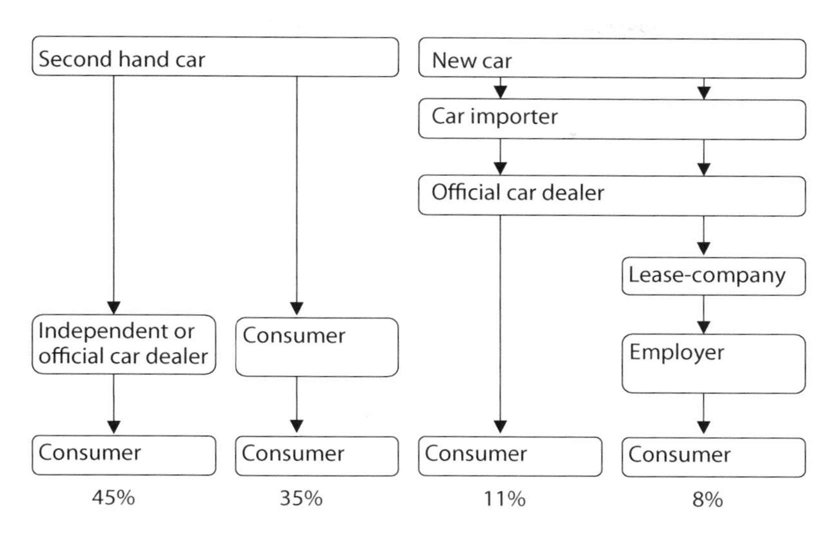

Figure 4.1. Ideal-typical routes for acquiring a car (percentages indicate approximate shares of total car sales).

car dealers (3,000 registered)[48]. The internet is by far the most used information source for finding a second-hand car. Typically, a consumer will search on a car vending site, and will also visit a local car dealer or independent car company for a test-drive. The rise in car vending sites has also resulted in an increase of consumer to consumer sales. These car sites have made it much easier for private consumer to sell and buy cars without third-party interference. Currently, one third of all cars in the Netherlands are sold directly from one private owner to another. This percentage, however, also includes cars that change from one owner to the next without complicated sale processes (inheritance, gifts or sales between family members and friends).

Approximately 20% of the cars sold are new cars. In the Netherlands by far the dominant route through which these new cars are sold is via 'official car dealers'. These dealers are highly structured by the corresponding car brands and as such differ from the independent car companies (which sell cars from varying car producers). Most automotive distribution models are based on the franchising model in which strict agreements between the car manufacturer and the local car dealers are in order. At the national level the territory of a dealer is determined. The local car dealers also receive training and support at the national level.

The average age of new car purchasers is higher when compared to used cars; the average age of new car purchasers is around 53 years old. This is also the result of the higher purchase price of new cars. Furthermore, the showroom has a much more prominent place in this sales route as

[48] Official car dealers are brand specific dealers who, as a general rule, are linked to one single manufacturer. Independent car companies sell cars from different car brands (in the Netherlands these are almost always used cars).

practically every consumer who purchases a new car visits at least one showroom.[49] Also, more than with second hand cars, up-to-date information about the car is provided to consumers in the showroom (predominantly brand oriented in-store marketing and leaflets).

The final route is the route via leasing companies. Leasing has a very prominent and specific place in the Netherlands. 32% of all new cars sold are sold to leasing companies. As such they have a significant influence on the automotive production-consumption chain. Lease cars are used primarily for business travel and commuting. Though lease cars are not necessarily new cars for the consumers who acquire the car, almost always they are bought new by the leasing company (the average car age of the lease fleet in 2006 was 22 months old). Leasing comes in different forms, with operational lease being the largest lease formula by far (74%). In operational lease the leasing company remains the owner of the car and the lessee, which is often a company from the private sector, receives the car for temporary use. It is common that the fleet owner of the employer makes available a limited number of cars from which the employee can make a choice. As such, the employee's car choice is structured more by the provision of the fleet owners and lease companies than the individual choice of employee.

In short, the descriptions indicate that the car purchasing practice is structured differently in each of the four routes. In each of the ideal type a different system of provision is present. The relevance of this typology is that the difference in the routes is also reflected in the search routines of the consumers, and in the consumption junctions between consumers and producers. This also has implications for the question how environmental information enters the automotive production-consumption chains. As the access points and the consumption junctions in the automotive chains are different, so is the environmental information provision. However, before the environmental information provision at these different consumption junctions will described, first we turn towards the purchasing practice.

4.2.2 Consumer's car purchasing process

In economics and marketing literature the purchase of a car is defined as a form of 'complex buying behaviour' (Reed *et al.*, 2004)[50]. Given that the car is a high-involvement product, the car-buying process is seen as a high-involvement process, leading to active search and use of information, deliberate evaluation of alternatives and a careful choice. Clearly the purchase of a car falls into the category of conspicuous consumption as discussed in the previous chapter. More specifically, the consumer information search itself usually includes both 'internal search' (retrieval of information based on previous searches and personal experiences) and 'external search' (accessing of different types of information sources) (Klein & Ford, 2003). Furthermore, research has highlighted that the car purchase can be seen as a two-stage process; in the first stage the vehicle class is decided

[49] In recent years car purchase through internet has received more and more attention. It is quite likely that in the near future more consumers will purchase a new vehicle through the internet. However, it is unlikely that many consumers will purchase a car without ever test-driving the vehicle.

[50] Complex buying behaviour means that consumers are highly involved in the purchase; the product itself is expensive, is bought infrequently, is perceived to be risky and is highly self-expressive.

on, based on costs and car capabilities, whereas in the second stage consumers undertake a more profound review of vehicles (Lane, 2005; Teisl *et al.*, 2007).

In the last two decades the automotive market has gone through a significant change in the way information is made available to consumers. The arrival of the 'information age' had an enormous influence on the purchasing process of complex products, including car purchasing. Traditionally, automotive dealers were seen as the dominant source of information, resulting in a situation of consumer-salesperson interfaces in which the salesperson 'led' the customer through the buying process (Reed *et al.*, 2004). Marketing research conducted in Germany by TNS Emnid (2004) showed a number of interesting developments in the automotive sector. The practice of information seeking has shifted in such a way that the majority of people purchasing a car make use of the internet as a source of information, making the internet one of the most dominant information sources.

Interestingly enough, another related development that can be witnessed is the decrease in customer ties, meaning that formerly fixed customer-supplier relations have become more and more fluid (CapGemini, 2010). Though important brand differences remain, in general, emotional attachment to a specific brand has lessened as a consequence of increased similarity and decreased quality differences in automobiles. The result is that in the past decade a strongly increased consumer empowerment has taken place. Car vending site Edmunds has summarised this development with its slogan: 'negotiate like a pro'[51]. Not only have car salesmen noticed that consumers enter the showroom armed with background information about the automotive sector, but the process itself has also changed; instead of visiting ten different showrooms, most consumers make a pre-selection of approximately three car types that they investigate intensively. As a consequence, the role of the salesperson likewise has shifted from leading to guiding, and from salesperson to advisor.

There is also a general trend in consumption towards more comfort and convenience. This trend, together with higher safety requirements, has led to an increased demand in size and luxury levels of cars (Van den Brink & Van Wee, 2001). As Shove (2003a) indicates, what starts out as an extra capacity or luxury can soon become normality, thus shifting consumer preferences and expectancies for automobile characteristics. As a consequence, the price of an average new car int Netherlands has risen from € 3,389 in 1970 to € 25,742 for petrol cars and € 27,396 for diesel cars in 2009. More importantly, the fuel consumption has not seen any substantial decrease in the past 25 years (Figure 4.2). The average decrease since 2001 has been around 1% a year (PBL, 2009). Only since 2010 fuel consumption has decreased more rapidly.

When looking at stated preferences most studies observe that environmental factors do not seem to play a major role in consumer car choices (Table 4.4). So, even though consumers mention sustainability issues as a major consumer concern (NIDO, 2002), the infamous attitude-action gap reveals that consumers' concern for environmental impact does not often translate into behavioural change (Lane, 2005). Furthermore, even though fuel consumption is mentioned as an important

[51] An important note here is that there might be a significant difference between second-hand car purchasers and new car purchasers. Furthermore, Lambert-Pandraud *et al.* (2005) have pointed out that older consumers, who constitute an important market segment, repurchase a brand more frequently when they buy a new car.

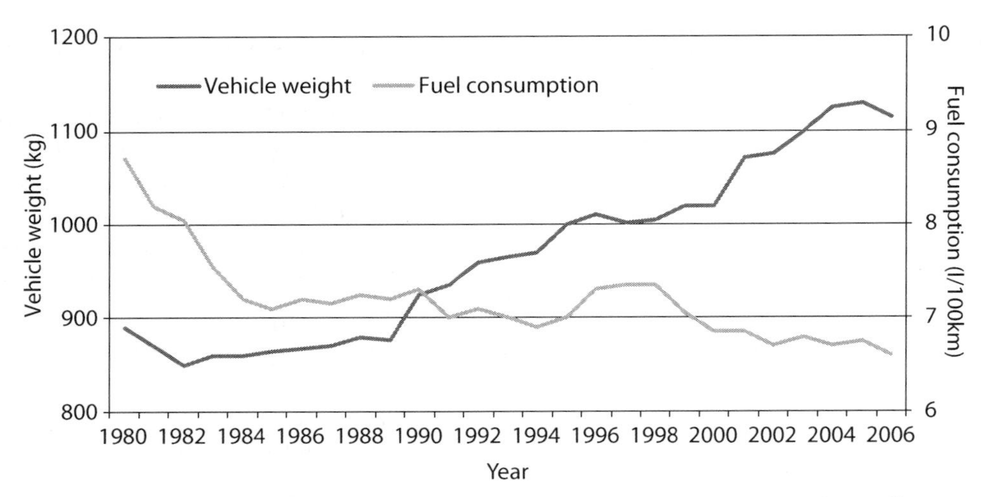

Figure 4.2. Average car fuel consumption and average car weight of the top 50 best-selling petrol cars (Bovag).

Table 4.4. Factors mentioned as being important in the purchasing decision (Lane, 2005).

Most important (10-30%)	**(5-10%)**	**Least important (<5%)**
Price	Performance and power	Depreciation
Fuel consumption	Image and style	Personal experience
Size and practicality	Brand name	Sales package
Reliability	Insurance costs	Dealership
Comfort	Engine size	Environment
Safety	Equipment levels	Vehicle emissions
Running costs		Road tax
Style and appearance		Recommendation
		Alternative fuel

factor, for most car buyers little effort is expended in comparisons of fuel consumption during the decision-making process (Boardman *et al.*, 2000; Lane, 2005).

One conclusion could be that people tend to be more concerned about status value and less about environmental performance then people would admit like to admit to themselves (Johansson-Stenman & Martinsson, 2006). This would fit well into the often-heard claim made by policy-makers and car producers that 'consumers are just not interested in environmentally friendly cars'. However, desk research by Lane (2005) sheds another light on this paradox. First of all, many buyers assume there are no major differences in fuel efficiency within the same vehicle class. By buying a new car, consumers automatically assume that it has good energy efficiency and is in compliance with strict environmental norms. It is also still widely believed that an environmental choice involves a certain sacrifice: in comfort, in performance or financially. Furthermore, because consumers' knowledge is incomplete, the environmental effects of car use are often confusing and complex for consumers. The relationship between fuel efficiency, CO_2-emissions and climate change is only very generally understood. Finally, the differences between local and global emissions are often mixed up (*ibid.*).

Different environmental information tools and taxation schemes have been developed over the years in order to help consumers take environmental aspects into account when purchasing a car. Indeed, these strategies were specifically designed to tackle exactly the above-mentioned problems and misconceptions. Before we investigate how these environmental tools have worked out in the practice of car purchasing, first the various types of information provision will be described.

4.3 Formats of environmental information provision in car purchasing

In this paragraph different formats of environmental information provision will be described. This means that not the subject of information (clean fuels or engines, emissions, waste), but the ways and methods of communicating about these subjects are placed in front. The three methods of communication which will be discussed here are: (1) environmental information tools, with a specific attention to the energy efficiency label; (2) face-to-face communication; and (3) general car marketing tools in which environmental aspects are emphasized (with specific attention to advertisements and in-store marketing). In the research most attention was directed to the development and analysis of environmental information tools, predominantly as these are the most commonly used methods of transferring environmental information.

4.3.1 Environmental information tools

In the introduction it was described that different informational strategies have been developed by different state and non-state actors with the aim to function as a consumer-policy tool. The development of three of these environmental information tools which have been implemented in the Netherlands will be described here. Most of the attention will go to the energy efficiency label as this label has been in effect for the longest time and has also been accompanied with varying financial incentives (Paragraph 4.3.4).

The first plans of the European Commission to formulate an overall fuel efficiency regulation for European cars were initiated in 1996. Next to a provider-oriented strategy – at that time based on voluntary agreements on emission targets with car industries – a consumer-oriented labelling directive was developed with the aim to directly target consumer's decision-making. Because no agreement was established on the exact format of the label, consensus was only formed on the minimum requirements: the energy efficiency label should at least contain the official fuel consumption and CO_2-emissions. Furthermore, the label should be present on or next to the car in the showroom, and the official fuel consumption and CO_2-emissions figures should be noted in car advertisements. The exact shape of the label was to be decided upon by the European member states.

The Dutch Ministry of Environment sought contact with various stakeholder representatives during the development phase of the energy efficiency label in the Netherlands. Present were representatives from the automobile sector, consumer organisations and environmental NGOs. Van den Burg (2007) describes how the proposed energy efficiency label in the Netherlands was clearly based on a number of assumptions of what the consumer wanted. First, it was believed that consumers would not understand the parameter 'gram CO_2/km', therefore a colour classification was proposed that was analogous to the European label on domestic appliances (Figure 4.3).

Next, the debate focused on the question if this classification should be done on the basis of absolute or relative fuel efficiency. It was argued that consumers, when considering purchasing a new car, have already decided upon the size of the car. As a fuel efficiency label would never convince a potential BMW-buyer to switch to a Fiat Panda it was decided that portraying relative fuel efficiency differences was more useful than absolute comparisons (*ibid.*). Furthermore, an absolute energy efficiency label would blur the differences between cars of the same car class: a small car would always do well, a big car never. It was decided that the formula, determining to which energy class a new car belongs to, should be based on length and width of the car, and its relative and absolute fuel efficiency. The 'average' car is between a C and a D-label and other cars would be compared to this standard[52]. An A-label means that the car is at least 20% more efficient than a regular car in that same car class, while a G-label means the car is at least 30% less efficient.

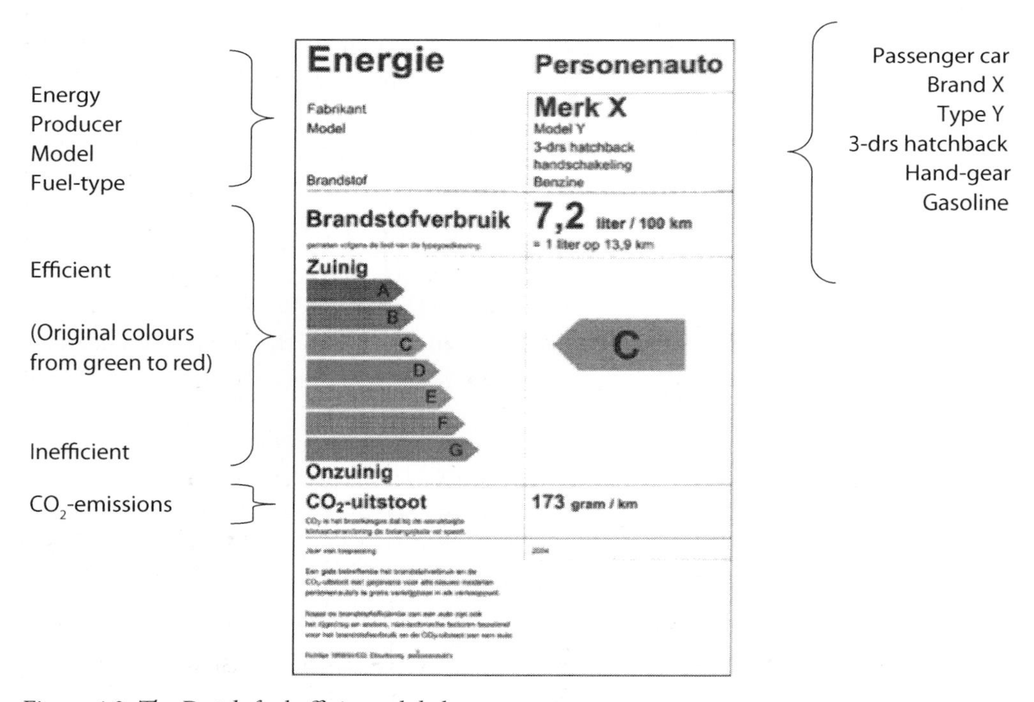

Figure 4.3. The Dutch fuel efficiency label.

52 Each year, the average is recalculated and car fuel efficiency categories are thus moving targets.

During the development phase the label met with severe cynicism from various parties who stated that a label would never change consumer purchase behaviour. For example, car industry representatives feel that car purchasing takes place more on the basis of emotional experiences than on rational choices. It is also likely that they did not like the idea of having a mandatory label standing next to the car in the showroom. After the introduction of a tax subsidy for A and B labelled cars in 2002 (based on a tax exemption on the purchase price of € 1000 and € 500 respectively for A and B labelled cars) the attitude changed towards the positive. However, after one year the tax exemption was stopped because a newly elected government cancelled the financial arrangement. Van den Burg (2006) describes how the attention from the varying stakeholders towards the energy efficiency label dropped to a minimum in the following years. The label was present in the showroom, but no organisation actively promoted the label. Even environmental NGOs undertook no activities to make the general public familiar with the label, as they felt that others were in a better position to do this (*ibid.*). Not until 2005 attention for the energy efficiency emerged again. In this time period a number of initiatives by governments, NGO's and corporations were undertaken. Local governments, drawing upon experiments from neighbouring countries, have started differentiating in parking tariffs where more polluting cars are charged heavier. Furthermore, actors such as the Dutch Automobile Association (ANWB), consumer organisations, and environmental NGOs took up the initiative to actively promote the energy efficiency label in their magazines and on their websites. Some of these actors even developed alternative environmental information tools by themselves. The ANWB has adopted the Eco Test, developed by its German sister ADAC, to influence decision-making of potential car-purchasers (Textbox 4.1). One reason for the development of the Eco Test is that the fuel efficiency label is seen as too limited as there are more car emissions than merely CO_2. Another reason is that 'despite a harmonised fuel labelling directive, unfortunately no harmonised calculation method for comparing cars inside a vehicle category [exists] because member states could not get to an agreement in Brussels for a harmonised relative system' (Van West, 2004). It was deemed confusing for consumers that some member states have the same format of labels with different coloured arrows for equivalent cars based on different calculation systems, while other member states have different labels.

Textbox 4.1. Background on Eco Test.

The Eco Test has been developed by the German Automobile Association (ADAC) in 2002, with funding from the global FIA Foundation. The main aim of the Eco test is to provide consumer information on aspects of the environmental performance of popular car models in Europe. Further aims are to increase sensitivity of consumers to ecological aspects, and to influence industry developments through consumer behaviour. Pollutants (CO, HC, NOx, PM) and CO_2-emissions are measured in a specially designed test. The pollutants are measured absolutely, while CO_2-emissions are measured relatively, depending on the vehicle class. These two values are added together on an equal basis resulting in a number (from 0 to 100) and a star rating (from 1 to 5 stars).

Recently, environmental organisations changed their consumer-oriented tactics by explicitly targeting new car purchases. One of the more influential Dutch environmental NGO, the Netherlands Society for Nature and Environment, initiated an Eco Top 10 of most energy efficient appliances (Textbox 4.2). The Eco Top 10 is part of a larger campaign to provide information to consumers about the environmental performance of products.

Textbox 4.2. Background on Eco Top 10.

The Eco Top 10 offers an outline of various energy efficient household appliances, such as televisions, refrigerators and cars (the measurements were done by a consultancy firm). It is comparable to Eco Top 10s developed in Germany, Switzerland, Austria and France. The Eco Top 10 is part of a Dutch climate program (HIER) stressing the necessity to implement adaptation projects and initiatives to climate change. The campaign involves 40 (inter) national charity organisations, government organisations and companies who have formed a collective communication campaign. Like the Eco Test the cars are divided into different car classes. Information is given not only about environmental indicators, but also about general characteristics such as storage space, purchase price etc. Furthermore, a special feature is an indication of fuel costs (in € per 15,000 km) of the Top 10 cars which is compared with the same indication for a very inefficient car in the same class. At first the Eco Top 10 was only available on a specifically designed website. Later on, the ANWB and the WWF adopted the Top 10 and actively supported it.

4.3.2 Face to face communication

The second format of environmental information provision is face to face contact at the consumption junctions. In all of the four purchase routes some form of personal contact takes place: between private consumers themselves, between consumer and car salesman, or between consumer and the fleet owner/employer. Unfortunately, little is known about the exact role of face to face communication in the provision of environmental information. However, there are indications that this role can be quite influential. This applies specifically to the role of salesmen in communicating about the environmental qualities of products. The purchase of expensive products is associated with an extensive search for information. An important source of information that consumers use is the information presented by car salesman (Boardman *et al.*, 2000). The car salesman can offer expert advice about the product and service qualities of a car. An important difference between face to face communication and other potential information channels is that the information is presented interactively in a reciprocal process of communication between purchaser and salesman (Spaargaren *et al.*, 1995). In contrast to the passive information provided by environmental information tools, a dialogue between these two actors can develop. In this dialogue a new and challenging role for car salesmen is expected in which they have a function as intermediaries between car producers and consumers. However, the way that salesmen take up this role depends on a lot of factors. Are the salesmen willing to provide environmental information, pro-actively or when asked, to potential car purchasers? Secondly, is the car salesman capable of informing consumers about the environmental aspects of automobility, in general, or one type of car, specifically? Both of these aspects (willingness and capability) relates to the question if the provision of environmental information has become an integral task of the car salesman. As the communicational tasks and knowledge of car salesmen is for a large part determined higher up in the chain, a significant role is set aside for the car companies and car importer who provide the training courses for car salesmen.

Though the role of car salesmen is important, a paradigm shift is necessary in their role in the provision of environmental information (see also Raad Verkeer en Waterstaat *et al.*, 2008). The society for nature and the environment conducted a research in 2005 about the availability of dust particulate filters in new diesel cars, and the information supplied by car dealers. Two aspects with respect to information provision were evaluated: (1) the pro-activity of the salesmen in discussing environmental aspects of diesel cars; (2) the knowledge of the car salesman about environmental aspects of diesel cars. This research was conducted shortly after a government subsidy for dust particulate filters was introduced. The research revealed that only 20% of the car dealers actively mentioned the subsidy, and only 34% of the car salesmen had adequate knowledge about the financial arrangement. Furthermore, a significant portion of the dealers had insufficient knowledge about the environmental aspects of diesel cars, in general, or about the function of the diesel particulate filter in particular.

A similar research has been conducted in 2001 by the Dutch Ministry of Environment with the aim to evaluate the newly introduced energy efficiency label. By way of mystery shopping the knowledge and motivation of the salesmen were investigated. The mystery shopper was asked specifically to shop for energy-efficient cars. When asked which engine size the salesmen would recommend to the mystery shopper (knowing the mystery shopper favoured a fuel efficient car) no salesman recommended the smallest engine size. Common reactions were that driving style was more influential, or that engine size has no significant influence on fuel efficiency (Stienstra & Jansen, 2001).

While it can be assumed that currently these specific examples have improved, the environmental aspects of consumer cars has received much more attention in recent years, the examples do show the significance of face to face communication between consumer and car salesmen in the showroom.

4.3.3 Advertisements and in-store presentation

The final format to be discussed is non-verbal communication about the environmental aspects of (environmentally friendly) automobiles, as provided by the car manufacturers. More specifically, this section focuses on the role of advertisements and the role of in-store presentation in the provision of environmental information[53]. Advertising is often seen from a producer point of a view, i.e. the goal of advertising is to persuade consumers to recognize and prefer a particular product. Advertising reflects what the advertiser believes the consumer wants and what it is available of providing (Ferguson *et al.*, 2003). In this respect it is interesting to see that environmental

[53] This format applies especially to new car sales as used cars are seldom advertised by car manufacturers.

information in the automotive chain has for a long time been disclosed or limited to the car company's environmental reports. Peugeot, for example, developed the dust particulate filter as far back as 1999, a time when the impact of dust particulates on local air quality received little public attention. Also BMW, Mercedes and Volvo distributed their diesel cars with this technology years before the Dutch government acknowledge the issue. These car producers put little effort in advertising the environmental characteristics of their diesel engines. This can partly be explained by the market failures of many green cars. For a long period of time environmental aspects were not seen as selling points: 'Why do people purchase a diesel car? That is not because the car has the cleanest diesel technology, but because the car has 180 HP. That is what a diesel rider gets a kick out of. Whether or not a car has a particulate filter is not one of his concerns' (interview Versteege, Toyota).

This viewpoint about the consumer has for a long time been reflected as well in car advertising (Figure 4.4). A review of television, magazine and newspaper advertisements in 2001 concludes: 'although manufacturers are making headway in developing alternative fuels and minimising effects on the environment, this is not coming through in the adverts' (Bristow, 2001, p. 26). The most common themes used in car advertisements are the affordability, desirability, comfort, responses and handling (linked to speed and power) (*ibid.*).

Recent years, however, has seen a massive increase in the provision of environmental information in car advertisements in the Netherlands. How can this be explained? Why has this

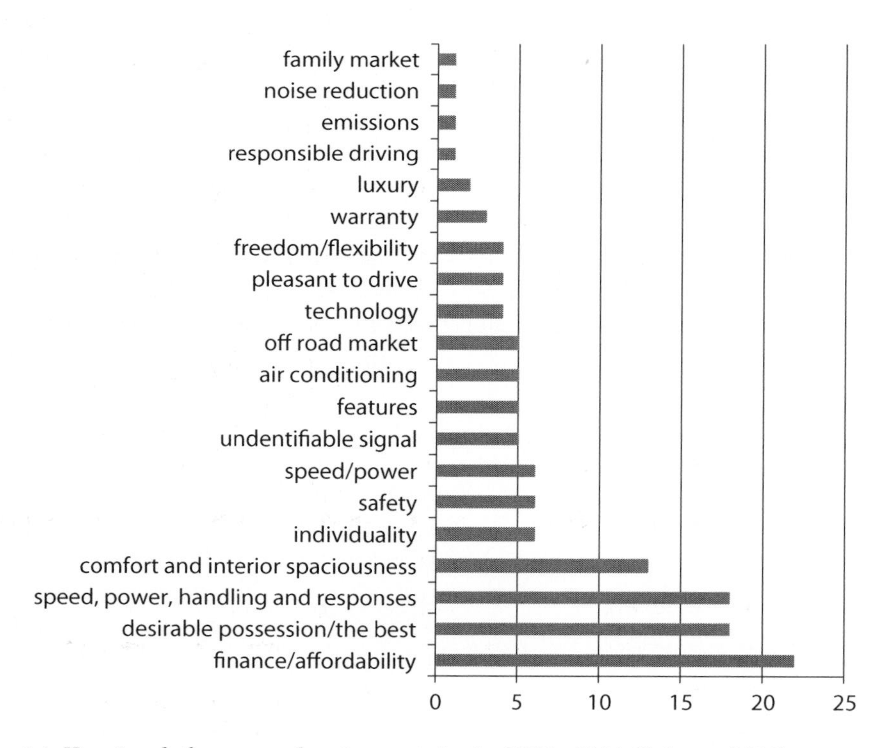

Figure 4.4. Key signals from car advertisements in the UK in 2001 (Bristow, 2001).

sudden and dramatic shift in the position of environmental aspects in advertising occurred? Attention to global warming and high gasoline prices may be seen as facilitating conditions. Furthermore, governmental tax subsidies provided car companies a tangible and direct marketing opportunity to promote the environmental aspects of their vehicles. Another explanation may be found in the core characteristics of the automotive sector. This sector is characterised by an enormous competitive market in which an increasingly reduced number of regime players operate. This competitiveness is strengthened by a production process which requires enormous investments leading to equally high break-even points (Orsato & Clegg, 2005; Nieuwenhuis & Wells, 1997). Due to this competitiveness, car manufacturers are in constant search for new market opportunities (and also are inclined to follow the steps taken by competitors to improve the ecological characteristics of the car). Another important characteristic is related to that of the object of the car itself. It has already been explained how material goods act as sources of social identity and bearers of social meaning (Lury, 1996). Automobiles are material goods for which this social phenomenon is especially true[54]. One could say that automobiles are the best marketed product worldwide. It is no coincidence that car manufacturers are the primary users of television advertising. The brand, as a set of relations between products or services (Lury, 2004), functions as the prime communication tool between the consumer and the manufacturer. The already mentioned topic of political consumerism indicates that consumers 'talk back' to producers and that consumers have specific opinions about a brand, the core activities of the brand, and the products that are affiliated with the brand. As brand and product images are crucial, brand (and product) positioning have a prominent place in the automotive industry. In fact, car manufacturers take great care of their brand and product position[55]. As environmental aspects have received higher attention in societal debates, environmental quality is increasingly seen as a positive asset to the brand and product image and is consequently emphasized in advertising. More than ever previously disclosed information about environmental characteristics has therefore been put to the forefront in car advertisements (sometimes in direct relation with the economic benefit of fuel-efficient driving or environmental tax subsidies).

Whether this turn must be seen within the light of an ecological modernisation where the display of environmental criteria is representative of the development of an ecological rationality within the car industry, or must be seen as the plain outcome of a highly opportunistic business, for now, we leave in the open. Nevertheless, car manufacturers' effort to put a spotlight on the environmental characteristics of cars has had a side-effect as well. Greenwashing or not, the transparency of an environmental claim is not always clear. BMW, for example, was 'surprised to find out that it had 129 car models with a green car label'[56]. Lexus promotes its hybrid SUV with

[54] According to an automotive survey, when asked why they bought a Toyota Prius, 57% of Prius owners said because it 'makes a statement about me' (Cap Gemini, 2007).

[55] Interesting in this respect is an example provided by a Toyota sales trainer. He mentioned that a representative of Toyota Netherlands would personally deliver the 100,000th Prius sold in Europe. However, before the delivery, first the person who had bought the Prius was 'checked' on representativeness of the Prius. This was to prevent that the prize-winner was a 'tree-hugging environmentalist' and damage the product position.

[56] These 129 models were based on the Dutch relative energy efficiency label. Also BMW considered a C-label a green label class instead of a yellow label class (the colour which is used to depict a C-label on the official energy efficiency label).

the slogan 'low emissions, zero guilt'. Saab has made the claim that their BioPower vehicle 'finally makes roads greener'. Saab had to withdraw this claim after Friends of the Earth Belgium filed a complaint for misleading advertising. These examples raise the question of what makes a high-quality and honest green claim, that is, from a consumer point of view. This is likely to depend on trust in automotive production-consumption chains and the mediator of the information. While currently, at varying governance levels, measures are being taken to reduce the number of insufficiently grounded green car claims (Sweden has gone so far as to completely banish the words 'environment', 'green', 'sustainable', etc. in car advertisements!); car manufacturers themselves have also used other measures to circumvent the risk of being charged with greenwashing and to provide consumers with objective and trustworthy claims. Among others, Daihatsu and Toyota actively use the aforementioned Eco Top 10 in promotional campaigns. The WWF logo, and the Eco Top 10 were all used in advertisements not only to indicate the environmental quality of the car but also as a tool for trust in the environmental claim.

While most of the attention with regard to non-verbal communication has so far been directed at car advertising in magazines and television, it is certainly not limited to these forms. Also the presentation of environmentally friendly cars and the display of environmental car-characteristics in the showroom itself, are influential forms of non-verbal communication. Products and the presentation of the environmental qualities by the use of environmental information tools are part of the situated setting of the consumption junction. In car purchasing this is most evident in the way a showroom, through its design, influences the consumer choice. Showroom designer Van der Schoot (interview) indicates that in general the design of a showroom can influence the recognisability of a product. However, he notes that currently in most showrooms little of the information used in national advertising campaigns can be found back in the showroom itself. This means that the content of a national message is not similarly reflected at the consumption junction. Van der Schoot believes that when a customer enters a showroom he or she should enter a world of information, amusement and experience. These elements should create an atmosphere that reflects the information and emotions raised by national advertising. A concrete example is given how showroom design differ between premium car brands and mass market car brands. A premium car is positioned on the basis of technological innovation, status, quality, design and exclusivity. This can be reflected in a showroom by displaying less cars with more attention. Each model is highlighted with audio-visual tools such as animations, images, etc. The mass market cars can position itself on the basis of price, volume, comfort. In this design the communication is much more directed linked with daily-life experiences and the versatility of use.

4.3.4 Monetarisation

In ecological modernisation theory, the economising of environmental flows is one of the prime conditions for ecological restructuring and strengthening the environmental rationality in society (Mol, 2005). In recent years this monetarisation of environmental flows is also strongly present in the automotive production-consumption chains. This is perhaps not surprising as in general in environmental policy an increasingly large role is attributed to economic policy instruments. Congestion charges, road pricing, tax subsidies, and CO_2 levies are different forms of monetarisation that have been introduced over the last decade in the field of mobility. The classic policy assumption behind economic instruments is that companies and consumers will make the choice that provides the greatest utility. This claim seems to be supported by the fact that price-differences between conventional and sustainable products are mentioned by consumers as one of the key reason why sustainable products are not purchased more often.

The monetarisation of the automotive sector has gone through quite a number of changes throughout the years. The analysis of the development of the energy efficiency label in the previous section already indicated how influential the political context can be. Table 4.5 reveals those shifts in the purchase taxes throughout the years after the introduction of a subsidy for energy-efficient cars in 2002. In 2002 a subsidy was introduced for the purchase of A- and B-labelled new cars which was directly abolished the year after. Between 2002 and 2006 no subsidy was available for conventional energy-efficient cars. However, a noteworthy additional development in the consumer-oriented strategy of the Dutch government was the introduction of a complete tax exemption for hybrid cars in 2004. This subsidy was specifically designed to facilitate the introduction of hybrid cars in the Netherlands. The tax exemption to compensate the higher purchase price of hybrid cars meant a decrease of approximately € 9,000 on the purchase price. This massive amount was reduced to € 6,000 in 2006 because of the exceedingly high costs for the national government and because of pressure from European car manufacturers who had no hybrid cars on the market.

Induced by the public alarm on poor local air quality, in June 2005 a subsidy for dust particulate filters was introduced. Consumers purchasing a new diesel car with a particulate filter received a € 600 reduction on the purchase tax. Again, this subsidy was calculated in the total purchase price (for existing cars there was also a € 400 subsidy available for consumers wanting to build in a 'retro-fit' particulate filter).

It has already been noted that from 2005 onwards the attention for the energy efficiency label had resurfaced. Very influential was the decision of the Dutch government to link the label to

Table 4.5. Dutch taxation schemes for new cars from 2002 onwards.

	De- or increase in car purchase tax in different energy classes						
	A	**B**	**C**	**D**	**E**	**F**	**G**
In 2002	- € 1000	- € 500	-	-	-	-	-
From 2004 to July 2006							
• Hybrid car	- € 9,000	-	-	-	-	-	-
• Non-hybrid car	-	-	-	-	-	-	-
From July 2006 to February 2008							
• Hybrid car	- € 6,000	- € 3,000	-	-	-	-	-
• Non-hybrid car	- € 1000	- € 500	-	+ € 135	+ € 270	+ € 405	+ € 540
From February 2008-2010							
• Hybrid car	- € 6,400	- € 3,200	-	-	-	-	-
• Non-hybrid car	- € 1,400	- € 700	-	+ € 400	+ € 800	+ € 1,200	+ € 1,600

fiscal instruments again. Furthermore, rising fuel prices and more attention for global warming led to a renewed public interest in the label. Table 4.5 shows how in 2006 the purchase tax of all new cars was made dependent on the label of the new vehicle; the financial stimulation for fuel efficient cars was reintroduced and complemented by a levy for fuel inefficient cars. In 2007 the newly appointed Dutch government strengthened the ecological monetarisation of the automotive sector with a range of new measures that went into effect in 2008 (see also Van den Brink *et al.*, 2007). Firstly, as Table 4.5 shows, the bonus-malus arrangement of the purchase tax was increased. The most significant effect is that energy-inefficient cars are charged more heavily. Secondly, a heavily debated CO_2-charge was introduced for cars that exceed a CO_2-treshold[57].

Next to these fiscal measures which directly influence the purchase price, other fiscal arrangements were introduced which influenced the use phase. For fuel efficient cars the vehicle excise duty was halved in 2008 and completely discarded in 2010.

Very influential as well are the fiscal measures in the leasing branch. In the Netherlands consumers who make use of a company lease car are charged a monthly payment based on the net catalogue price. As a lease car is seen as an indirect form of payment from the employer to the employee, an amount is added to the wages of the employee. As income taxes have to be paid over this virtual payment the net wage will be lower. For a lease car of € 30,000 this means that € 550 is added to the monthly pay check which, depending on the income, means the net payment is about € 200 per month. In 2008 very fuel-efficient cars received a substantial reduction on this monthly charge (the same criteria apply as for the reduction in VED). For very fuel efficient cars the fiscal charge was reduced to 14% while for other lease cars the charge was increased from 22% to 25% (Table 4.6). Very fuel efficient cars therefore received almost halve the fiscal charge in comparison to regular vehicles.

This section has showed that environmental monetarisation has been introduced to the fullest in the Dutch automotive production-consumption chains. Two major concerns, local air quality and climate change, have been directly translated into a monetary bonus/malus system for car purchasing. While used cars are still almost completely left out of the monetarisation of car

Table 4.6. CO_2-borders for yearly taxation on lease cars.

0%	Electric vehicles	
	Diesel	**Other fuels**
14%	<95 gram CO_2/km	<110 gram CO_2/km
20%	96-116 gram CO_2/km	110-140 gram CO_2.km
25%	>117 gram CO_2/km	>141 gram CO_2/km
35%	cars older than 15 year old (over real value)	

[57] For diesel cars this threshold is 240 gram CO_2 and for petrol cars this threshold is 200 gram CO_2 (e.g. a petrol car which emits 229 gram CO_2 gets an addition charge of € 3,190 at the moment of purchase).

purchasing, the incorporation of the leasing branch is a noteworthy new development. However, the excessive account of the renewed monetarisation (from 2008 onwards) is also a clear indication of the complexity of the current fiscal structure. The visibility of each incentive can be questioned as the consumer pays the total consumer price (Table 4.7).

Table 4.7. Governmental taxations linked to the purchase of a new car.

Ford Mondeo 2.0 TDCI	
Net price	€ 22,815
Purchase tax	+ € 9,650
Dust particulate charge	- € 900
Energy label D	+ € 400
CO_2 charge	-
Diesel charge	+ € 307
V.A.T.	+ € 4,335
Consumer price	€ 36,507

4.3.5 Towards a framework of environmental information flows

So far, on a general level the provision of environmental information and monetarisation of four car purchasing routes have been analysed. As it is beyond the limitations of a single case study to investigate all the changes in environmental information provision within these four routes, in the upcoming empirical sections one of the described routes, namely new car purchasing, will be investigated in more detail. New car purchasing is chosen because the provisioning of green cars through mainstream consumer channels is a fairly recent phenomenon. The quantities of green cars on offer are therefore much higher in the purchase of a new car. Furthermore, this case study puts an emphasis on the role of environmental information and monetarisation in the selling and buying of cars; the previous sections clearly pointed out that the far majority of information flows and fiscal incentives focus on new cars. The present study is limited predominantly to the analysis of one specific site of the consumption junction, namely the car showroom. The reason for choosing this specific site is the fact that every consumer who buys a new car will visit one or more showrooms. Finally, prior research has indicated that consumers' experience in the showroom, the technical information that can be acquired there, and the role of the car salesmen are among the most influential factors in car purchasing (Boardman *et al.*, 2000).

Surprisingly enough, only limited research has been conducted that connects car purchasing decision-making with the consumption junction (e.g. the showroom) and its wider connection in the systems of provision, the automotive industry. However, investigating the access and provision points in the automotive production-consumption chain is of crucial importance to know (1) how the practice of new car purchasing is contextually structured; and (2) how environmental information and monetarisation influence the purchase decision; that is not only on an abstract national level but also at the level of the consumption junction itself. This means that analysing the provision of sustainable cars, not only as technological artefacts, but also with regard to different governance strategies that directly or indirectly target consumer decision processes, is a key variable worth researching. Furthermore, the sections in this paragraph show that informational strategies, as one specific governance strategy aiming to influence the acquisition of sustainable innovations, deserve more attention. The elements already discussed in this chapter show that many variables have to be taken into account in assessing the provision of environmental information. Three sources of environmental information have been described in the previous chapter: environmental information tools (the energy efficiency label, the Eco Test and the Eco Top 10), face-to-face communication, and advertising and in-store marketing. When the provision of these sources is combined with consumer information search in car purchasing, a framework of information flows in new car purchasing can be formulated (Figure 4.5). The figure shows how car purchasing is contextually structured by the system of provision. In new car purchasing the national importer and the local car dealer play a significant role in providing information to consumers. Through advertising, in-store structuration, and conversations with car salesmen, consumers access different types of information. Next to information from car manufacturers, consumers use a lot of different other information sources when purchasing a new car. On average, consumers consult four to five information sources, including information provided by national automobile clubs, consumer's associations, car magazines and of course their social network. Figure 4.5 also shows the different sources of the three environmental information tools; the energy label which originates from the

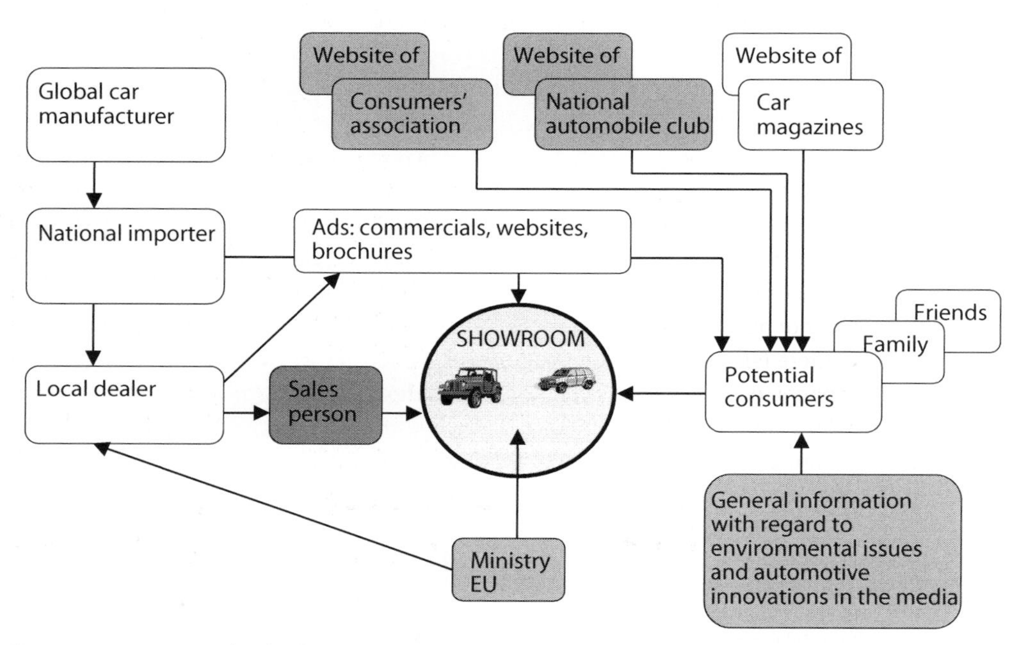

Figure 4.5. Framework of information flows in new car purchasing.

Ministry of Environment, the Eco Top 10 which originates from the Netherlands Society for Nature and Environment, and the Eco Test which originates from the Dutch Automobile Club ANWB.

Figure 4.5 shows that environmental information can be provided to consumers through many different sources. Whether this information is actually accessed by consumers depends on different variables, a number of which have been described in this chapter already. Based on the information presented in this chapter the key variables of importance in consumer assessment of the different information formats can be identified. Therefore, the next sections will be structured on the basis of the following questions:

- What are the most relevant sites or locations where environmental information flows for the buying and selling of new cars are organised?
- How do citizen-consumers assess and make use of environmental information flows during the practice of new car purchasing?
- How are the three environmental information tools that are provided by different societal actors evaluated by citizen-consumers in terms of comprehensibility, trust, and 'attractiveness'?[58]
- In what ways do representatives from the systems of provision (car manufacturers and salespersons in particular) develop and use flows of environmental information during the sales process?
- How can the location and presentation of green cars at the different sites of the consumption junction influence and facilitate the sales process?

Via these variables we can investigate consumer concerns, consumer search strategies and consumer decisions in direct relation with provider strategies of car manufacturers, and environmental information tools developed by societal actors aiming to influence potential purchasers. In the next paragraph it is explained how the abovementioned variables have been investigated.

4.4 Methodology

The majority of the empirical analyses is based upon a focus group research conducted with car purchasers and car salesmen. The research has been conducted in the Netherlands between February 2006 and February 2007. In groups of eight to ten participants a discussion was facilitated through open questions and group assignments. The choice for focus group methodology was made because of its specific attributes. In everyday life (including professional life of car salesmen) the normative order of behaviours and opinions is rarely articulated. In focus groups an attempt is made to bring these taken for granted assumptions to the foreground (Bloor *et al.*, 2001). It allows the participants to question and rethink that what is normally taken for granted in order to discuss (future) possibilities and options. On the one hand the focus group provides important research information. On the other hand they can provide information that helps us to formulate better survey questions and enables us to contextualize at least some survey questions with the help of everyday life experiences as reported by the respondents.

[58] The aspects of comprehensibility, trust and attractiveness are specifically analysed as research indicates that these are influential factors in determining if and how consumers pay attention to eco-labels (See also Thøgersen, 2000, 2002, 2005; Van den Burg, 2006).

The focus groups were conducted with Toyota car salesmen on the one hand, and Toyota purchasers on the other. The outline and topics in both focus groups were kept as similar as possible to enable comparisons between the provider focus group and the consumer focus group. Toyota was selected because they can be considered as one of the frontrunners in the mainstream automotive sector. The salesmen focus group consisted of seven men and one woman coming from different car dealers in the Netherlands. The focus group took place at Louwman and Parqui in September 2006, the national distributor of Toyota, where the training of all Toyota salesmen takes place. Secondly, in the consumer focus group ten individuals, six men and four women participated. These participants registered on the basis of an invitation letter send to a group of 200 car purchasers. Importantly, all participants had recently bought a new Toyota which makes them 'experts of practice' and able to talk about their own personal experiences. The selection of participants was not based on environmental characteristics of the new car, but based on a representative spread of the Toyota car types sold in that same time period (summer 2006). The average age of the group was approximately 50 years old which makes it comparable to the general age of new Toyota purchasers in the Netherlands[59].

In both focus groups the purchasing process was discussed, including the information search routes. Also, the three information tools were introduced: the energy efficiency label, based on the European Union labelling directive; the FIA Ecotest developed by the German automobile club ADAC and provided in the Netherlands by its Dutch sister ANWB; and the ECO Top 10 developed by the Netherlands Society for Nature and the Environment. The principle was to analyse the different perspectives on these three tools in order to investigate the current information provision during the car purchase, and also to abstract general characteristics which might help future developments of environmental information provision in the automotive sector. The focus groups also provided information about the knowledge and willingness of car salesmen to provide environmental information (either with or without the use of environmental information tools). The influence of in-store presentation was to a large extent only discussed with the car salesmen, the focus group methodology and the topic made the issue largely unsuitable for discussing it with consumers.

The focus group research was supported by two qualitative internet surveys. The quantitative data is derived from two digital consumer surveys conducted in collaboration with Milieu Centraal[60]. In February 2006 the first survey was conducted about car use and car purchase (n=883, 69% response rate). One year later the second survey on environmental labelling was posted (n=896, 50% response rate). The respondents were representative with regard to sex, age, education and region. This data was used primarily to make comparisons with the outcomes of the focus groups and place these outcomes in a wider context.

[59] The participants present had purchased the following car models: three times the Toyota Aygo (A Label, 109 grams), three times the Toyota Yaris (B label, 134 grams), two times the Toyota Corolla Verso (C Label, 178 grams), one time the Toyota Avensis (C Label, 172 grams) and one time the Toyota Prius (A label, 104 grams). Between brackets is the designated energy label and the absolute CO_2-emissions, the CO_2-emissions are estimates as the exact emissions depend on the specific motor size and accessories in the car.

[60] Milieu Centraal is NGO which aims to provide consumers with practical and reliable information about the environment and energy use in daily life.

4.5 Results 1: an analysis at the consumption junction

In this section first the car purchasing practice is described. Because space limitations prohibit a complete depiction of the individual story lines the focus is predominantly on the relation with environmental aspects in car purchasing. Next, the provision and evaluation of environmental information tools that are currently available inside and outside the showroom are discussed. In succession three aspects of this evaluation (availability, comprehensibility, and trust) will be described. Finally, albeit explorative, the role of in-store presentation in the positioning of green cars and their environmental dimensions is discussed[61].

4.5.1 The purchasing practice and the role of environmental factors

The consumer focus group revealed that the choice for a specific car type was most of the times inspired by very individual stories and backgrounds. Most consumers relied on positive experiences of previous cars and on experiences within their social networks. The focus group also showed that car consumption can be seen as a continual process instead of one-point-in-time purchase. When asked how consumers came to their car of choice often the participants started out years back in time. Furthermore, consumers continue to keep an eye on automotive developments long after the actual car purchase. Finally, the car as a product, and the accompanying car services at the dealer or garage, are discussed long before and long after the purchase.

When discussing the practice of car purchasing, consumers often had specific ideas of what was significant when buying a car. Important is that most purchasers were loyal to the brand, even though in every purchase car types of competing brands were compared, most participants were Toyota customers for years. Next to brand and dealer loyalty, factors such as price, comfort, safety, reliability, space and appearance were mentioned as key decision features. This does not mean that environmental aspects play no role. More than a few consumers expressed that they took sustainability, fuel efficiency and emissions into account. Not surprisingly, this was expressed more often by consumers who had purchased a relatively fuel-efficient car. An interesting example is from one participant who made a deliberate trade-off by purchasing a small and fuel-efficient new car (Toyota Aygo), which he used for daily commuting, to extend the life-span of his big family car just long enough that the car will last until the moment that his children will have left the parental home. By making this trade-off his daily fuel use would drop, and he wouldn't have to buy a big car in the near future.

The environmental subsidies were mentioned in a number of occasions. One participant noted that the environmental subsidy for her car steered her attention mentally. A Toyota Prius purchaser remarked that the energy efficiency of his car worked in two ways: the Prius was not only fuel efficient and environmental friendly, these benefits were also even subsidised by other

[61] The clear analytical distinction between the three formats of environmental information is not always similarly present in the results sections. The same applies for the distinction between environmental information and monetarisation. The focus group participants automatically mingled the different information formats when discussing the role of environmental aspects and environmental information in car purchasing.

car purchasers. But, in general the subsidies were mostly seen as facilitating and not determining factors in the car choice.

The car purchasing practice was also discussed with the car salesmen. In addition to the above they pointed out the influence of much less tangible aspects such as the total car experience in the showroom and during the test drive, and the indirect role of advertisement. With regard to environmental aspects, the car salesmen indicated that overall interest in fact has increased in the last ten years. Firstly, in the automotive industry itself; when the catalytic converter was introduced the industry thought it was ridiculous: it made the car more expensive, the performance would go down, the car industry was strongly opposed to it and found it unacceptable. Now this attitude has completely changed to the point that 'everybody who works in the branch just wants to go to very environmentally friendly cars'. Secondly, regarding consumer interest the salesmen perceived that it has increased over the years, for instance, more and more questions in the showroom are environmentally related. The standard response of the car salesmen when confronted with questions about the environmental characteristics and fuel consumption of a car is to grab the specific car brochure and explain the details with the help of this brochure.

Furthermore, in principle, the salesmen perceived environmental information provision a tool which definitely has potential. For one, the salesmen experienced that the current energy efficiency label is recognized by large numbers of customers. The salesmen made a comparison with the Euro NCAP crash test to indicate how consumer demand can influence production design.

> *The NCAP crash test now has five stars. Well, now every manufacturer wants to be able to attain those five stars. You should be able to do the same with the environment: environmental stars.*

Nevertheless, the salesmen remain sober and sceptical about (information concerning) environmental factors being enough to convince consumers to buy sustainable cars. Overall environmental aspects were considered foremost as additional factors; as a bonus.

> *Horsepower is important, but if you give the customer a bonus [read: environment] with which he can sell this horsepower to his employer or his social surrounding, you can pull him in on that point alone. He can say: I have a new car and it's clean as well.*

> *Regarding sustainability and cars there are two things that are important: first the car must be able to perform the same as other cars, and second, it must either not be more expensive or else it must be clear that the extra money can be earned back.*

The salesmen were positive about the fiscal measures attached to the energy efficiency label. Especially the massive tax subsidy for hybrid cars and the € 1000 for small A-labelled cars was considered to be a big influence in the purchase decision. They mentioned a direct increase in subsidized cars after the introduction of the tax subsidy in July 2006. However, the salesmen remained critical about the subsidy's transformative effects because for more expensive cars the financial incentives, negative or positive, were considered to be negligible.

4.5.2 Assessment of environmental information tools

In this section the assessment of the three environmental information tools will be discussed. In the conceptual framework it was suggested that different aspects can play an important role in the assessment of the information tools. Three aspects have been discussed in the focus group: the location of the environmental information tools at the consumption junction, the comprehensibility of the tools and trust in the tools.

The first aspect to be discussed is the location and presentation of the environmental information tools in the different sites of the consumption junction. It is not only the visual indicators of the information tools which determine if consumers appropriate the information presented by these tools. The location of the information itself also plays an important role in the amount of influence this information has. Already mentioned was that the energy efficiency label, positioned next to or on new cars in the showroom, is very well-known to new car purchasers. Another aspect which became clear was that information tools not present in the showroom were either unknown or hardly known by both the salesmen and the consumers. During the research period the Eco Top 10 and the ADAC Eco Test were available only on the website of the corresponding provider; later on the Eco Top 10 was more broadly used, for example in advertisements of different car companies. Even though the ANWB, the national Automobile Association, is commonly consulted during the search practice this has clearly not resulted in a spreading awareness about the Eco Test. If the greater awareness of the energy efficiency label is directly related to its position in the showroom is hard to determine. However, the question is relevant as almost all environmental information targets consumers outside of the showroom. Of course there has been much more support and attention to the energy efficiency label than to the Eco Top 10 and the Eco Test. One the one hand this relates to the fact that the energy efficiency label is in effect the longest and on the other hand to the fact that the governmental fiscal strategies are primarily based upon the energy classes presented on this label. Nevertheless, when the results from the consumer surveys are compared with the results of the consumer focus group, an interesting aspect is the major difference in recognition of the label. While only 42% of the respondents of the survey knew about the energy label (Koens & Nijhuis, 2006), in comparison all consumers of the focus group had knowledge about this label. Therefore there are strong indications that the presence of an environmental information tool inside the showroom is important for consumer awareness of the tool.

The second aspect discussed was the comprehensibility of the environmental information tools. Access to environmental information is also to an important degree dependent on the question if an information tool is understood correctly. In the focus groups the consumers and salesmen were asked to elaborate on the clarity and attractiveness of the information tools in smaller subgroups and to report their opinions back in a plenary session.

In both focus groups most of the attention focused on the energy efficiency label, which is the most known and influential environmental information tool. In the consumer focus group the first reactions on the energy efficiency label were generally very positive. All participants were aware of the existence of the energy label and of its presence in the showroom. Most consumers recognized the energy label because of the similarity with the energy label for domestic appliances. Furthermore, the tax subsidy was mentioned on several occasions as a factor in the purchase decision. Although the participants knew that category A was best and G was worst they found it was very subjective as there was debate about the exact meaning of these letters. They did not know exactly that the A category represents a car which is 'at least 20% more fuel efficient' but used it more as a heuristic to determine if the car is 'environmentally okay'. It was also apparent that none of the participants was aware of the fact that the energy efficiency label represents relative energy efficiency. This means that none of the purchasers had understood precisely how this energy efficiency label works, and that it was also not explained to them. When the design of the relative energy classes was brought to attention in the consumer focus group the positive feelings of the consumers altered to a certain extent. First, the consumers felt that the relativity of the classes is not clearly communicated on the label itself. Second, some consumer felt almost deceived by the confusing message of the label.

> *Without foreknowledge this is selective information.*

The energy efficiency label was also discussed in the large-scale consumer surveys. In this survey there was massive support for the energy efficiency label and the taxation scheme: almost 80% of the survey respondents supported these schemes for new cars. After explaining how the energy label must be interpreted, the dilemma of an absolute or a relative energy label was also presented in the survey. The mixed results are indicative of the complexity of the dilemma: 44% of the survey-respondents thought the current relative energy label was preferable, 32% found the current label lacking in clarity, while 21% had no clear preference.

The energy efficiency label and the Dutch implementation with relative energy classes also elicited heavy emotions from the car salesmen. First of all, the salesmen strongly disagree with the choice to measure solely energy efficiency in the label.

> *The energy label is not an environmental label, however in the heads of the customer ... they are.*

> *The only thing they measure is CO_2, not nitrogen and particulates, while those are much more essential.*

Secondly, the salesmen found the relativity of the energy label very difficult to communicate to consumers. The rationale behind the Dutch implementation of the energy efficiency label (relative fuel efficiency differences are more useful than absolute comparisons), was found to be completely unacceptable. An important reason for these heavy reactions are the hard to grasp measurement methods. Also, these measurements are made more strict each year which means that a vehicle can receive a B label in one year, and C label in the next year.

> *In the Netherlands there are cars that run 1 on 8 and are sold with a subsidy of € 500 to € 1000 from the government while being terribly harmful to the environment.*

> *Consider the Corolla hatchback and Corolla station wagon. The station wagon uses more fuel, has a higher CO_2-emission than the hatchback but gets a better label. The customer really gives up at that point.*

But, even to us it is abracadabra.

That is the conclusion I think. You talk around the label and continue with other subjects.

These quotes indicate that the energy efficiency label, during the research period, was often disregarded by the salesmen during the sales process. Furthermore, the sales trainer expressed that during training courses he advices his salesmen not to focus customers' attention to the energy efficiency label at all!

The two other environmental information tools where discussed much more briefly as there was little experience of use amongst consumers and salesmen. In the consumer focus group the reactions towards the ADAC Eco Test were far from positive. For instance, there was a lot of confusion related to the distinction between CO_2-points and pollution-points. Furthermore, the long numeric list was found to be far from attractive.

I found the table very unclear; in the end I understood the aspect of the more stars the better, but the rest is all for experts. Well, as a normal consumer I don't know what all those points mean.

The salesmen's reactions were also far from enthusiastic. Though it was praised that emissions which have an effect on local air quality were also included in the test, the way the Eco Test is presented was deemed completely unsuccessful:

But, does it really look like this? Do people get to see it like this? That is just hopeless.

Nobody is going to read this.

From the reactions of both consumers and salesmen, one can conclude that the Eco Test clearly is lacking in communicative value. This is an interesting finding as the Eco Test was developed precisely because the lack of a harmonised calculation method for comparing cars with the energy efficiency label was deemed confusing for consumers.

With regard to the Eco Top 10 there was a difference in reactions between car purchasers and car salesmen. Participants of the consumer focus group responded positively to the Eco top 10. Most found it helpful that the Top 10 provided not only environmental information on CO_2-emissions, but also information about fuel-costs, price, and volume. Furthermore, the presentation of the Eco Top 10 was considered easy to understand. While more than one consumer mentioned that in the next purchase he/she might consider using the tool, the true effectiveness of this tool remains to be questioned. First, none of the consumers had used this tool during their previous car search and second, these kind of intended actions are notoriously unreliable. Finally, due to its design of a Top 10, only a very limited number of cars are represented in this tool.

The salesmen reactions to the test were simple and clear: they were sceptical about it and found the test disputable.

4.5.3 Mediating environmental information

As the extensive literature on eco-labelling has shown, not only the location and communicative value are important characteristics of environmental information flows. Trust relations between provider and consumer are one of the crucial aspects when environmental claims are concerned. As Thøgersen (2000, 2002, 2005) has shown, belief in the presented information is one of the major factors in the decision to purchase eco-labelled products, or not. Analytically three types of reliability can be discerned: trust in product reliability, trust in the presented information, and trust in the source of the information itself.

Confusion and mistrust in environmental information tools and its sources can occur when there is competing and contradictory information about the environmental performance of a product. This occurs when one environmental tool indicates a specific environmental rating for one vehicle type which does not correspond with another tool. This could happen when different standards are used (1) for the measurement of the environmental impact; (2) for vehicle classification; or (3) test ratings. All the three environmental information tools discussed in this chapter differ from each other in at least one of these aspects. For example, all the tools make use of a different measurement method. Furthermore, the Eco Test is the only tool which measures local air pollutants next to energy efficiency[62].

The point of contradictory information was also highlighted in both focus groups; a common outcome was the feeling that there should be only one clear information tool, with one measuring method to prevent confusion. Both consumers and salesmen expressed great concern about the increasing amount of labels and tests in daily life. Again, the salesmen used safety as an example of how contradicting information can influence trustworthiness:

> *The Kia Carnival was presented in America as the safest car available. That has been communicated here in various magazines. However, half a year later the Euro NCAP organisation puts the Carnival against the wall with only two stars. If you have the same thing with environmental labels, then the customer gets disoriented.*

To keep contradictions in environmental information provision to a minimum, the salesmen felt that the source of the information should be the same in all cases. The idea is to 'keep it simple' and to use only one type of information system on which different kind of policies can be based, such as price mechanisms and parking regulations. However, how the label should look like, and who should provide it was an unsolved matter. Some salesmen pointed to the highest government (either European Union or national government) as it has both the data available and seems to some as the most trustworthy agent. Other salesmen pointed again to the Euro NCAP as an example of a new organisation which has now become an influential institution in itself. Surprisingly, information coming from the car industry itself and from environmental organisations was

[62] A Toyota Aygo, for example, is listed in the Top 10 of most fuel-efficient cars, also has an A energy label, but receives only a three star ranking at the Eco Test (out of the maximum of five stars).

considered to be biased and untrustworthy. They especially considered their own position to be universally infamous because research had shown they were often seen by customers as unreliable.

In the consumer focus group a similar discussion followed on what could be considered a trustworthy source. The only agreement was that the car producers were considered not to be very trustworthy in making environmental claims. Furthermore, there was a great concern for information overload because of an excess of environmental information tools. However, this could also be the result of the focus group set-up where all automotive labels were discussed.

One important issue with regard to product reliability are potential hesitations of consumers in acquiring new (and often 'unproven') technologies. This aspect of risk-taking in purchasing innovations is of course a matter which has been recognized as far back as the beginning of Rogers's theory of innovation diffusion. Rogers's well-known division of consumers in different adopter groups describes the varying willingness and abilities to adopt an innovation. Indeed, one of the characteristics of the so-called 'innovators' group is their greater willingness and propensity to take risk, indicating that innovation adoption naturally involves risk-taking. The concern for risks was also cited within the consumer focus group by a participant who recently had bought a new Toyota Prius. He wanted the full eight year guarantee to be on the safe side and therefore refrained from buying second-hand.

Within the car salesmen focus group the aspect of trust in hybrid technology elicited a heavy debate between those who disregarded the influence of technological reliability and those who felt it was critical to address this point:

> *There are still a lot of people who question if hybrid technology doesn't break down sooner.*
>
> *Come on! The technology has been on the market for nine years.*
>
> *No, she is right. It is a battery and a mobile phone battery only lasts two years.*
>
> *That is why we give them an eight-year guarantee, over 160.000 km, period!*
>
> *That is what we know, not what the customer knows.*
>
> *And, also customers ask: what happens after those eight years?*

With this section on trust the analysis of the assessment of environmental information tools is concluded. In the last section a more explorative turn is taken when the potential for different showroom settings is discussed.

4.5.4 In-store structuration based on potential provider strategies

In the conceptual framework it was already suggested that different contextual factors influence the embedding of ecological rationality in the practice of car purchasing. So far two formats of environmental information provision have been discussed: environmental information tools and face to face communication with car salesmen. The third format is in-store structuration of the green car in the showroom. In this final and explorative section we will highlight the third point by focusing on the construction and evaluation of possible provider strategies for sustainable car purchasing. The construction of different strategies and the accompanying showroom settings springs from the premise that the consumer does not exist, and that different consumers must be addressed in different ways because they are motivated by different factors, experience different barriers for behaviour and are affected in different ways by policy (Anable, 2005). Different types of consumers are motivated by different consumer concerns and respond differently to provider strategies. Acknowledging the multiple faces of consumers in car purchasing means that different types of informational strategies need to be constructed. These strategies can come into existence in the consumption junction by the development of specific structural settings. For showrooms this means that a showroom setting or theme is developed with a specific sustainable consumer strategy in mind. These strategies resolve around the question how sustainable cars are presented in the showroom and which leitmotif is used to convince and seduce different types of consumers to purchase sustainable cars. The aim of these vehicle images is that the utilitarian functions of the car (fuel efficiency) are connected with symbolic meanings (see chapter three).

In the focus group sessions we have presented various showroom settings or themes to the car salesmen to investigate how they see possible connections between environmental aspects with general consumer preferences in car purchasing. The purpose of these themes is to evaluate possible formats which try to embed environmental performance of cars in a way that gives environmental quality more personal significance to potential car-purchasers[63]. To accentuate these themes to consumers we also focused on how changes in showroom design can accentuate these themes. According to show room designer Van der Schoot (2006), currently, car dealers express little vision in showroom design. Van der Schoot believes the showroom must become a place where information and experience come together, where the theme is emphasized with a specific showroom setting, and communication centres on consumers' life-world experiences.

The showroom settings were the outcomes of the focus group assignment with car salesmen were the salesmen were asked to investigate the link between environmental aspects and the consumer's everyday life-world concerns and interest in car purchasing.

Theme 1. High-tech showroom (outcome of focus group assignment).

In this futuristic setting technologically pioneering comes to the forefront. The main message is that sustainability becomes connected to gadgets and innovations. The presented cars are high-tech, powerful and clean, which is communicated and reiterated in the showroom setting. An openwork car or engine can be used to visibly present the new technology in action. Furthermore, interactive touch screen or speech guided computers can be used to present information about the clean

[63] These themes did not come out of the blue as these where constructed on the basis of the main messages used in the green car advertisements. Depending on the car class and the type of car a different message in car advertisements is present which link the environmental performance of the car with other consumer preferences.

technologies. The theme is supported by drawings or paintings with images of future sustainable transport and a clean environment, while the material used in the showroom is state-of-the-art.

According to the car salesmen especially engineers and businessmen like to talk about such new technologies. These consumer groups see these innovations as conversation pieces and admire the intricate high-tech gadgets in the car. Especially hybrid cars, plug-in hybrid technology and new generation electric cars form the innovations fitting into this theme. The Toyota Prius brochures already reflect many of the items discussed in the showroom setting: a strong emphasis on the future, progress and innovation is equalled with ecological aspects.

Theme 2. Economic showroom (outcome of focus group assignment).

The focus is on the connection between high fuel efficiency and the possibility of driving a car inexpensively. In contrast to the next theme in this showroom calculating the precise car costs plays a much bigger role. Comparisons between regular cars and fuel efficient cars are shown to make the savings in fuel costs clear. This can be done by images of big and smaller petrol pumps. Prepaid constructions can be presented to enable young consumers to purchase these new and fuel efficient cars instead of second hand and possibly polluting car.

For the car salesmen this was the easiest link to make. They clearly saw how a connection between the environment and economics can be made, especially since fuel prices have become a major consumer concern in car purchasing. This theme obviously targets the more rational calculating consumer. Materialistic consumption is seen as important but only by getting the best deal. Economy is clearly the first argument while the environment is seen as an extra bonus. Low and medium priced cars with a low emission internal combustion engine fit best into this theme.

Theme 3. Trendy showroom (outcome of focus group assignment).

In this theme pragmatic idealism forms the key message. In contrast to the prior theme more emphasis is put on conspicuous consumption involving aspects of fun and youthful lifestyles. The cars presented in this showroom are mostly appreciated because of the design and the trendy urban lifestyle it reflects. Pragmatic idealism is reflected in the belief that it is completely feasible to continue a fast and fun-filled life while taking the environment into consideration. Many trendy elements (novelties in sports, fun, and music) are eminent as an overall package of experiences in the showroom. This theme links with hedonistic and variety-seeking consumers (most often to be found in young, highly educated, self-conscious urban dwellers) with a focus on lifestyle, shopping and distinction. Small, low emission internal combustion engine cars fit most closely into this theme.

The car salesmen found this theme the most difficult to visualize. The link between trendy lifestyles and the environment was found to be very hard to make. To quote one salesman: 'It is quite possible to make a trendy car that is less damaging to the environment, no problem at all, but you cannot sell something that is environmentally friendly as trendy. That is impossible.' He pointed out that the car as an aesthetic, identity-forming, multipurpose artefact sets the boundaries from which to build a sustainable car, not *vice versa*.

Theme 4. Sustainable showroom (interview Luuk van der Schoot).

The main theme is sustainability in combination to personal and family health. The showroom itself is made by sustainable design, while the showroom is filled with plants and natural wild-life sounds to elicit a green experience. Instead of a system of provision that is structured around one car brand, a wide variety of different sustainable cars and products are available in this showroom: this means a shift towards a car supermarket. In theory it is possible for a retailer to start a thematic showroom for new cars in which sustainability is the main message. The showroom is located in close vicinity of other retail-activities, and even the showroom itself can supply other non-automotive products that are attractive to the aimed lifestyle group and fit within the theme. Combined with an increased showroom experience, this would stimulate more frequent visits from potential buyers, e.g. young families.

This showroom fits best with consumers which can be described as involved and responsible. They feel responsible for a good personal health and worry about their children's health situation. Concern for nature and environment are seen as a constituting element of this responsible lifestyle. Bio-fuelled cars fit most naturally into this theme as they are organically powered cars. However, as mentioned in the theme description different types of environmentally friendly cars could be brought together as well. This way a consumer can be sure always to buy green when purchasing at this showroom.

4.6 Results 2: an analysis of the Dutch automobile market

Whereas Paragraph 4.5 provided an in-depth analysis of consumer-oriented strategies at the level of the showroom, next we concentrate on an overall impact analysis of the environmental information tools and taxation schemes with an assessment of the Dutch automotive market. The analysis is based on an overview of sales of hybrid cars and developments in market shares of A-labelled and B-labelled cars from 2001 to 2010, both for the private car market as for the lease car market. This chronological analysis presented hereafter makes it possible to observe shifts taking place in the sales of green cars while it simultaneously provides a general assessment of the influence of the Dutch consumer-oriented strategies in the greening of new cars.

The sales figures of new cars per energy class between 2001 and 2010 provide interesting information in the light of the Dutch consumer-oriented strategies (Figure 4.6). In general, the number of new cars sold with an A-label was relatively small until 2008. More specifically, we can see that there was a sales peak of A-labelled cars in 2002, the year that the first subsidy for energy-efficient cars was introduced. Though still small in relative terms, the number of sold A-labelled cars increased almost tenfold in one year. There is also a clear decrease in sales of A-labelled cars since the abolition of the subsidy in 2002.

We mentioned earlier that in 2005 a renewed interest in the energy efficiency label could be witnessed. This is also reflected in the number of cars sold[64]. The tax subsidy of € 1000 (from July 2006 onwards) has had a further positive effect on the sales of A-labelled cars. Car sales show

[64] The increase in the number of A- and B-labelled cars in 2005 must also be partly attributed to a more lenient calculation method (PBL 2009).

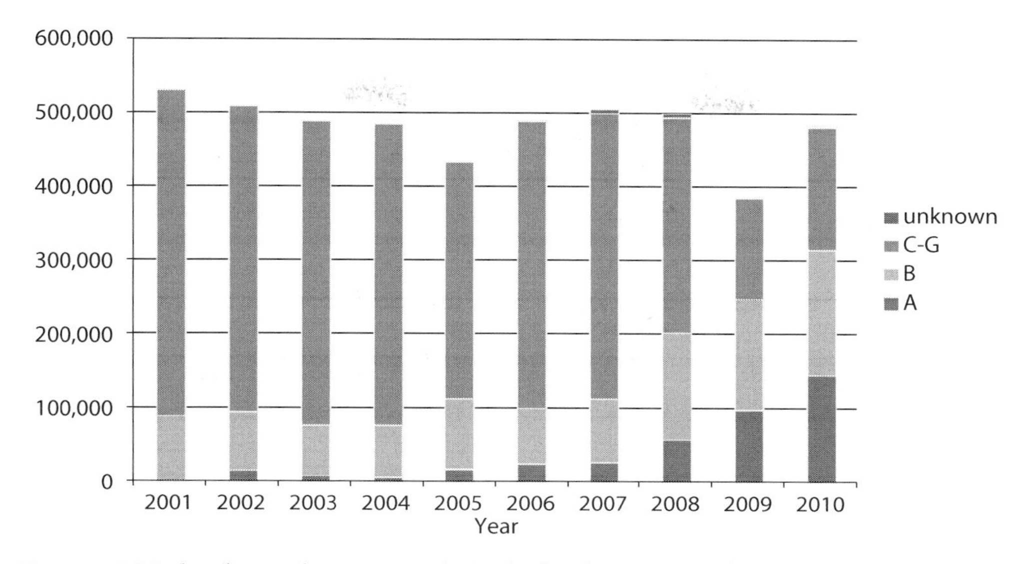

Figure 4.6. Market shares of new cars in the Netherlands per energy class (Bovag, 2012).

a more substantial increase in the number of A- and B-labelled cars since 2008. The amount of A-labelled new cars increased from 5% in 2007 to approximately 30% in 2010. Similarly, the percentage of B-labelled cars has increased from 17% in 2007 to 35% in 2010. One explanation for this outcome is that the taxation scheme introduced in 2008 not only provides a higher subsidy for energy-efficient cars, it simultaneously places a significantly stronger levy on energy inefficient cars. Though one should be careful to claim a direct and causal relationship, it is likely that the new fiscal measures which were implemented in 2008 played an important role in the stark increase of A-and B-labelled cars.

However, other developments such as the economic crisis are likely to play a role as well. In addition, Ecorys (2011a) explains the new trend of 'downsizing' in the sales of new cars. The market segment of the two smallest car classes has increased from 33% in 2005 to 50% in 2010. Small and compact cars have therefore become increasingly popular when compared to medium or large sized cars.

While the sales of energy-efficient vehicles provides a clear indication of the influence of price mechanisms, this effect can also be illustrated by taking a closer look at the sales of the Toyota Prius, the most sold hybrid car in the Netherlands (Table 4.7). We can see that there was virtually a non-existent market for this car before 2004. After the introduction of the special tax subsidy to stimulate the introduction of hybrid electric vehicles the sales increased rapidly. The introduction of a new type of Prius formed a contributing factor in the increase in hybrid car sales. The new Prius had a much more attractive design and was marketed not as an environmentally friendly car but intentionally as an innovative and high-tech car (interview Versteege, 2006). Interesting to note is that the hybrid car sales dropped directly after the tax subsidy was reduced from € 9,000 to € 6,000 in the year 2006. The price difference between the Prius and comparable cars apparently

Table 4.7. Sales of Toyota Prius in the Netherlands (Bovag).

2002	2003	2004	2005	2006	2007	2008	2009	2010
63	18	1,107	2,736	2,388	2,229	6,415	8,302	7,832

had become too large for consumers. In a reaction to these decreasing results Toyota Netherlands reduced the price of the Prius by € 2,000 in February 2007.

More noteworthy is the enormous increase in Prius sales in 2008. This is primarily the result of two additional fiscal measures which explicitly target the car leasing market. As discussed, in 2008 fuel-efficient cars (maximum of 110 g CO_2 per km) received a substantial reduction on the monthly charge which lease car drivers have to pay as taxes. In addition, the vehicle excise duty for these highly efficient cars was halved, both for lease car drivers and for private car owners. As hybrid cars, at the time, were one of the few car models to meet these criteria, consumer interest in these cars has greatly increased.

The share of HEVs in the total amount of new car sales increased from approximately 0.5% in 2007, to 3% in 2010. However, for newly bought company cars the share of HEV in 2010 was approximately 6%. As Figure 4.7 shows, HEV are increasingly adopted as a company car, especially in comparison to private cars. Even more interesting, the fleet of company cars has become more fuel efficient than the private car fleet. While in 2007 the CO_2-emissions of company cars and private cars were comparable, by 2010 the average company car emitted 7% less CO_2 per kilometre than the average private car (Ecorys, 2011a).

Increasing attention for corporate social responsibility, together with the economic crisis and the aforementioned fiscal policies have drastically changed the character of the company car (*ibid.*).

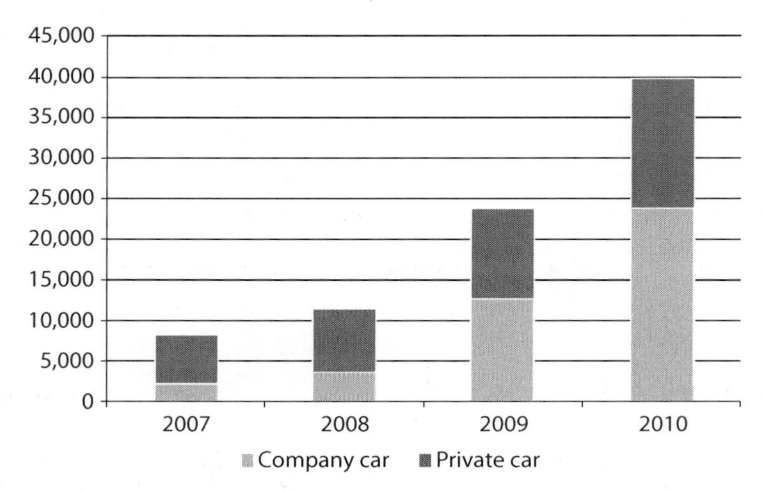

Figure 4.7. Hybrid car fleet in the Netherlands, 2007-2010 (Bovag, 2012).

Interesting to note is that the importance attached to green leasing is primarily a demand-driven development. A survey conducted amongst 2,000 European fleet managers revealed that the environment is on the top of the agenda of fleet managers (see Appendix A). Under the heading of corporate social responsibility, green procurement has become an important business concern. One of the outcomes of green procurement in car leasing was that many employees were stimulated or even restricted by their employers to lease an A, B or C-labelled car only.

4.7 Review of the results and lessons for consumer-oriented policies

This chapter began by stating that an increasing number of consumer-oriented strategies are being initiated that (aim to) make use of consumers as agents of change. These strategies are based on ideas of political modernisation in which typical command-and-control policies and legislation are complemented by so-called new environmental policy instruments. In this regard, Mol speaks of 'informational governance' as the production, the processing, the use and the flow of – as well as the access to and the control over – information which is increasingly becoming vital in environmental governance (Mol, 2008). Both the in-depth analysis at the level of the showroom as the analysis of car sales provide interesting input for discussion about the use of environmental information in the practice of car purchasing and the effectiveness of Dutch consumer-oriented policies.

The focus groups show that there is a clear need for help in reducing the complexity of the environmental information in the car-buying process. Much environmental information is abstract and hard to understand, whereas consumers should be stimulated and encouraged to compare cars at the level of sustainability. While one would expect consumer knowledge about the fuel economy label to have increased since 2006 (the year of the focus group research), more recent studies do not support this. A stated preference research conducted by the Netherlands Assessment Agency shows that knowledge of the label is predominantly limited to owners of A-labelled cars (PBL, 2009).

On the basis of the results from the focus groups we can distillate three general conditions that promote the successful use of environmental information in the purchase of new cars. First, environmental information that is actively provided in the search process of consumers is more likely to be accessed. This means that these tools should be actively provided to potential purchasers in the places that matter, such as the showroom, websites of car producers, automobile associations and car magazines. The increase in environmental information at these consumption junctions which has taken place in the last couple of years can therefore be seen as a positive development. Second, consumer interest increases when this information is communicated in an attractive and comprehensible way. Regarding the comprehensibility, the focus groups outcomes indicated that without clear explanation the energy efficiency label is hard to understand for consumers. Nowhere on the label is the meaning of A-G explained, let alone the precise meaning of the relative energy classes. One could argue that the latter is unnecessary as consumers will most likely search for cars within one specific car class. However, in the focus group and in an evaluative research conducted by Stienstra and Jansen in 2001 it turned out that most dealers display cars with a small engine which influences consumer's perception of the energy efficiency of a specific car model. Furthermore, numerical visualisations of CO_2-emissions on the label most of the times have no significant meaning for consumers. This critique holds even more truth for the ADAC Eco Test whose long numerical lists and it's unclear measurement methods simply puzzle both consumers and salesmen. Third, it is important that environmental information be provided and/ or supported by a 'trustworthy' source.

Next to these general conditions that can influence interest in the environmental information tools, contextual factors play a determining role in how consumers interpret these information tools. The other information formats through which environmental information can be communicated, namely face to face communication, advertising and in-store presentation have been shown to influence the way environmental information is provided and accessed. The way the information tools are presented in the showroom, and the role of the salesmen that work there, obviously can play an important role in stimulating the use of the environmental information tools. Even though increased consumer empowerment has changed the role of the car salesmen to that of an advisory function, as consumers are increasingly knowledgeable about car characteristics, these salesmen could still play an important role during the sales process, certainly with regard to less tangible information such as the environmental qualities of a car. The car salesmen were without a doubt knowledgeable enough about the energy efficiency label. The car salesmen in the focus group also were not unfavourable of providing environmental information or using environmental information tools during the purchase process (if the latter are made more comprehensible). These aspects seem to suggest that throughout the years some positive changes have taken place in the automotive industry. While in 2001, the year in which the energy efficiency label was introduced, car manufacturers strongly opposed the idea of environmental information provision, by 2006 the car salesmen felt that environmental aspects of cars are important elements in the practice of car purchasing. In addition, in comparison to 2001 the capabilities and willingness of car salesmen to provide environmental information about car characteristics has improved.

However, at least in 2006, the salesmen saw sustainability primarily as a bonus and not as a stand-alone selling-point. So the strategy of the car salesmen in the showroom can best be described as passive information provision; once a consumer has shown interest in environmental aspects the topic is addressed and explained. Furthermore, little information in the showroom is available that points consumers to environmental aspects. The energy label, which is present because it is compulsory, clearly has not bridged this gap. As the energy efficiency label is seen as being too complex, the car salesmen have explicitly expressed that the information provided by this label is ignored and qualified to consumers as being unimportant. Also salesmen have no real incentive to direct consumer's attention to the label. On the contrary, because profit margins on new cars are so slim, salesmen are stimulated to sell as much extra's as possible, including bigger engines. The need to change behaviour is therefore present both at the consumer and the provider side.

Though disclosure of environmental information in the practice of car purchasing could improve, the car market analysis indicates that the energy efficiency label and the accompanied fiscal measures are likely to have been influential in stimulating market demand for energy-efficient cars. On the one hand, the Dutch taxation scheme which aimed to influence consumer car choice can be seen as a catalyst for change. Especially for hybrid cars, car sales show an almost direct relationship with the introduction and height of the tax subsidies. Also, for the smaller car segment a subsidy of € 1,400 or € 700, the current subsidy for A- en B-labelled cars, can be a substantial and decisive factor. On the other hand, the influence of the tax subsidy or penalty in the bigger car segments is limited, not in the least because consumers in this car segment are less price-sensitive. Furthermore, the fact that the final catalogue price is a concoction of many different components also makes it easy for car dealers to make the levies less visible (MMG Advies 2008). In addition, the relationship between the different label classes (A-G) and the size of the car (mini, compact, executive, etc.) is important. While the goal of the Dutch version of the fuel efficiency label, with its classes based on relative performance, was to provide an action perspective for each car size, in practice by far the majority of A- and B-labelled cars were to be found among the smaller car sizes. The Netherlands Assessment Agency therefore concludes there is a mismatch between the supply of A- and B-labelled cars on offer and consumer preferences for medium- to large-sized vehicles (PBL, 2009).

On the basis of an European-wide analysis of the fuel efficiency label, ADAC (2005) concludes that the main source of CO_2 reductions of new cars are technical developments. Ecorys (2011b) comes to the same conclusion by stating that autonomous technological improvements in new cars have reduced the CO_2-emissions of cars in the Netherlands by 5% a year between 2007 and 2010. The informational and fiscal strategies have further reduced these emissions with an additional 1.3% (*ibid.*).

When the energy efficiency of Dutch new cars is placed in an European perspective another interesting insight can be gained. Whereas the CO_2-emissions of new cars in the Netherlands in 2001 were above the European average, by 2010 the CO_2-emissions were so substantially reduced that they were below the European average (Figure 4.8). So, while the energy efficiency of European cars is improving as a whole, the Dutch fuel efficiency is improving in an even faster rate. It can be no coincidence that this faster rate kicked off in 2008, the year that the new fiscal policies were implemented. It is more than likely that the consumer-oriented strategies in new car purchasing which were discussed in this chapter played a crucial role in this development.

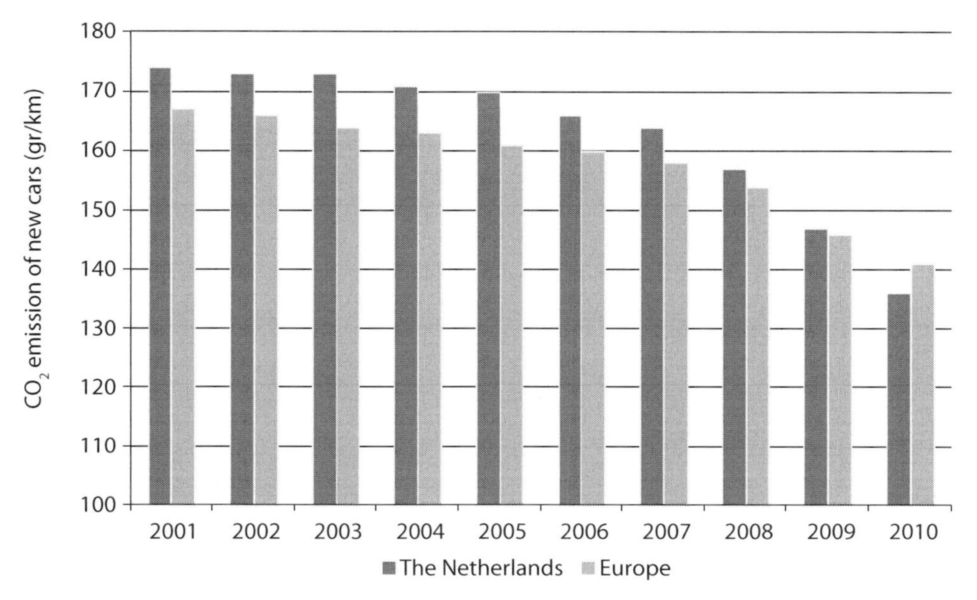

Figure 4.8. Average CO_2-emissions (gram/km) of new cars in the Netherlands and in Europe (Bovag, 2012).

4.8 Conclusions and discussion

4.8.1 The ecological modernisation of automotive production-consumption chains

The amount of environmental information being produced, and the availability and accessibility of this environmental information have, especially in the Western world, increased to a great extent over the last decades. The quantitative and qualitative increase of environmental information has changed environmental reform to such an extent that scholars even denote that a new mode of environmental governance has materialized (Mol, 2005). While environmental information has always been important as an enabling condition for environmental reform, recently though environmental information has become a constituting and transformative factor (*ibid.*). It is through the monitoring of environmental flows that ecological concerns become tangible, i.e. monitoring enables the incorporation of ecological concerns into political decision-making processes and industrial design (Van den Burg, 2006). More recent is the change that has taken place in the form and function of environmental monitoring and information. Not only is environmental information used as a tool by governments and industries, environmental monitoring and environmental information are more and more used as tools to inform citizen-consumers (*ibid.*). At different points in the production-consumption chains environmental information is made visible. At the access points where producers and consumers meet, environmental information is displayed in numerous different ways: on products, on information sheets, on eco-labels or on websites. Also environmental NGOs have found new ways of using environmental monitoring and information to influence production-consumption chains. By using (previously disclosed) information about the environmental performance of car manufacturers and their products, NGOs can exert influence on both ends of the production-consumption chain[65].

In this chapter we investigated the role of consumer-oriented informational and fiscal strategies in the practice of new car purchasing. Reverting back to the question posted in the introduction, is the increase in environmental information and monetarisation which were portrayed in this chapter a sign of an increasing ecological rationality in the practice of car purchasing? Are we witnessing an ecological transformation which points to the beginning of the large-scale diffusion of greener vehicles?

According to the theory of ecological modernisation, the ecological transformation of the automotive production-consumption chains (more specifically in this case study the car purchasing practice), can be traced by three consequential steps. First, through environmental monitoring the relevant environmental flows are made visible and tangible. Information about the environmental characteristics of a car can be used by different actors to integrate environmental concerns into various decision-making processes, whether the user is a representative of a car manufacturer (e.g.

[65] The case of Transport and Environment provides an interesting example. In the report 'How clean is your car brand?' T&E monitored and revealed the lack of progress in carbon dioxide reductions by the global car industry. While it was fairly common knowledge that car manufacturers would not be able to meet the voluntary agreements made with the European Union, the performance of individual brands was unknown as they were not revealed by the car manufacturing associations (ACEA, JAMA, and KAMA). Transport and Environment were the first to track the progress of individual car brands in reducing emissions, thereby bringing the environmental performance of manufacturers to the forefront and into the spotlight.

Toyota Netherlands where environmental monitoring is used to track the national average of CO_2-emissions of Toyota cars), the manager of a lease company (e.g. Leaseplan where environmental monitoring is used to identify a company's fleet carbon footprint and to provide advisory measures to clients), or an individual consumer (who uses environmental information tools to make sure a fuel-efficient car is purchased). Second, monetarisation is the process where monetary values are being placed on environmental aspects to stimulate and facilitate the incorporation of these aspects in decision-making processes. Finally, substitution is the process where the regular products are gradually replaced by the products with better environmental qualities.

In general, we see that the practice of car purchasing has changed considerably in the last decade. Consumers today are increasingly knowledgeable about (environmentally relevant) vehicle-characteristics and about the practice of car purchasing. This is the result of a strong consumer empowerment which has taken place in the last decade as information has become more accessible. The internet has become one of the most dominant sources of information, and in response the role of car salesmen has shifted from leading to guiding, and from a selling role to advising.

An important outcome of this trend is that the information age has disclosed information about the environmental impact of cars and about alternative vehicles that was until recently inaccessible to consumers. Though mostly initiated by the Dutch government, an increase in the provision of environmental information and fiscal instruments in the practice of car purchasing can be witnessed. In contrast to European-wide conclusions made by ADAC (2005) and TNO (2006) – which mentioned a lack of green car advertisements, and a lack of consumer interests in green cars – in the Netherlands we see different developments. Consumers are asking more and more environmentally related questions in the showroom and car advertisements are actively promoting the environmental aspects of new cars.

In addition, the car salesmen made the comparison between environmental information tools and the NCAP measurement of car safety. The focus groups show the importance currently being attached to car safety, the result of an emancipation process of the social dimension of automotive production-consumption chains which was initiated in the late 1960s. By comparing the environmental dimension with the social dimension the car salesmen reveal a gradual emancipation of ecological concerns in the automotive production-consumption chain.

4.8.2 Societal embedding of environmental information

The role of environmental information provision as a tool to influence the car purchase practice has received much attention over the last five years. For example, the transport advisory council indicates that it is crucial that consumers, at the moment of purchase, have sufficient access to environmental information (Raad Verkeer en Waterstaat *et al.*, 2008). However, this same advisory council also notes that the policies aiming to stimulate the purchase of fuel efficient vehicles (education, information provision, training, and fiscal measures) have achieved only limited success (*ibid.*). Similarly, in the car market analysis it was mentioned how various reports indicate that the consumer-oriented strategies only had a limited effect on stimulating fuel efficiency in the automotive sector (Ecorys, 2011b).

Two important points can be made about this conclusion. First, this chapter has pointed out the importance of the social context in which environmental information is provided. The social shaping of access indicates that one cannot assume that consumers are empowered with information solely by increasing the amount of environmental information (Van den Burg, 2006). We showed that it is worthwhile to analyse specifically what environmental information is provided, via which structures of provisions and how they are meaningful supported at important sites of the consumption junction. For instance, the focus groups show that current information tools have not successfully assisted in reducing the complexity of the environmental information. Furthermore, as this research shows, these processes do not end once a label is introduced; in the actual practice of purchasing a car, the meaning of the energy efficiency label is also contested by car-salesmen.

Furthermore, whether or not environmental information tools exert influence depends on the embedding in automotive production-consumption chains. The notion of embedding refers to the number of societal actors who use the label, and the number of instances in which the label is used because, by doing so, the label is reproduced and justified. In the first years of its existence, all legal requirements were fulfilled but not many additional organisations used it (Nijhuis & Van den Burg, 2009). Therefore the energy efficiency label was present in the showroom, but exerted little influence. In recent years, the label can be found more often, for example on automotive websites, but it is also more clearly linked to other policy measures, particularly financial incentives.

The second point is that the impact of consumer-oriented strategies and the correlating consumer purchase practice is greater than the direct environmental change. The success of a label not only depends on the question whether or not it is taken into consideration by consumers. To understand the impact of a label one should investigate how it is developed, how consumers' interests are articulated, the extent to which the label is contested and the ways in which it is embedded in society. Once a label is successfully embedded, it exerts power through various mechanisms. The fuel efficiency label has put a spotlight on the issue of fuel efficiency in the automotive sector. As such, labelling also exerts influence prior to consumer choice by directing corporations to develop and market different products. Therefore the strict distinction which evaluative reports often make between autonomous improvement in fuel efficiency of new cars due to technological developments, the additional effect of consumer-oriented strategies, and the changing car preferences of consumers is a dubious one as these aspects are clearly interrelated. The case of the European fuel efficiency label has shown that the introduction of the label not only means that consumers make different choices; it also means that producers make different choices. A structural effect of labelling was that car manufacturers brought fuel-efficient models on the Dutch market that had not been for sale beforehand (Nijhuis & Van den Burg, 2009; Van den Burg, 2006).

4.8.3 The system of provision structures consumer-oriented strategies

Another point for discussion is that there seems to be an unequal development in the ecological modernisation within the four systems of provision. In Paragraph 4.2, four different routes for acquiring a new or a used car have been portrayed. In this chapter we have focused on the practice of new car purchasing only.

However, each of the four described routes for acquiring a car is structured by a different system of provision with different participating actors. In line with the social practices approach it is suggested that the structuration of the system of provision heavily influences the way information is provided and accessed by consumers. In general one could say that the systems of provision for new cars and lease cars are much more structured when compared to the systems of provision of used cars, whether these used cars are sold by private consumers or not. It is also clear that the spread of environmental monitoring and monetarisation is not the same in the four different routes for acquiring a car that have been described in this chapter. While gradually environmental information about new cars is becoming more widely available, for used cars this information was until recently almost completely absent. Though used car websites provide very detailed information about car characteristics which allow consumers to specify one's car search, in 2006 environmental aspects were not one of these characteristics to choose from. Only since a few years these options have been made available. Furthermore, for used-car dealers environmental aspects are currently not high on the agenda. Therefore, car salesmen are in general very ill informed about the environmental aspects of used cars; also no information tools are available to help consumers or car salesmen to point out the environmental aspects of the potential car. Lacking a clear system of provision, needless to say there is little or no information available for consumer-to-consumer sales of used cars. Only an intensive search on car review websites will provide the potential buyer with information about the environmental characteristics of the used car.

Finally, for lease cars, the system of provision is completely different. The provision of lease cars is highly structured by lease companies and the fleet owners of employers. In the last five years a major transformation has taken place in the leasing branch. In the Netherlands attention and interest of the leasing branch for environmental aspects of car leasing is quickly increasing. Lease companies through various mechanisms are undertaking environmental initiatives. Through monitoring leasing companies make information available about the environmental performance of actors and car fleets. Analyses of emissions and fuel consumption is made available to fleet owners and individual car drivers, and in a significant number of cases this is complemented with advise on the possibilities of emission reduction, both at the level of the company's fleet and at the level of the individual driver (see Appendix A). Thus, brought back to its essence these lease companies have taken up a facilitating and advisory role in promoting green leasing. The provision of lease cars is also structured to a large extent by the company and its fleet owner. The company's environmental goals, expressed in green procurement as part of the corporate social responsibility programme, can influence the types of car on offer for the lease driver.

Finally, the difference in ecological modernisation between the four different ideal-typically routes are also reflected in the fiscal incentives provided by the Dutch national government. Currently, no direct financial incentives are present which stimulate consumers to take environmental considerations into account during the purchase of a second-hand car. The massive subsidies that have been introduced for the purchase of new cars are therefore in sharp contrast with the lack of incentives for used cars.

4.8.4 Changing consumer attitudes?

The improvement in the eco-efficiency of consumer vehicles in the Netherlands, surging roughly from 2010 onwards, had taken many by surprise. The commonly held perspective has always been that 'consumers are just not interested in eco-efficient vehicles'. A more fundamental question is therefore whether or not this recent phenomenon truly reflects a structural change in the consumer demand for eco-efficient vehicles (see also Dijk *et al.*, 2012).

The dominant view among policy-makers and industry representatives has always been that environmental considerations play a very minor role in decisions on car purchasing. Type, price, colour, distinction and so on are all considered to be more important. On the one hand, this is confirmed by the consumer focus group, where consumers first mentioned these aspects as being most important. Only a small niche exists for consumers who purchase environmentally friendly cars mainly because of environmental performance of these cars. On the other hand, we see that through various mechanisms an ecological rationality has slowly but gradually permeated the practice of new car purchasing.

Whether or not a structural shift in consumer preferences for alternative vehicles will take place also depends on the ways that car producers and environmental policies are able to connect to the everyday life world perspective of car purchasers. Research by Dijk (2011) indicates that car consumers consist roughly of three sub-groups with regard to engine preferences: one group for which engine price is most important (35%), one group for which engine size is most important (60%), and a green car segment (5%). It is likely that the strong focus of the Dutch government on fiscal strategies has appealed to the first consumer group thereby broadening the market of fuel efficient vehicles from the green car segment only, to the more price aware consumer.

However, these fiscal strategies have also placed a strong economic burden on the Dutch government[66]. To maintain the strong environmental improvement of the last few years in the future a more structural connection must be made with the values and preferences of the car purchasers as it is unlikely that the massive subsidies will be maintained. Environmental information tools and fiscal measures are consumer-oriented strategies which aim to influence car purchase decisions at a rather rational and cognitive level. Chapter 3, however, showed the enormous importance of symbolic meanings attached to the car. Car companies, as the main providers of green cars, are crucial for the success of that incorporation process of environmental aspects and the life world of car consumers.

[66] In 2010 the reduced governmental income due to the taxation strategies for green new cars was approximately 500 million Euro.

Chapter 5. Towards a practice-based approach of everyday mobility

The analysis of why people travel, and whether they should travel in the way they currently are, is to interrogate a complex set of social practices, social practices that involve old and emerging technologies that reconstruct notions of proximity and distance, closeness and farness, stasis and movement, the body and the other.

John Urry (Mobility and proximity, 2002)

5.1 Introduction

In this theoretical chapter the aim is to establish an account of a practice-based approach to everyday mobility. This implies that the social practices model, as described in Chapter 3, is conceptually elaborated and specified for everyday mobility in order to understand the key dynamics of social behaviours and (environmental) change in this consumption domain. The different components of the social practices model need to be adequately translated in order to be useful for the analysis of practices of mobility.

As discussed before, there is not one specific practice theory which automatically candidates as the best approach to the study of mobility. Moreover, notwithstanding a few exceptions, so far a practice-approach is almost completely absent from the domain of mobility. Therefore, in this chapter the intention is to establish a practice approach to mobility by answering three sets of questions derived from the practice based approach presented in Chapter 3. First, we ask which core elements are implicated in a practice to be labelled a mobility practice. Which types of mobility practices can we discern and what are their major characteristics? Second, we explore how practices relate to the socio-material structures which enable and constrain mobility-routines in everyday life. What contextual factors do mobility practices have in common? Thirdly, we try to establish which variables help explain the individual variation at the life-style side of the social practices model. How do we analyse the diverging ways in which citizen-consumers conduct their mobility practices? What kind of mobility strategies and patterns do citizen-consumers employ to realize their projects and plans? These questions indicate that the central theme of this chapter is focused on understanding how mobility is constructed, not in the sense of analysing the socio-economic drivers behind the increase of (auto)mobility, but in the sense of what elements are essential for being mobile and which strategies human agents employ to effect movement.

In the next paragraph 'mobility practices' are explored by diving deeper into the 'social bases' of everyday mobility. As mobility generally is considered to derive from 'demand' in the sense of being undertaken by human agents in order to be able to perform certain activities at a certain destination, it is important to investigate the role of physical co-presence in mobility. Furthermore, in this paragraph the basic characteristics of mobility practices are introduced and elaborated. Although there are many types of mobility practices, it is argued that they all share the same four dimensions: a temporal dimension, a spatial dimension, a material/technological dimension, and an experience dimension.

In Paragraph 5.3 the concept of 'motility', as it has been developed by Kaufmann (2002) is explored. Motility refers to the potential of human agents to be or become mobile. This notion is useful to conceptualize the human agents' side of the social practices model as applied to the domain of mobility. It defines the structural dimension of mobility practices not (only) at the system-of-provision side of the model, in terms of available transport networks and technological infrastructures (place and location accessibility) but discusses the rules and resources of mobility practices specifically in terms of human agent's access to mobilities (portfolio), and their skills and cognitive appropriation which are all needed to participate in specific mobility practices. By focusing on the character of human agents' motility it is possible to better understand the differences between individual human agents in performing 'similar' mobility practices. The chapter is concluded with a first exploration and conceptualisation of practitioners and their mobility portfolios, a theme to be taken up in more detail in Chapter 6.

5.2 Practices of mobility

5.2.1 Co-presence: the social base of travel

Before describing the different elements of a practice-based approach it is important to consider the question why people travel in the first place. While it is easy to understand that people travel to perform activities at certain locations, it is vital to consider that these activities are often socially inferred. Therefore, in this section we will focus on the 'social bases' of travel in order to better understand why people meet and come together at specific locations. The word 'social', in this respect, should be considered in its broadest sense. The purpose of grocery shopping is more than just acquiring the goods you need as fast and convenient as possible. Similarly, the purpose of non-daily shopping for clothes and furniture has not merely a utilitarian function but has many social elements as well. Hence, the term 'fun-shopping' which has been introduced to indicate that the 'purpose' of this type of shopping is as much about undertaking a leisure activity (relaxing, looking around, experiencing new stores, and being together with family or friends) as it is about acquiring a new product.

According to John Urry (2002), geographers and transport experts have insufficiently focused on the social bases of corporeal travel. Transport researchers, so Urry claims, have treated mobility as a black box in which the demand for movement is taken for granted. Understanding the connectivity of social life should not begin with the types and forms of transport as they are mostly a means to socially patterned activities (Urry, 2003b, p. 156). Therefore, Urry asks the question why people travel physically in the first place: 'why bother with the risks, uncertainties and frustrations of corporeal movement' (Urry, 2002, p. 256). The answer for the desire to travel has to be found in the significance of corporeal co-presence (*ibid.*). This is in line with the sociology of Anthony Giddens who states that 'in the course of their daily activities individuals encounter each other in situated contexts of interaction – interaction with others who are physically co-present[67]' (Giddens, 1984, p. 64). So even though social life, due to the importance of internet and mobile communication, is increasingly becoming virtual and mediated by network, physical co-presence is still a central, constitutive feature of social life (Urry, 2003b; Collins, 2004).

Giddens uses a multitude of concepts, many of them derived from Goffman's typology of interaction, to discuss the meaning of co-presence. Gatherings refer to assemblages of people in contexts of co-presence. Context in this respect includes on the one hand the strips of time-space within which the gathering takes place, and on the other hand the physical environment (or 'locale') of interaction with its associated norms and values (Giddens, 1984, p. 71). Social occasions are formalized contexts in which gatherings occur; they can be said to provide the structuring social contexts of gatherings. So, while gatherings may have a very loose form with barely any social interaction, social occasions typically have a specific pattern of conduct (the work day at the office is a good example of a social occasion). Simultaneously, the same physical space can be the site of several social occasions. During gatherings and social occasions interaction between individuals may occur. This interaction comes in the form of focused and unfocused interaction. It is especially focused interaction which is important, not only because individuals coordinate their activities through facial expression and talk, but also because it introduces an enclosure of those involved from those not involved (*ibid.*, p. 72). A 'unit' of focused interaction is called an encounter. It are these encounters, sustained by talk, as Giddens explains, which form the guiding threads of social interaction.

Giddens adds to the work of Goffman that most encounters occur as routines with a rather complex structure. What might seem at first sight to be an insignificant interaction is actually very substantial but hidden from view by routinization: 'what is striking about the interaction skills that actors display in the production and reproduction of encounters is their anchoring in practical consciousness' (*ibid.*, p. 75). That is, most of the time all participants of a face-to-face encounter make use of the same collectively shared rules indicating the appropriateness of talk and behaviour in a specific context. These interaction skills are taken for granted and used in an 'automatic' way. When investigating these interactions in detail, they turn out to be guided by a rather complex configuration of rules and resources which actors draw upon to make the practice happen. The smooth, routine-like reproduction of social practices is essential for creating and maintaining trust of people in other people, in technologies, and more general their socio-physical surrounding.

Like Giddens, Urry (2002, 2003b) draws upon Goffman to indicate that travel is rooted in the demand for physical proximity to other people, places and events. Social life (of work, family, education and leisure) thus involves continuous processes of shifting between being present with

[67] Both Urry and Giddens make extensive use of the term co-presence as it is originally employed by Goffman. The full conditions of co-presence are found whenever agents sense that they are close enough to be perceived in whatever they are doing, including their experiencing of others, and close enough to be perceived in this sensing of being perceived (Gofmann, 1963, in Giddens, 1984).

certain people, places and events and with being distant (Urry, 2007, p. 47). Urry defines five social bases of travel which generate the necessity for proximity (Urry, 2003b, p. 163). First, travel for legal, economic and familial obligations involves proximity to people for formal obligations such as travel to work, weddings, court, etc. Second, social obligations involve less formal meetings with family and friends which are associated with spending quality time. This can also involve travelling to specific locations away from the normal patterns of everyday work and family life. Third, object obligations involve travel in which persons come together for a specific object, for instance, to work on an object or text which has a specific location. Fourth, co-presence for place obligations involves travel to sense a place directly, to experience a leisure place (*ibid.*). While the first three social bases of travel are face-to-face, this social base is 'face-the-place' (Urry, 2002). For instance, while cities have always been places of work and habitation, increasingly cities are also perceived, and indeed marketed, as centres of consumption: 'they are places in which advertising, shopping, and entertainment are incorporated into every aspect of urban life' (Hannigan, 1998, p. 65). Mommaas (2000) emphasizes the strategic dual character of the media and leisure industry in shaping the experiences of places. Both are mediators of experiences, the media because it digitalizes and broadcasts the experiences, and the leisure industry because it attracts people physically to the location of experiences. As such, there is an increasing integration between the global media-industry and the local physical world of leisure (*ibid.*). Finally, event obligations are about experiencing a particular event. As these events occur at a specific moment in time they are also labelled 'face-the-moment' (Urry, 2002).

Face-to-face encounters are important because they contribute to the establishment and confirmation of trust. Social talk is an important part of establishing trust, and this can seemingly be conducted without being co-present. However, during face-to-face encounters individuals can literally look each other in the eye while discussing a variety of topics and trying to correct possible misunderstandings (Urry, 2007, p. 236). Indeed, some conversations and decisions are only possible, or at least deemed appropriate, in physical co-presence[68]: 'trust is something that gets worked at and involves a joint performance by those in such co-present conversations' (Urry, 2002, p. 260). It is this joint performance which Giddens emphasizes when he notes that the interaction skills that actors display in the production and reproduction of encounters are anchored in practical consciousness. In contrast to face-to-face encounters, face-less commitments and communications are characterized as being more functional and task-oriented, less rich and multi-faceted. They are in some respects less effective in establishing long-term trust relations (Urry, 2003, p. 170). In a similar vein, Randall Collins in his theory of Interaction Rituals states that co-presence and mutual focus are key factors in bringing about a process of collective effervescence. People travel along interaction ritual chains in order to meet, look each other in the eye, get 'excited' and become energized with 'emotional energy' (Collins, 2004; Spaargaren, 2011).

The social bases of travel must be considered as highly relevant for the study of sustainable mobility. For instance they are part of the social context in which mobility innovations have to be embedded. The importance of this becomes clear when looking at, for instance, the question to

[68] For example, while it can be considered appropriate to notify a person by phone of a positive outcome of a job interview, reversely, it is often considered highly inappropriate to tell the same person he/she is fired via this same medium.

what extent virtual travel can become a substitute for physical travel. Geels and Smit (2000) have labelled the neglect of social aspects in technological trajectories a form of 'narrow functional thinking' which helps explain the numerous failed technology forecasts in the mobility sector. They ask the question why so many images about technology futures, such as the envisioned impacts of ICT on traffic and transportation, were wrong. Famous examples are the promise of a reduction of business travel due to teleconferencing, a reduction of commuting due to teleworking, and a reduction of shopping travel due to virtual or internet shopping. Geels and Smit note that purely functional thinking about shopping activities in which virtual shopping substitutes for physical shopping neglects the social aspects that are involved, such as elderly shoppers who like to talk with retailers, neighbours and acquaintances they encounter in the shop (*ibid.*). Illustrative too in this respect is research with regard to teleworking in the Netherlands (Walrave & De Bie, 2005). While 77% of the non-teleworkers is interested in teleworking, many of them feel that the job is not feasible for teleworking because they need frequent face-to-face contact with colleagues (38%) or face-to-face contact with customers (28%). Furthermore, while according to 50% of the teleworkers the productivity of their work has increased due to working from a distance of the office, 60% of the teleworkers claim that the alternative work-type has a negative effect on their promotional opportunities. Both teleworkers and non-teleworkers strongly belief that teleworking has a negative effect on the social contacts with colleagues and that one becomes less involved in corporate activities, and one experiences a lack of company information. While managers of non-teleworking companies fear a lack of control and high ICT-costs, managers of teleworking companies are much more positive about these aspects (*ibid.*). So, considering the abovementioned role of trust in face-to-face meetings, it is therefore no coincidence that lack of possibilities for trust-performances is one of the greatest obstacles for the introduction of teleworking in non-teleworking companies. As Geels and Smit (2000, p. 876) remark: 'the process of the societal embedding of teleworking does not proceed automatically and smoothly. Instead, all kinds of practices need to be adjusted and changed'.

Because of the social bases of travel the prophesized death of distance (see Cairncross, 1997; Makimoto & Manners, 1997) remains improbable. These remarks are not suggesting that virtual travel and virtual co-presence can never simulate (to use Urry's terms) some forms of physical travel. On the contrary, activities such as teleshopping, telecommuting and teleconferencing have increased tremendously in recent years and might indeed on the long run substitute for certain trips. Indeed, the main message of the research by Walrave and De Bie (2005) is that many teleworkers are very enthusiastic about the possibilities of teleworking. In an increasingly networked society there are clearly significant opportunities for virtual travel to 'simulate' some social bases for physical proximity, while simultaneously there are many other social bases for which virtual proximity is no substitute (Urry, 2002, p. 269). Instead of 'eliminating' distance, it is more likely that transport and communication technologies are becoming 'travel partners'. According to Urry (2007, p. 179) this leads to processes in which new information and communication technologies co-evolve with extensive forms of physical travel.

The specific mobility practices and their characteristics, which we have shown to be based on patterns of social occasions and encounters, are described in the next section.

5.2.2 Defining mobility practices

In this section the specific mobility practices will be described and selected. This is a matter which deserves special attention because these practices will be used as the unit of analysis and for the formulation of a citizen-consumer oriented environmental policy in the domain of mobility. However, as the paragraphs on practice-theory in chapter three showed, practices are not a clearly defined unit of analysis. To circumvent this problem Spaargaren *et al.* (2002) have formulated two general requirements for selecting sets of social practices which can be said to be relevant for environment and climate policies. The first criterion is that social practices should be environmentally relevant. That is, the formulated social practice should be policy-relevant because of the environmental impact that is implicated in the social practice. While every social practice to a more or lesser extent has an environmental impact, it is more worthwhile to focus on the 'larger fishes in the pond' than on the smaller ones. The relevance for environmental policy is also related to the question whether or not the social practice can be targeted with (existing) environmental policy instruments. The second criterion, formulated by Spaargaren *et al.* (2002), is that social practices should be relevant for and recognizable by citizen-consumers. The practices have an everyday character in the sense that they constitute familiar repetitive activities of day-to-day social life. Furthermore, citizen-consumers recognize and acknowledge that the social practice generates environmental pressure, that this pressure is related to their own consumption behaviour and that there are means and methods available to help reduce the environmental pressure.

While there has been very limited research conducted on mobility from the specific viewpoint of practices, the majority of practice approaches to mobility are based on transport modalities and the networks surrounding these modalities. Deriving their analysis predominantly from science and technology studies, especially actor-network theory, many of these practice approaches focus on the interaction between mobility behaviour and mobility technologies. For example, Peters (1999) illustrates this aspect by looking at the travel practice of cycling in different contextual settings. He explains that although the bicycle itself and its (physical) way of being used have not gone through major changes, the historical changes in the networks surrounding the bicycle have significantly altered the practice of cycling and the meaning associated with cycling. Simultaneously, Peters describes that specific cycling policies, including the design of cycling infrastructures, provide a contextual setting which highly influences the practice of cycling[69]. Similar to Peters, Urry (2007) describes the multiple kinds of movement (such as the different modes of 'doing walking', car driving, etc.) and the systems that move human actors. In sum, the majority of practice approaches to mobility focus on the systems of mobility and the way that a specific modality and its user interact.

---

[69] A famous example is 'bicycle-city' Houten which was constructed (literally) with a spatial and infrastructural design in which precedence is explicitly given to slower modes of transport. The bicycle infrastructure forms the frame for major facilities such as schools and shops while the connections by car are largely restricted to a central ring around the town. Peters indicates that the construction of Houten as a bicycle city is a clear product of the 1970s during which the modal shift from car use to other means of transport was highly promoted (see Peters, 1999, p. 37). More recent examples of innovative biking infrastructures are to be found in the many city-biking-projects which have been developed to deal with urban pollution and congestion problems while making the city accessible to bikers and biking tourists in particular.

Quite a different practice-based approach is provided by Stock & Duhamel (2005) who have developed a typology based on the 'geographical code of practice'. This code takes into account the conditions in which the movement occurs and the qualities of the geographical place involved. In focusing on the geographical code Stock and Duhamel emphasize the influence of the character of places on the experience and conduct of different mobility practices (Table 5.1).

For example, tourism mobility is always based on the first two elements of the code in the sense that it always involves a non-daily activity and a personal choice of movement, while all the other elements may vary (familiar/unfamiliar, far-away/near, non-exotic/exotic). The fundamental difference between tourism mobility and business trips to these authors is the distinction between obligation and choice to go on the journey. The five elements of which the geographical code is constructed leads to 64 possible combinations of the conditions of practices in relation to qualities of a place (*ibid.*).

Distinguishing practices on these journey types is interesting because it corresponds with the social bases of travelling which we have discussed in Paragraph 5.2.1. Meeting relatives and friends, travelling from home to work, visiting places and events during leisure time, bringing kids to school, going on business trips to meet clients; these are just a few cases of social practices in daily life which involve mobility. These everyday journeys, as Pooley *et al.* (2005) argue, form an important part of the social fabric that constructs our daily life.

Therefore, based on the considerations mentioned above we suggest the following set of mobility practices to be considered in the context of environment and climate policies: commuting, business travel, home-school travel, shopping, leisure travel en visiting family/friends (Figure 5.1). In addition to this, the practices of shopping and leisure travel can further be divided into a set of less commonly recognized sub-practices. Leisure travel, for instance, is made up of event travel, day trips, etc. It is clear that these practices are not necessarily mutually exclusive, neither is mobility

Table 5.1. Examples of a geographical code of practices (Stock & Duhamel, 2005).

Example of practice	**Geographical code[1]**	**Type of practice**
A Londoner going on holiday in Morocco	non-daily/choice/unfamiliar/ far-away/exotic	tourist practice
A Londoner going to Marrakech for a conference	non-daily/obligation/ unfamiliar/far-away/exotic	business trip
A Brightonian going to work in London	daily/obligation/ familiar/near/ non-exotic	commuting
A Londoner going on to Brighton to stroll on the beach	daily/choice/ familiar/near/ non-exotic	leisure

[1] The distinction of daily/non-daily refers to habitual and non-habitual mobility; choice/obligation refers to the autonomy in the decision-making; familiar/unfamiliar refers to whether or not the place is regularly visited; far-away/near refers to the accessibility to the place; and non-exotic/exotic refers to differences in language, habits, food, etc. (Stock & Duhamel, 2005, pp. 64-65).

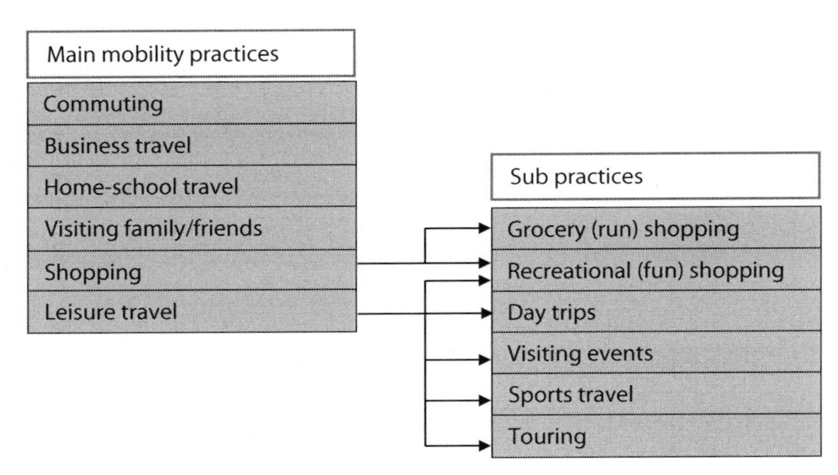

Figure 5.1. Main mobility practices.

limited to these practices. However, these six practices are the most commonly distinguished journeys in research, and are recognized by travellers all over.

Furthermore, this conceptualisation of mobility practices is interesting because it relates closely to the specific target group approach developed under the heading of mobility management. The European Platform on Mobility Management distinguishes a multitude of trip purposes which all have certain characteristics and provide specific opportunities and problems for the implementation of mobility management strategies: 'depending on their trip purpose, people visit different places, and origin/destination patterns may vary from "close" to "dispersed". The degree of freedom may differ from "limited" (e.g. in the case of commuting) to "high" (e.g. in case of leisure trips), and people are more or less flexible in their time management. Additionally, people will have different demands or preferences according to the properties of the chosen mode of transport for example if they need to carry goods or if they "only" want to enjoy the landscape' (www.epomm.org). This quote illustrates two relevant aspects. First, conceptualising mobility practices on the social bases of travel gives clear opportunities for mobility policies such as mobility management. Second, mobility practices are made up of a variety of dimensions. These dimensions will be further discussed in the next section.

5.2.3 Characterizing mobility practices: four dimensions of mobility

So far I have argued that mobility practices can best be conceived of as resulting from the participation of human agents in social occasions and social activities which take place at specific sites. Mobility is thus the result of situated human interaction at specific moments in time which take place at specific locations. The purpose of this section is to further define some of these core characteristics of mobility practices. The focus will be on the specific elements or dimensions which constitute mobility practices. Mobility practices can be described as the interplay between four different constitutive dimensions: the temporal, the spatial, the material and the symbolic dimension (Figure 5.2).

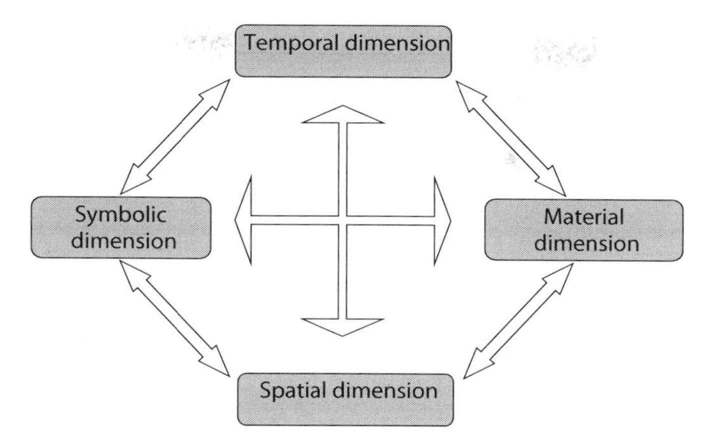

Figure 5.2. Four dimensions of mobility practices.

On the one hand movement implies that human agents negotiate space and time to create specific space-time paths. So, all mobility practices have a temporal and a spatial dimension. That is, the structuration of daily activities is connected with the specific space-time trajectories of human encounters. On the other hand, the concept of contextuality implies that social practices occur at determinate locations (with specific properties) in space and time. We will focus on structuration theory and time-geography to specify these concepts of space and time and how they relate to human agents, social practices and social structures. Next to temporal and spatial dimensions, mobility practices are dependent on the physical and material infrastructures of transport systems. So, the modal dimension of mobility practices refers to the use of a specific transport technology embedded in socio-material transport systems serving as systems of provision for mobility practices. Finally, the dimension of (lifestyle) experience is important to understand the instrumental and affective factors in the different mobility practices. Mobility practices cannot solely be understood by the temporal-spatial separation of locations and the different modalities that link them. To understand mobility practices it is important to know about the cultural aspects of mobility in modern society and the place and role assigned to being mobile. The four dimensions will briefly be introduced in what follows.

The importance attached in structuration theory to interaction in situations of co-presence in time and space makes it clear that mobility practices have both a temporal and a spatial dimension[70]. Indeed it is seemingly impossible to analyse mobility without considering the time-spatial patterns of mobility practices. The temporal dimension indicates that practices are

[70] The focus on timing and spacing correlates with classic time-geography which shares the focus on the routinized character of everyday life with the structuration theory. Time-geography is based on the premise that every action and event in a person's life has both spatial and temporal attributes (Pred, 1981). It is a geographical approach to understanding human behaviour in which time and space are resources which human agents draw upon to realize personal projects and plans. The possibilities to realize these projects and plans in time-space are shaped by various types of contextual constraints such as shop opening hours (Hägerstrand, 1970).

carried out over time: they have a certain duration or length of time, and they are undertaken at a certain moment in time. The temporal rhythm of opening and closing times of shops, companies, educational institutes and childcare facilities have a strong influence on the mobility patterns that people undertake. These temporal rhythms break the day into available time intervals or time windows (Schwanen & Dijst, 2003).

Furthermore, the majority of traffic jams can be understood as resulting from the temporal synchronicity of institutions and the collective impact of individual behaviour of human agents in the conduct of everyday mobility practices. Thus, traffic jams exist because too many people are on the road at the same time which leads to an 'inefficient' use (from a transport economic perspective) of the physical infrastructure. Also, mobility practices can have a distinct temporal patterning. Commuting evidently has the strongest temporal structuration with strong temporal peaks around half past seven in the morning and five o'clock in the evening during weekdays. In contrast, leisure related mobility practices have a different temporal structure, with less tight levels of synchronization[71]. Furthermore, leisure related mobility takes place more often in the evening hours during weekdays and during the weekend as a whole (see Harms, 2008).

Social occasions are made to happen at specific socio-spatial locations. The spatial dimension of mobility practices therefore refers to the place and space in which activities are undertaken, as well as the spatial distance travelled. Locations (or locales) are sites of interaction where people work, eat, sleep, visit friends and relatives, and are entertained. In everyday mobility the home is often the dominant location, it is the starting point of a trip and at the end of the day the home is generally speaking the final destination. An important point made by Giddens is that locations are not mere 'stopping points in space' (a criticism on traditional time-geographical approaches) but are also sites with a certain identity which influences the interactions that are occurring at the location. Giddens uses the term 'locale' to indicate the setting of interaction spaces (Giddens, 1984, p. 118). Locales not only contain material elements but also contain information, knowledge, rules, signs and images[72]. Thus, places and the practices conducted there are highly interwoven and co-constitutive (Urry, 2007). For example, hospitals, business parks, theme parks, and homes are all places between which people travel but which all exert a different contextual influence on the social occasions and often also on the modes of travel. This is related to the point that locales have a specific geographical location and a geographical spread which makes them more or less accessible. Especially in the Netherlands, with its strong planning tradition, more and more attention is being paid to the relation between spatial planning and mobility. Another significance of the spatial dimension becomes clear when we look at the distinction between different types of shopping practices. Run-shopping, which mostly consists of shopping for groceries, takes place only at a short distance from the home. Grocery shops are geographically dispersed as they are frequently visited by people. They are also much less centred when compared to shops for fun-shopping (clothing, multi-media) which are often clustered in city-centres.

---

[71] However, temporal synchronization is visible in the yearly peaks at airports and main roads connecting favourite holiday destinations in Europe.

[72] See Holloway and Hubbard (2001) for an overview of perspectives on the relation between people and place in everyday life.

The first two dimensions of mobility practices are thus related to the temporal and spatial aspects of more or less institutionalized practices and they refer to the creation of time-spatial paths by human agents. Peter Peters (2003) speaks of 'passages' to indicate the strong relation between the two dimensions. A passage indicates the creation of time-spatial orders in which space and time cannot be separated from each other. Peters stresses the point that the history of mobility must be seen as a sequence of mobility innovations (e.g. timetables, traffic information) which have been introduced to ascertain not only a faster, but especially a trouble-free and predictable journey (Peters, 2003, p. 221). Thus, the passages make routinization and trust in travelling and the transportation systems possible.

The third dimension of mobility practices refers to the material dimension. It is composed of the different modes of mobility which enable and constrain the possibilities for practitioners to be on the move. Indeed, being mobile is dependent on the means human agents have at their disposition when making use of different transport modalities. As knowledgeable and capable mobile actors, people have to know about the ways in which travelling by car, train or plane 'works' and about the many implicit and explicit rules that are implied in the successful use of technical objects and systems. While being labelled as the material dimension, it is important to keep in mind that a mode of mobility is reliant upon a whole set of socio-technical systems and infrastructures which are surrounding the material artefact and which help constitute it as modality. Each form of mobility, be it walking, cycling, car driving or train travel, presupposes the use of a mobility system (Geels, 2005c; Urry, 2007, 2004). 'The private car on leaving the assembly line is only a semi-finished product, since it requires a system of collective or public roads, signs, lights, etc. without which it becomes valueless' (Simonsen, 1973, in Otnes, 1988, p. 131). When people know how to drive a car, they are not just capable of driving the physical object but they know how to apply the rules of the traffic system, how to maintain the car, where and how to fuel it, where and how to buy a new one, etcetera. The fact of physical objects being interwoven into complex mobility systems also means that changes in the social practices of mobility entails not so much a case of substituting old technologies or technological artefacts with new ones; it might involve radical shifts from one kind of socio-technical system into another (Geels, 2005c; Hoogma *et al.*, 2002). These systems may function both as an inhibiting or facilitating factor in mobility transitions. However, because of the characteristics of mobility systems (or infra-systems in general) they have a tendency for incremental change and are less likely to undergo radical changes (Frantzeskaki & Loorbach, 2010).

The fourth dimension of mobility practices relates to the symbolic or cultural dimension of mobility. In Chapter 3 I have already outlined the various cultural and symbolic dimensions with which the automobile is associated. The car has not only become a dominant cultural symbol, it is also being associated with a specific type or modus of 'being on the move'. Urry (2004) talks about cocooning or dwelling the car to indicate that the car is increasingly becoming a home away from home in which car-drivers are skilled multi-taskers who can communicate with work or home, listen to music or simply enjoy a moment of privacy.

The symbolic dimension is not only related to the cultural framing of the transport modality but also includes the specific mobility experience of actors when being involved in the different mobility practices. Salomon & Mokhtarian describe that over two-thirds of the travellers disagree that the only important thing about travelling is arriving at the destination. Nearly half their respondents agree that getting there is half the fun. Also research has shown that travel experience may differ for different mobility practices. Generally speaking people like social and leisure journeys better than mandatory travel such as school and commuting journeys (Ory & Mokhtarian, 2005; Anable & Gatersleben, 2005). This means that travel experience in commuting is different from the travel experience when travelling to a leisure activity such as a large-scale event. As more and more trips are recreational by nature, this is reflected in the way people experience mobility in general.

In home-school trips children's safety is a major concern. Even though in most Western societies the number of traffic accidents with children has decreased for a number of years in a row, the social construction of risk perception together with the importance attached to safety concerns has a major influence on the modal choice in home-school trips. When parents have the feeling that an unguided bicycle trip to school is risky, they are hesitant to choose for this travel mode.

In commuting and business travel, traditionally, productivity of travel time is very important (Lyons & Urry, 2005). While most travel time in the car is currently unproductive time, the possibility to work and study in a train is argued to result in heightened productivity[73]. The trend of providing internet access in public transport systems in order to increase the possibilities of work and study, fits perfectly with the tendency of experiencing commuting and business travel as productive ways of being mobile.

In sum, the journey itself is therefore something that is worked upon in the sense that it consists of a multitude of activities which influences the way it is framed and experienced by the actors involved.

Herewith we have shortly introduced the four separate dimensions of mobility practices. However, what is most characteristic and informative of a practice-based approach to mobility is the integrated analysis of mobility conduct which does not give precedence to one specific aspect of mobility, but which investigates the core constitutive elements of mobility and their interrelations as a whole (Peters, 1999; Stock & Duhamel, 2005). These interrelations are also visible when the characteristics of the train system and the automobile system are compared. The train system is a system which is characterised by collective time-tables, fixed routes, clock-times, and public spaces in which it is more difficult to 'maintain' and 'repair' spatio-temporal orders of passages (to use Peters terminology) in a way that they fit specific lifestyles of end users. Furthermore, in order to reach one's destination with collective public transport it is often necessary to shift between different modes of transport at fixed nodes in the network. The system of automobility, in contrast, provides almost maximum (individualized) time-space flexibility: it has shifted clock time towards instantaneous time (Urry, 2007). Importantly, the automobile has made mobility patterns possible that seem hardly possible with non-automotive transport. So, the different systems of mobility do not only enable people to enact certain practices in a more efficient way, they simultaneously modify and define what those practices are about, how they are experienced and how they are configured and structured as 'normal practices' (Shove, 2002).

Most theoretical approaches to mobility tend to focus on one specific dimension, for example, in the sense that most mobility innovations are judged on the aspects of travel time and time

---

[73] Lyons and Urry (2005) agree that the specific journey also influences the productivity because of aspects such as the degree of crowding, availability of seating, ride quality, temperature, noise level, the degree to which the journey is familiar, reliable, and the duration and stage of the journey.

reduction. In contrast, a practice approach concerns an 'amalgam of issues, actors, and materials' (Peters, 1999, p. 42). In a similar vein, many policy strategies in the field of mobility which target mobility behaviour can be said to have a limited scope (Dijst, 1995). While policy strategies focusing on the use of mobility systems seem to imply that only the supply and demand is changed, in fact more is at stake. For example, Dijst (1995) points out that strategies aiming for a modal split (such as parking policies, and price mechanisms) not only have an effect on how people move, but also on the activity patterns of those people. That is, when people shift from car use to other modes of transport the pattern of time-spatially dispersed activities in a certain period of time changes as well, as do their ways of experiencing travel (*ibid.*). Thus, there is close connection between the spatio-temporal order, the system of mobility and its way of being used and experienced by practitioners. These interrelations in mobility practices are summarized in Table 5.2 in which the mobility practices are characterised.

Table 5.2. Characterisation of mobility practices.

Practices	Characterisation
Commuting	Commuting is the most habitually conducted mobility practice. It is undertaken very frequently and at fixed time periods resulting in high peaks in traffic around eight o'clock in the morning and around half past six in the evening. Also commuting journeys tend to be longer in time (30 minutes) and distance than other practices. The far majority of commuter trips are conducted by car (60%), though there is a relatively high percentage for public transport (10%) especially in the larger cities which have better access to public transport networks. Commuting is almost always a solitary activity and is the practice most associated with necessity. Because of the routinized character of commuting, the problems with congestion and the strong influence of the employers most policies have focused on this mobility practice. Employers increasingly consider their employee's mobility, and how it is conducted, as part of their responsibility and as a means of cost reduction.
Business travel	As business travel occurs during working hours it is the only practice which is not conducted in private time. Therefore the phrase '(travel) time is money' does indeed apply to business travel. Because of the lease-cars, the long distances travelled and the high value placed on flexibility, punctuality and representativeness, business travel traditionally is the practice with the highest car usage of all mobility practices. Because lease-cars make up approximately 40% of new car sales this practice is also important for the car purchasing practice. As with commuting, recently there is a massive increase in the number of private and public policy measures aiming to increase the number of green lease cars and the use non-automotive means of travel modes.

Table 5.2. Continued.

Home-school travel	This practice refers to journeys to and from primary school, secondary school and childcare facilities and is typically short-distance. How children and adolescents go to school depends for a large part on the children's ability, and the parent's approval, to travel to school independently. While the far majority of children (especially adolescents) still go to school by non-automotive transport there is an increase in car usage. Safety, and especially safety perception, is an enormous issue in this practice as children are one of the most vulnerable practitioners due to their fragility and their inexperience. Low safety (perception) in the vicinity of schools can lead to a downward spiral in which parents increasingly bring children to school by car, thereby adding to a disorderly and unsafe traffic situation, and simultaneously reducing their children's independency. There is increasing attention for this issue and schools, often by including traffic education, show an accommodating role in this.
Grocery shopping (run-shopping)	Grocery shopping involves main food shopping at the local shops and supermarkets, and the purchase of everyday household products. This type of shopping takes place very often, entails little time, and is done closely to home (with an average distanced travelled of 2.3 km). On average, each year people make around 230 trips to and from shops (this number includes fun-shopping). The quality and size of the supermarket (which has a market share of 70% in this practice) strongly determines the pull to the local shopping centre. However, the way that grocery shopping is conducted, that is on a weekly or on a daily basis, also highly influences the modal share. Grocery shopping conducted on a daily basis is done most often by walking and cycling. Regardless of the distance to the shops, most people who conduct grocery shopping on a weekly basis do their shopping by car. This is obviously related to the large quantities of shopping goods purchased (see AVV, 2006a; Mackett, 2003).
Recreational shopping (fun-shopping)	Fun-shopping occurs much less frequently than grocery shopping and has much more of a recreational character. The activity itself (having fun, relaxing, looking around, being together with family or friends) is often as important as the purchase of new products. Time is therefore much less of importance when compared to run-shopping. In this practice trendy and fashion-sensitive supply of products and services is very important. Large department and fashion stores, located traditionally at the city centre, are the most dominant pull factors. In recent years new retail formulas focusing on multimedia and sport have become attractors as well. Finally, because fun-shops are often concentrated in city centres these shops are more accessible by public transport. As a result fun-shopping has a relatively high share of public transport use. The choice of destination is often more flexible and may differ per shopping trip. Currently, there is only a limited knowledge about shopping trips and the impact of transport policy on shopping (see Harms, 2006; TTR, 2002; Van Beynen de Hoog & Brookhuis, 2005).
Touring	Touring concerns travelling (walking, cycling, touring by car/motorcycle) for its experience and intrinsic benefits: physical exercise, adventure-seeking, enjoying the scenery, relaxation, getting a fresh air (see Ory & Mokhtarian, 2005).

Table 5.2. Continued.

Day-trip	This leisure practice refers to going to theme parks and other recreational locations in which people enjoy a special day out. Spending leisure time together and having fun is considered very important. It differs from visiting large-scale events in that it is not a temporarily organised event and therefore the location is more used to high traffic flows. Often attractions are located far from city centres because of the space required. Also, increasingly municipalities and the larger entertainment parks recognise the importance of maintaining high levels of accessibility and also in promoting non-automotive mobility. With regard to the latter it is suggested that the journey itself can (and should) be part of the leisure experience (see KPVV, 2006; KiM, 2008).
Visiting large-scale event	An event is an organised recreational happening aimed at the general public which, due to its size and peak traffic flows, temporally disrupts the normal ways of life in the surrounding area (a single event may attract millions of visitors in a few days which overexert the available infrastructure). Events most often occur on holidays and/or weekends in the centre of large cities and they attract people from a wide range (on average the distance travelled is 30 km). Events are a joyful experience for the visitors which incite feelings of expectations and willingness to do and learn new things (see CROW, 2007; Ecorys, 2006).
Travel to sport activity	This practice refers to trips to either participate in or watch sport activities. Next to home-school travel and grocery shopping, this practice is conducted most closely at home. This practice is most often conducted during the weekdays after seven o'clock and during weekends at the end of the morning.
Visiting family and friends	This practice is primarily about being close with friends and family. Visiting family and friends accounts for 15% in number of trips and for 21% in distance travelled. Half of these journeys are conducted during weekends, especially the longer ones (during the weekends the distance travelled for social travel is at least twice as high as during weekdays). Because of the relatively long distance travelled and the low accessibility of many residential areas by public transport, often the car is used as a means of transport. Also this practice is more than other practices conducted by partners or by the whole family resulting in a high (car) occupancy rate.

5.3 Conceptualizing practitioners in the conduct of mobility practices

In the remainder of this chapter we will go deeper into how practitioners conduct their mobility practices and which elements are necessary to be able to participate in mobility practices. According to the methodology suggested by Giddens in the context of his structuration theory, social practices can be analysed from two different angles. First, in the mode of 'institutional analysis', the focus is on the rules and resources as chronically reproduced features of social systems (Giddens, 1984, p. 288, 375). In this type of analysis attention for the structural properties of social systems is favoured over the skills, awareness and actions of individual actors. Visualized in Spaargaren's social practices model this means that practices are analysed from the right side of the model (Figure 5.3). When pursuing an institutional analysis of the reproduction of social practices, the researcher is bracketing the motives, values and interests of the situated agents and their lifestyles as they are involved in the reproduction of the practices. These 'actor-related elements and questions' are the central focus of the second mode or modality of researching social practices: the analysis of strategic conduct. In this mode the focus is on how actors reflexively monitor what they do and draw upon rules and resources in the constitution of interaction (*ibid.*, p. 288, 373). This means that in the latter primacy is given to the discursive and practical consciousness of human agents, their motives and their strategies within specific contexts. Similar, this means that in terms of the social practices model, practices are analysed from the left side. Though methodologically the institutionalized properties of the practices are bracketed or taken as a given in the analysis of strategic conduct, Giddens reminds us that these properties are still subject to the reproductive capacities of human agents. *Vice versa*, human agents remain subject to the (changes in) institutionalized properties and their influence on human agents' ability to participate in social activities (see Appendix B).

In order to be mobile, practitioners need to acquire and put to use various modes of mobility. Part of this competence refers to the point that throughout their lives practitioners become acquainted with various systems of mobility. What is interesting about the conduct of human agents in mobility practices is that on the one hand it can be characterized by a clearly existing inertia; this refers to the difficulty that policy-makers experience in trying to influence people's everyday mobility patterns in order to reduce congestion, to improve traffic safety, and to reduce environmental impacts. Thus, practitioners seem to be persistently resistant to change once a particular habit has been established. On the other hand, throughout their lives people continuously give shape to their mobility patterns, often in a changing infrastructural context. The development of (social and technological) innovations continuously makes new modes of mobility available to practitioners, thereby increasing the possibilities to be mobile. In a historical analysis Pooley *et al.* (2005) describe how in various UK cities citizens have, apparently without much difficulty, throughout their life-course integrated various modalities into their mobility patterns. So, human agent's conduct in

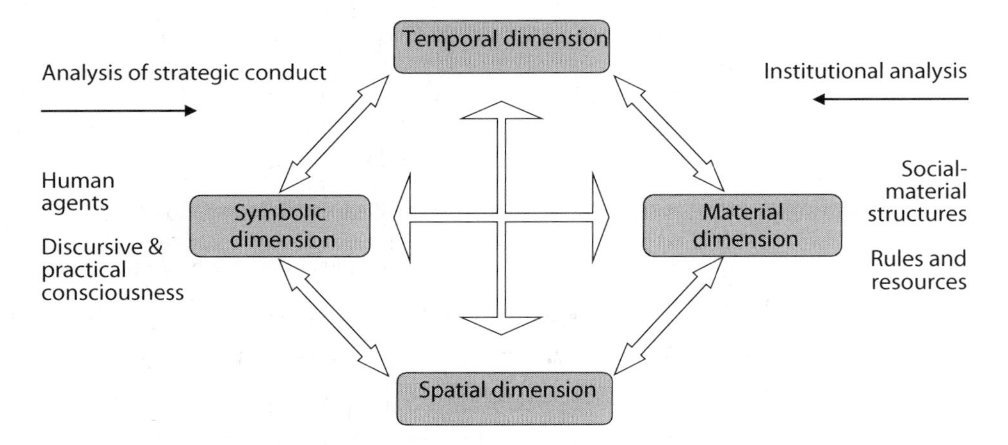

Figure 5.3. Institutional analysis and the analysis of strategic conduct.

mobility practices is both inert due to the routinized character of daily mobility, while simultaneously human agents adopt and adapt to the new mobility options (made) available to them.

An important question is how the variation in the conduct of human agents in mobility practices can be conceptualized. Especially the material dimension, involving how people travel, is relevant because herein lie important possibilities for the reduction in the environmental impact of mobility practices. Partially this conceptualization clearly involves the ownership of various 'mobility tools' by human agents such as cars, bicycles and public transport passes. Beige & Axhausen (2008) have indicated that through the ownership of mobility tools people to a certain extent commit themselves to a specific type of travel behaviour. Due to the habitual nature of mobility patterns and because of the investment costs put into acquiring these mobility tools, the specific set of mobility tools available to practitioners can have a substantial influence on how people travel. In their analysis they have identified how the ownership of automobiles and various public transport passes shifts during the life course. However, while the concept of ownership of mobility tools is interesting, it does not quite grasp the complexity in the available modalities and the conduct of mobility practices. Firstly, as has been described, for instance by Jeremy Rifkin in 'The age of access' (2001), access to modalities is broader than the mere ownership of mobility tools. The widespread existence of car-leasing and the more recent development in the collective use of private transport modes such as car-sharing and bike-sharing, increasingly blur the boundary between mobility ownership and mobility services. Instead of ownership of mobility tools, the concept of 'access' better accounts for all the different modalities that human agents may put to use in order to be mobile. Secondly, as we described in Chapter 3, the conduct of practices involves more than the access to different modalities. It is also related to the knowledge and competences of human agents, and the evaluation of the different transport modalities.

A much more rich concept to understand the logic of human agent's conduct in mobility practices has been develop by Kaufmann (see Flamm & Kaufmann, 2006; Kaufmann, 2002, 2004; Kaufmann *et al.*, 2004). His notion of 'motility', which focuses on human agent's potential to be mobile, encompasses and integrates the various elements which are important to be able to participate in practices. Motility is defined as 'how an individual or a group takes possession of the realm of possibilities for mobility and builds on it to develop personal projects' (Flamm & Kaufmann, 2006, p. 168). The motility or mobility potential of human agents is constructed out of three components: (1) access to different forms of mobility; (2) the competence and knowledge to recognize and make use of these forms of mobility; and (3) the cognitive appropriation which describes how human agents assess and evaluate (perceived) access and skills. In short, motility therefore is about the ways that human agents access and appropriate the capacity for mobility (Kaufmann, 2004).

In order to broaden the scope of ownership of mobility tools, Flamm & Kaufmann (2006) use the term 'personal portfolio of access rights' to indicate all the mobility products and services available to individuals or households. This mobility portfolio includes the privately owned automobiles and light vehicles (bicycles, scooters, etc.), and access to mobility services provided by mobility providers (through public transport passes, membership(s) of new mobility services such as car-sharing, bike-sharing, etc.). Evidently, access is also heavily influenced by the specific mobility context such as the spatial morphology (population density, number of parking spaces, etc.) and the available public transport networks within a geographical area.

Competence refers to the physical ability, the acquired skills and the organizational capacity of human agents to make use of the available modes of mobility. For example, as indicated in Chapter 4, the lack of a bicycle culture and the lack of cycling skills is an important barrier for (female) immigrants in the Netherlands to use a bicycle. Similarly, Kaufmann indicates that the act of travelling involves a complex set of cognitive and physical activities which not only involves the obvious skills of driving a vehicle, but also the skills of planning and timing a trip with a specific modality. Especially when one has the choice between different modalities and/or when one is unfamiliar with the destination of the trip the organization of a journey can be a highly complex decision process in which different modalities and uncertainties need to be weighed. In particular multi-modal trips can be difficult to organise for inexperienced multi-modal travellers, while these trips are also surrounded with uncertainty in the travel schedule in cases of delays. Especially the last 'chain' in a multi-modal trip is considered the weakest link as it is surrounded with high levels of uncertainty. While, the supply of new mobility services could strengthen the links in multi-modal trips, most travellers have no knowledge about, or access to, these mobility innovations. Also, research on multimodal traveller information has repeatedly revealed that most travellers do not consider their modal choice for the majority of the trips (see also Kenyon & Lyons, 2003).

The final component of motility refers to the human agent's strategies, values and representations of the various systems of mobility. This cognitive appropriation is based on the social and individual representation and the evaluation of the extent in which various modes of mobility suit the realization of everyday activities in the domain of mobility. Because this process of appropriation involves a normative assessment of the functional and symbolic suitability of various forms of mobility, it is the most subjective component of motility. Cognitive appropriation is mainly dependent upon one question: 'which criteria are people likely to apply when evaluating a particular means of transportation to satisfy their way of organizing daily life, taking into account their overall lifestyle?' (Flamm & Kaufmann, 2006, p. 178). The criteria's used to assess the functional suitability of a trip predominantly refer to aspects such as travel time, flexibility, accessibility, reliability, travel costs, comfort and the capacity to convey goods. While it is evident that some modalities are better suited to fulfil these criteria, more important is the question how certain habits in the use of modalities are constructed. Flamm & Kaufmann indicate that personal experiences are very important because of the way that skills and cognitive appropriation are constructed. Skills are to a large extent specific to a certain travel mode and these skills need to be maintained with regular conduct. The authors indicate that on the individual level the learning-process of acquiring skills and cognitive appropriation of travel modes is often a self-reinforcing process. The more a human agent becomes acquainted with a travel mode, the more skilful the agent becomes at travelling with this travel mode and the more this means of transport is appreciated. This relation is also confirmed by Harms (2008) who indicates that there is a positive relation between the use and the evaluation of a mode of transport. Nevertheless, the functional characteristics of public transport are assessed much more negatively than the car and the bicycle, even by frequent public transport users.

Why is this conceptualization of human agents on the basis of motility such a promising approach? Clearly, motility allows us to conceptualize differences between groups of human agents in the conduct of mobility practices on the basis of meaningful characteristics. One of the interesting aspects of the portfolio of access rights (and of the concept of motility in general) is that the choice for a specific (component of a) portfolio is part of an individual's or a household mobility strategy. Flamm & Kaufmann (2006, p. 185) see this strategy in the light of controlling uncertainty. What these authors mean is that certain access rights are not purchased because of a desire to use them immediately, but more to be able to deal with uncertain or exceptional situations. Comparable with a savings account 'excess motility' may be acquired only or primarily to be used in times of need. In addition, it is clear that there a major differences in the motility of human agents: 'Depending on context, individual actors, groups and institutions differ in access, competence and appropriation, and have thus at their disposal different motility options' (Kaufmann *et al.*, 2004, p. 754). As mobility serves human agents to accomplish personal projects and plan, motility is an important factor in social positioning. The authors use the concept of motility therefore as much to explain and conceptualize social stratification as they do to understand spatial mobility (in that sense they focus both on the vertical and horizontal dimension of social position). Social and spatial mobility, they argue, are highly interrelated and because of it motility can be seen as a (new) form of capital, comparable to economic, cultural and social capital. Because of this argumentation motility is highly relevant for studying social in- and exclusion in relation to mobility.

Thus, motility influences human agent's (possibilities) to participate in mobility practices. On the one hand we intend to use the notion of motility, especially the concept of portfolio of access rights, to understand which mobility portfolios are prone to more sustainable mobility practices. Can we distinguish greener or less green mobility portfolios on the basis of accessibility to sustainable mobility innovations? Are green mobility portfolios in any way related to green attitudes and lifestyles? And, when mobility innovations become part of the mobility portfolio of human agents, thereby increasing the possibilities of the green conduct in mobility practices, does this indeed lead to more sustainable mobility practices? These specific questions will be addressed in full detail in chapter seven.

On the other hand we aim to use the concept of motility to better understand whether changes in the mobility portfolio of human agents lead to greener conduct in mobility practices. While Kaufmann analyses motility from the viewpoint of individual human agents in relation to a specific geographical area with a specific supply of transport modalities, we want to add the focus on the role of mobility providers and stakeholders in shaping human agent's motility. This addition not only makes the rather static concept of motility much more dynamic, it also provides information on how the system of provision has a structuring influence on how motility is constituted. Under the heading of mobility management increasingly stakeholders from the public and private sector (such as employers, lease companies, and governmental agencies) aim to influence the mobility choice of human agents. Often in these cases mobility innovations are 'made' accessible to human agents. While it is unproblematic to assume that in theory the motility of human agents may increase in such a changing context, the question is whether this newly acquired green mobility portfolio automatically leads to more sustainable mobility practices. For example, Weert Canzler, on the basis of an innovative car-sharing project, concludes that: 'the tendency of road users to consciously weigh the various transport possibilities does not automatically grow with the range of those services' (Canzler, 2008, p. 113). This corresponds with one of the theoretical starting-points of the CONTRAST-project, namely that merely increasing the supply of sustainable socio-technical innovations does not guarantee that sustainable consumption practices will emerge. As important is the question if and how sustainable innovations become integrated into mobility practices. This question will be addressed in the next chapter.

Chapter 6.
Understanding innovation in mobility practices

Contrary to the original expectation, the opening of Britain's early railways had a more immediate impact on the pattern of passenger travel than it did on the goods traffic. With very few exceptions, whenever a new line was opened to traffic there was a spectacular increase in the number of persons travelling along the route served compared with the numbers previously using the road. After the opening of the Newcastle and Carlisle Railway throughout its entire length on 18 June 1838 eleven times as many persons travelled by trains as had previously gone by coach.

Bagwell, P.S., The transport revolution (1988, p. 107)

6.1 Introduction

In the previous chapters the social practices approach for the domain of everyday mobility was conceptualised. Based on theories of practice, we have described how everyday life consists of a constellation of interwoven social practices. However, so far the (dimensions of) mobility practices were described in a rather static way while one of the foremost aims of this thesis is to analyse and understand environmentally relevant dynamics within and between these practices. Therefore, in this chapter the attention will shift towards innovation in mobility practices. On a conceptual level we will describe how changes in mobility practices may come about. What are the processes that keep current mobility patterns stable or produce shifts in mobility patterns? What are potential leverage points for deliberate intervention towards more sustainable mobility patterns? And what is the impact of socio-technical innovations on these mobility practices?

In describing innovation in mobility practices we will further relate to the perspective of socio-technological transitions. A transition, defined as a shift from one socio-technical regime to another, is a co-evolutionary and multi-actor process which takes place on various levels of structuration. The key idea of the multi-level perspective is that transitions and systems changes can best be understood by the interaction taking place between the niche, regime and landscape level. On the level of the niche societal or technological innovations (attempt to) break through by taking advantage of the windows of opportunity in the socio-technical regime. If these niche-innovations have 'sufficient momentum' adjustments in the socio-technical regime may occur or the whole socio-technical regime may change. The specific transformation that takes place in the socio-technical regime depends on the timing and the specific nature of the multi-level interaction (Geels & Schot, 2007). As a policy approach, the logic of transition management is to create the (legal, financial, organizational) space for experimenting and up-scaling which is needed to facilitate a potential regime shift. Strategic niche management, designed to assist the introduction and diffusion of sustainable innovations through protected societal experiments, and guiding visions about (future) system innovations therefore play an important role in policy based on transition management.

However, while the transition perspective is strongly focused on emerging technological trajectories and transition pathways, we will pay specific attention to the everyday life changes these trajectories and pathways entail. As Geels & Schot (2007, p. 413) acknowledge, agency is always present in the multi-level perspective and in multi-level interactions, however, the role of agency is not always made explicit. In this chapter the socio-technical regime is therefore complemented with a practice-centred view on regimes. In this viewpoint the regime is described as the whole of implicit and explicit rules and patterns of thought that give direction to the practical actions of human agents in their practices and in turn are shaped by and reconfirmed through these actions (Loeber, 2003). From a practice point of view, what happens when socio-technical innovations become embedded in mobility practices? Which elements of mobility practices change and in what manner? These questions, and their answers, remain underemphasized in the current transition literature.

To illustrate how transition dynamics take shape in everyday mobility, in this chapter three cases studies in the practice of commuting are presented. In each case study influential actors in the system of provision of commuting attempted to orchestrate a shift in the mobility practices of a designated group of commuters. These three case studies are examples of initiatives which can be labelled under the heading of mobility or demand management. Like the greening of consumer cars, mobility management is one of the guiding visions for the transition to sustainable mobility (Kemp & Rotmans, 2004; Urry, 2009). Mobility management aims to facilitate shifts in mobility practices (e.g. towards multimodality and teleworking, etc.), not by creating major infrastructural changes, but by alterations in the institutions in which mobility practices are embedded. While the case studies are interesting in their own right, the main contribution is that they will help us understand the dynamics of change and stability in mobility practices. In addition, they will shed a different light on the appearance of regime shifts and the role of practitioners in socio-technical transitions.

In the next paragraph we will first draw upon the concepts of deroutinization and reroutinization to explain the nature of innovation processes in individual habits and collective practices. In Paragraph 6.3 the principles of mobility management are described in greater detail. Thereafter, the three cases of mobility management in the practice of commuting are described. In the concluding paragraph the focus of attention shifts towards the theoretical implications of the case studies for the analysis and understanding of transition dynamics in mobility practices.

6.2 Processes of change: deroutinization and reroutization[74]

The routine character of situated social practices is a crucial factor in the conduct and reproduction of social practices. In the conduct of everyday life, human agents make use of a practical consciousness in order to avoid the obligation to reflexively consider the wide range of possible behavioural options. Only when the taken for granted and trusted instruments are no longer considered sufficient or do not provide a solution for a new situation, groups of actors individually or in an organised way shift from an automatic mode of decision-making to a reflexive mode of decision-making. As routinization is key to relative stability in the conduct in social practices,

[74] See also Chapter 4 in Spaargaren *et al.* (2007).

the concept of 'de-routinization' is key to alterations in the conduct of social practices. Giddens loosely defines de-routinization as 'any influence that acts to counter the grip of the taken-for-granted character of day-to-day interaction' (Giddens, 1997, p. 220). Furthermore, Giddens states that 'any influences which corrode or place in question traditional practices carry with them the likelihood of accelerating change' (*ibid.*). During processes of de-routinization, whereby the mode of practical consciousness is (temporarily) shifted to a mode of discursive consciousness, the taken-for-granted ways of doing things are challenged and put to question. De-routinization may lead to a trial period during which human agents actively 'experiment' with new ways of conducting the practices of everyday life. Re-routinization occurs when, throughout a transitional phase, destabilized routines are reconstructed into new, more or less stable, routines. So, processes of de- and reroutinization may eventually lead to a different set of rules and resources which may lead to a different shape and content of how the social practice is conducted.

What is important about the processes of de- and re-routinization is whether the alterations take place at the level of individual habits or at the level of collectively shared routines; a point which we also stressed in chapter three. Deroutinization at an individual level can mean that an actor decides to change parts of his or her lifestyle by refraining from certain practices and starting to participate in other practices. When focusing on the individual actor and his or her lifestyle, an important source of de-routinization is the occurrence of what Giddens terms 'fatal moments' in the personal life of the actor. This fatal moment often refers to changes in the behavioural context of the individual human agent. Examples of fatal moments in the domain of everyday mobility are residential relocations and major construction works on the daily commute trajectory. In both cases the routinely conducted mobility behaviour ceases to function and needs to be re-examined. These situations provide 'windows of opportunity' in the sense that people need to re-evaluate existing habits and are more open for experimenting with different transport options. Especially if human agents move to a new residential area which has congestion problems, good access to public transportation and few or expensive parking places, planned and organised intervention measures can have a strong influence on shaping new routines. Nevertheless, while trigger moments draw travellers into a state of discursive consciousness, in many cases this does not lead to alterations. That is, the practitioner returns back to its old habit, either directly or after deliberate experimenting with alternatives.

The abovementioned fatal moments can be labelled as negative trigger moments because the routines are (suddenly) disrupted and have become dysfunctional. On the other hand, planned and organised forms of de-routinization can be labelled as positive trigger moments. Positive triggers happen when groups of participants in social practices are 'confronted' with sustainable alternatives in the organization and configuration of the social practice (Spaargaren *et al.*, 2007). Positive trigger moments may involve a day of driving around in a rental car which runs on alternative fuels, or a stay at a car-free holiday park. In both situations there is also a (temporary) de-routinization of mobility habits as citizen-consumers come into contact with socio-technical innovations; however, this may occur without them actively being in search of alternatives for existing routines. Positive trigger moments may also be part of deliberate intervention policies, initiated by companies and/or governmental agencies, in order to influence car use habits. Examples are promotional activities such as temporary free travel cards to experiment with public transport (see also Thøgersen & Möller, 2008) or the purchase of electric cars for the pool of company cars.

Though the innovations presented to practitioners may spark an interest, there is no acute need for practitioners to reconsider the habit. It is especially in situations where the context of routines is disrupted (negative trigger moment) that orchestrated behavioural change (positive trigger moments) has the highest change of success[75].

So, through processes of de- and re-routinization individual human agents may change their mobility routines. However, alterations on the level of collectively shared social practices are of a different kind and character. Here the collectively shared notions and meanings of everyday conduct in a specific consumption domain or social practice are put to question. Considering the practice-centred view on regimes, an alteration on the level of social practices implies a (fundamental) shift in what is considered as the normal way of doing things. Thus, in order for a regime shift to occur, a shift in the rules and patterns of thought that give direction to the actions of human agents in their mobility practices has to take place. To illustrate, in the domain of mobility there are collectively shared representations of how best to organise one's mobility practices. In this light Freudendal-Pedersen (2005) talks about 'structural stories' which contain the arguments people commonly use to legitimize their everyday actions and decisions in the domain of mobility. These arguments are generally regarded by human agents as universal truths, thus expressing a certain standard or collective convention in society. Examples of existing structural stories are: 'when one has children one needs a car', or 'one cannot rely on public transport services, there are always delays' (*ibid.*).

These structural stories, or patterns of thought, reconfirm that the car-regime is the dominant socio-technical regime in mobility. De- and re-routinization on the level of social practices are therefore of a much more fundamental nature than they are on the level of individual habits because it encompasses alterations in the general conduct of social practices, including the lines of argumentation behind this conduct.

6.3 Reconfiguring commuting practices: three cases of mobility management

Any shift in the car-based regime requires that reconfigurations in mobility practices occur. The case studies in this chapter focus on initiatives to orchestrate alterations in commuting practices, predominantly involving a shift from single car use towards other modes of transport and telecommuting. In their book on strategic niche management Hoogma *et al.* (2002, p. 50) distinguish three basic approaches to facilitate such a shift in the car-based regime towards alternative transport modes, aside from large-scale investments in infrastructure and technologies. Firstly, initiatives may be implemented which make public transport more flexible and better adapted to consumer's needs. Measures in this category consist mostly of improvements in ICT, providing better and more adequate (multi-modal) travel information, and services that connect various public and private transport systems together such as mobility cards. Secondly, one can distinguish initiatives which make new transport options available to the traveller. Examples in this category are product-to-service systems, such as car-sharing and bike-sharing, but also the

[75] Bamberg (2006) describes a policy intervention which uses residential relocation as the starting point for triggering habitual change in the mode of transport by providing information on available public transport options in combination with a free one-week pass.

implementation of van-pools and shuttle-busses. Though the transport modes as such are not always new, the novelty in this approach resides in the fact that commuters make collective use of private means of transport. Finally, initiatives may be undertaken that alter the decision context made in commuting practices. Examples are financial incentives, changes in a company's formal and informal rules towards e-working, parking policies, etc.

The abovementioned types of strategies, especially in Europe, have become known under the heading of mobility or demand management. Mobility management, brought down to its essence, is defined as the organisation of smart travelling (KpVV, 2007). Mobility management is the whole of activities undertaken by governmental bodies, companies, and consultants, focused on the stimulation of conscious behavioural choices of employees (SER, 2006). The choice in this sense refers to the question if and when the trip is made and by which transport mode[76]. In the Netherlands, mobility management has received a large impulse after the advice of the Social Economic Board in 2006 (SER, 2006). Projections of accessibility problems for the economic core regions prompted the Board to point out the necessity of demand oriented policy in addition to increasing network capacity. In the advice of the SER the responsibility for a sustainable mobility system is laid down at every stakeholder involved: governmental bodies, employers, transport companies, and employees[77].

Central to mobility management are the travel demands of the everyday traveller and the way in which he gives expression to these demands (VenW, 2002). Mobility management involves the whole of measures undertaken to influence the behavioural choice of the traveller with the aim to direct these demands away from single car use to other travel modes, other destinations and/or travel times. What is particularly interesting about the mobility management approach for this thesis is that it centres on the question how existing mobility routines may be (deliberately) altered and new routines may be formed.

Though mobility management is a relatively new branch in the field of mobility, the separate cornerstones of which mobility management is made of are not (examples are the introduction of teleworking, the use of price-mechanism, facilitating multi-modal travelling). Nevertheless, mobility management is novel in its integrated approach to mobility. First, mobility management does not start out from the perspective of one specific stakeholder but aims to incorporate all the relevant actors involved, most notably travellers, employers, and mobility providers. Not surprisingly, mobility management aims for a win-win situation for all the stakeholders involved. For example, (large) companies, who are often the initiators of the mobility measures, aim for a better company image, higher employee satisfaction, reduced parking and travel costs, and a

---

[76] Mobility management distinguishes itself from traffic management in three ways. First, mobility management usually focuses on pre-trip travel choice, while traffic management focuses on on-trip travel choice. Second, mobility management targets all travellers, while traffic management targets road users. Third, mobility management is always related to behavioural choices in contrast to traffic management which can also focus on optimizing traffic flows without any choice of the car driver involved (e.g. by way of controlling traffic lights).

[77] Based on the advice of the SER a TaskForce Mobility Management (TFMM) was initiated in 2007. The TFMM was called to life by the Dutch employers branches (VNO NCW and the MKB) by request of the Dutch cabinet. The TFMM had as a target to come up with a whole package of concrete measures to reduce the amount of car kilometres and the excretion of environmentally harmful emissions during the rush hours by 5% by 2011.

more accessible location. Second, the integrated concept of mobility management is visible in the strategies pursued which focus not on one specific mobility measure but on a cluster of strategies and measures.

To understand stability and change in the conduct of mobility practices in the following sections three case studies of mobility management are presented. Though these case studies provide just a few examples of the numerous mobility management initiatives currently undertaken, the primary aim is to show the patterns of innovation processes in the practice of commuting. Furthermore, the three case studies shed light on the barriers and opportunities for mobility management as one of the transition management pathways.

The first case study describes mobility management as a temporary measure to reduce congestion during a period of road construction works. It provides an insight into the processes of de- and reroutinization in a short time-period due to a combination of various context-changes (negative and positive trigger moments). This case is an example of non-structural changes in the everyday routines of a group of commuters. The empirical material is derived from evaluation studies of mobility management during road construction conducted by Rijkswaterstaat.

The second case study provides an example of stability and continuity of mobility practices despite the implementation of mobility management measures and efforts to change routines. In this case the management of a lease company introduced various sustainable mobility options to the employees albeit with little effect on mobility patterns. The empirical material of this case is based on interviews with eight employees of a car leasing company.

The third and final case describes an integrated approach of a company mobility management plan where the provision of mobility innovations was interwoven with a novel approach to working ('time and place independent working'). The material for this final case is based on document analysis (available written material, presentations, etc.). The case will show that an integral approach, which focuses not only on the supply of mobility innovations but also on the social contexts of commuting and work, can be successful in facilitating structural changes in commuting practices.

6.3.1 Mobility management during road construction works

The use of mobility management as a measure to reduce congestion during road maintenance and construction works was initiated by Rijkswaterstaat (the Dutch Highway Agency) in 2006. That year many overdue maintenance works on the public highways and bridges were carried out. This was succeeded by a new round of thirty maintenance and construction works in 2008, which will be in effect until 2015 at the least. These maintenance works are on such a large scale that a significant part of the road network has been en will be negatively affected. Road maintenance and construction works in most circumstances lead to periods of reduced network capacity as lanes are either narrowed or taken out of order completely. Consequently, without the utilization of compensating measures the road user will be confronted with longer and unreliable travel times on the Dutch highways. Considering the ambition of Rijkswaterstaat to be a 'public oriented network manager', efforts are undertaken to keep the nuisance for road users to a minimum. To deal with the severely reduced accessibility, especially during rush hours, temporary mobility management measures are taken with the aim to stimulate road users to find alternatives. Rijkswaterstaat distinguishes four categories of behavioural alternatives to travelling by car during rush hours: travel time shift (travelling outside of the rush hour periods), travel route shift (travelling via another route); modal shift (travelling by public transport or by bike), and telecommuting.

Due to its high frequency nature, the practice of commuting is one of the most habitually undertaken mobility practices. Therefore, commuters are unlikely to make unprompted alterations in their everyday commuting trips. Road construction works can provide an important impetus for the search of alternatives for car commuting. That is, the extra travel time alone may provide enough momentum for commuters to make use of alternatives. By providing incentives for behavioural change, Rijkwaterstaat makes the alternatives even more attractive. A wide variety of measures has been implemented over the years to (temporarily) increase the attractiveness of travel alternatives. The provision of free (or reduced cost) public transport passes stimulates a modal shift form car use to public transport use during commuting trips. The measure 'rush hour avoidance' uses financial incentives to encourage road users to avoid the rush hours of the specific highway under construction.

To evaluate the effectiveness of mobility management measures during road construction works Rijkswaterstaat conducts *ex post* evaluation studies amongst road users and participants of mobility management measures. Below the results of two studies are presented. Though these projects differ in scope, both in the instruments employed to influence the road users, as in the context of the road construction works (travel time delays, specific location of the road network, availability of alternatives), the projects have in common that the scope is of a temporary nature.

The A4/A10 South pass in 2006

The maintenance and expansion of the A4/A10 in the Amsterdam area in 2006 was one of the first cases where mobility management measures during road construction works were applied on a large scale. The goal of the implemented measures was to reduce traffic volume on the A4 and A10 with approximately 10% during the rush hour periods for a period of two months (approximately 9,000 vehicles). Because the region involved contains an important business district, the mobility measures where developed and implemented in a public-private collaboration which included the residing corporations. Every company with more than ten employees was actively approached with the offer to participate.

The A4/A10 South pass, the most noteworthy measure of the project, provided free public transport for 30,000 commuters in the two-month period of road construction works (AVV, 2006b). The South pass was accompanied with a personal travel advice. In the personal travel advice a comparison was made of the travel times and travel costs of the commuting trip by car and by alternative transportation modes. Finally, temporary shuttle busses were used to improve the connection from the train stations to the work location.

In order to measure the effects of the travel pass two online surveys were conducted under travellers who were in possession of an A4/A10 South pass (2,283 travellers completed both surveys). The South pass was used at least once by approximately half of the pass holders while the other half had never used the pass. More specifically, 24% were everyday users of the pass (4-5 times a week), 22% were regular users of the pass (1-3 times a week).

As Figure 6.1 shows, the modal split of the pass holders changed drastically during the period of road construction. While 61% of respondents used the car for commuting trips before the road works, during the works this was reduced to 42% (*ibid.*). On the other hand train usage by the pass holders doubled, while bus, tram and metro usage increased by 60%. Obviously, public transport usage was much higher among South pass users than with the non-users.

Because of the rise in public transport use, which on average lead to 3,200 less cars during rush hours, the South pass significantly helped to reach the predetermined goal to reduce the traffic volume by 9,000 vehicles. Nevertheless, the majority of the commuters who changed their modal use during the road construction works returned to their old routines afterwards (see Figure 6.1). Shorter travel time by car, not in the least facilitated by an extra lane on the A4/A10, and the wind-up of the free travel card were the main arguments.

Rush hour avoidance A6

The conditions of the road works on the highway A6 were relatively similar to the previous project. The works took place in 2008 on a major transport artery in the Amsterdam region (the bridge between the cities of Almere and Amsterdam) and consisted of maintenance works and the broadening of the A6 with an extra lane. An important difference with previous project though was that the construction works lasted for a full year instead of a couple of months. Two main mobility management measures were implemented to avoid an increase in rush hour congestion: a free public transport travel card and a monetary reward for rush hour avoidance[78]. As the results of the free travel card are similar to the South pass, the focus will be purely on the financial reward system.

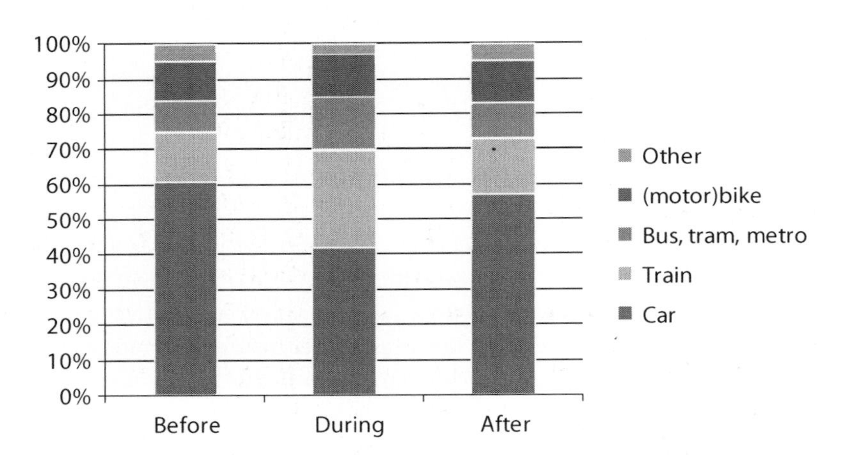

Figure 6.1. Modal split of A4/A10 South pass holders (AVV, 2006b, n=2,283).

[78] Commuters were not allowed to participate in both projects at the same time.

As a policy approach the rewarding scheme of rush hour avoidance can be seen as the counterpart of congestion charging. By providing a positive incentive car users are prompted to search for alternatives (travel routes, travel time, modal shift, telecommuting). The idea of the scheme is that participants receive a small reward (in the order of four to six euro) each time a rush hour period for a designated location is avoided (see Bliemer *et al.*, 2009 for a synthesis study of rush hour avoidance rewarding schemes in the Netherlands). Prior to the construction works, cameras were installed which were able to detect license plates in order to contact potential participants. Car users who travelled frequently over the bridge on the A6 were invited to participate in the rush hour avoidance project. In total almost 3,000 commuters participated in the project. The financial reward for rush hour avoidance A6 was four euro for the morning rush hour, with an additional two euro if the participant avoided the bridge the whole day.

In order to know the behavioural response of the commuters an evaluation study was conducted amongst 700 participants of the rush hour avoidance scheme (RWS-DVS, 2009). Before the road construction works the participants travelled on average 2.1 times a week during rush hours over the bridge. During the project period this was reduced with 40% to 1.3 times a week. Interestingly, the behavioural response of the participants of rush hour avoidance during the construction works is quite different from the response of the participants of the free public transport card. As Figure 6.2 shows, the majority of the alternatives chosen are travelling outside of the rush hour period (16%) and travelling via another route (9%). Though there is a clear increase in public transport use, the car is still the dominant transport mode in the commuting trips. Nevertheless, as was the case with the South pass, after the completion of the project with the resulting increase in road capacity, most of the commuters returned to their prior routines.

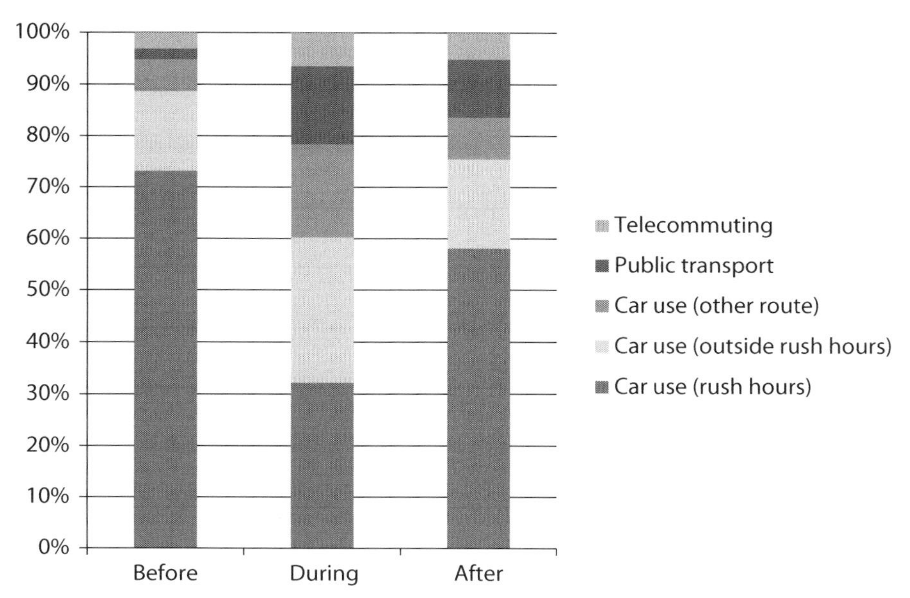

Figure 6.2. Travel choices of rush hour avoidance A6 participants, based on number of trips (RWS-DVS, 2009, n=711).

Lessons learned

The two cases of mobility management during road maintenance and construction works provide a good illustration of the processes of de- and reroutinization in a short time-period due to a combination of negative and positive trigger moments. As the evaluation studies show, contextual changes may lead to a variety of alterations in mobility practices. Clearly, the capacity reduction during road works provide a negative trigger moment for car drivers on that specific road (and often also on the surrounding road network), especially for commuters who travel during rush hours. It is important to note that this negative trigger alone may be impulse enough for some commuters to alter their travel patterns. As the evaluation studies show, also non-users of mobility management measures may alter their behaviour. However, by actively providing or stimulating alternatives to car driving during rush hours, a positive trigger is provided which complements the negative trigger moment. This combination of positive and negative trigger moments may lead to a phase of experimenting with alternative ways of commuting. On an individual level the following line of thought, derived from the introduction of the South pass, is an example of how the phase of discursive consciousness may work out (AVV, 2006b, p. 12, translated by author)[79]:

- Awareness: 'I will get hopelessly stuck in traffic jams.'
- Interest: 'You offer alternatives? Which are they?'
- Desire: 'Not in traffic jams and faster as well, I would like to try that.'
- Action: 'There is a bus around the corner, and then the train at 8.05 AM departing from platform 2. What a hassle. There better be a traffic jam.'
- Evaluation: 'Travelling by public transport is not as bad as I always thought it would be.'

The participants of rush hour avoidance A6 reduced the number of commuting trips during rush hours with 40%. In addition, one quarter of the South Pass holders travelled each day from home to work by public transport, and one fifth of the pass holders used public transport once to three times a week. Rijkswaterstaat concludes that the program 'rush hour avoidance A6' was able to tempt a group which does not easily change behaviour to alternative ways of travelling (RWS-DVS, 2009). That is, in comparison with car drivers who did not participate in the program, the participants travelled more often via alternative routes, travelled more often outside of rush hours, and worked more often from home. Especially interesting is the fact that 10% of the participants of 'rush hours avoidance A6' experimented with telecommuting for the first time. Amongst the non-participants this was only 3%. Rush hour avoidance therefore has the additional effect that car drivers try out new ways of commuting. Moreover, the evaluation study reveals that approximately 30% of the participants think more positively of the alternative they chose most often.

Thus, the mobility management measures had the effect that new ways of doing things were introduced in the mobility practices of these groups of commuters thereby increasing the mobility portfolio of the practitioners. Considering the four dimensions of mobility practices discussed in the previous chapter, the participants of rush hour avoidance predominantly made alterations in the temporal dimension of mobility practices (travel time shift), or the spatial dimension of mobility

[79] Evidently Rijkswaterstaat used the marketing philosophy AIDA (awareness, interest, desire, action) for the introduction of the South Pass.

practices (travel route shift). Also, the mobility management measures are shown to have an effect on various dimensions at the same time. Practitioners who experimented with telecommuting or public transport use had different time-spatial patterns, used different transport modes and often changed their opinion about travel modes altogether.

Nevertheless, as soon as the construction works were finished both the negative trigger (reduced accessibility) and the positive trigger (free public transport passes/financial incentives) were discontinued. As a result of that, the far majority of the car users returned to their old routine of commuting by car. As with the case of the South Pass, only a small proportion of the commuters (approximately 7%) did indeed continue with the newly acquired routine of public transport use. Whether one considers six hundred extra train travellers from prior pass holders disappointing or not depends on the eye of the beholder. More important is the lesson that though contextual changes may lead to alterations in mobility practices, these practices will mostly only be undertaken in a novel way when these context changes are of a more or less permanent character.

6.3.2 Company level mobility management: stability in mobility practices

In the second case study the focus of attention shifts to the level of one specific company where, as a part of corporate social responsibility strategy, various mobility management measures were introduced. The company which introduced the measures, namely Arval, is one of the leading car leasing companies in the Netherlands. In an objective benchmark analysis of Dutch lease car companies the customer service of Arval was awarded as the best (Heliview, 2008). Especially customer contact and quality of service are considered the key characteristics of the company. The reason to implement a sustainability program for Arval's own employees was two-fold. First, the same benchmark report indicated that the service of Arval on corporate social responsible fleet ownership could be improved. That is, the supply of information on green fleet management and proactive advice on environmental solutions were considered points of improvement (Heliview, 2008). For the board of directors of Arval green procurement demands of their customers provided an direct impetus for corporate social responsibility, both internally and externally (Bakker, 2008). The second reason was the personal motivation of several board members to realize environmental improvements within their own circle of influence.

The environmental program consisted of measures to improve environmental performance of the operational performance of the office (waste management, lighting), measures with regard to energy efficiency of the company cars (energy efficient driving, ABC-energy label policy), and measures to facilitate alternatives to the car (introduction of a mobility card). Because a survey amongst the employees of Arval revealed that environmental awareness and commitment were generally low, twenty persons were asked to act as sustainability ambassadors. Their task was to bring attention to the abovementioned measures and to promote environmental awareness within the company. Because this case study is on mobility management in the practice of commuting we will limit the focus in this paragraph to the introduction of a mobility card.

Every employee of Arval with a lease car was given a mobility card (the NS Business Card) which aims to facilitate multimodal travelling and travelling by public transport. One of the main advantages of the card is that every train trip can be pre-booked and is automatically placed on a monthly invoice which is send to the company, thus reducing the administrative burden for the employees. Furthermore, 'door-to-door services' such as a taxi, a public transport bike, and a parking place near train station, can be billed on the mobility card. Next to promoting sustainable mobility, the aim of the introduction of the mobility card into the company was to get the employees acquainted with the workings of the card. As customers of Arval increasingly requested the mobility card as an add-on to the lease car it had become one of the company products.

To monitor and evaluate the effects of the introduction of the mobility card on the commuting practice, interviews with eight staff members were conducted in December 2008. It is important to note that the mobility card was not specifically introduced for commuting, but was also applicable for the other mobility practices (especially business travel). The aim of the interviews was to examine what the impact of this mobility management measure was on the employee's motility (mobility portfolio, skills, and cognitive appropriation) and actual travel behaviour. How did the introduction of the mobility card affect the knowledge and cognitive appropriation of public transport? What was the impact of the enlargement of commuter's mobility portfolio on the actual travel behaviour? Which factors explain the use or non-use of the mobility card? In order to answer these questions eight structured interviews, preceded by a short questionnaire, with owners of the mobility card were undertaken, half of them were users of the card while the other half were non-users.

Based on the interviews with the employees there seemed to be a clear distinction between travellers who had used the mobility card and those who had not. As one of the directors put it:

> *There is sort of a dichotomy within this micro-society: users and non-users of the mobility card. Especially the non-users in the fieldwork I have challenged by saying: You have to sell this product to your customers. Therefore I believe that you have to make use of it. You have to experience for yourself in order to be able to sell it properly. Even if there is not a train station nearby where you live, there is always, once a month or once every two months, a trajectory where you can say, I will park my car here and go further by train.*
>
> Interview Bakker (2008)

Furthermore, it became clear that all the users of the mobility card were very positive about the introduction of the mobility card. While both the users and the non-users of the mobility card are highly frequent car drivers, the mobility card users took the opportunity to experiment with public transport modes. In that sense the provision of the mobility card can be seen as a clear case of a positive trigger moment which prompted at least part of the employees to reconsider their travel routines.

> *Without the card I would probably have taken the car just out of habit. Now when you have the card you think: well let's give it a try, and the experience is quite positive. In any case, it has made me alert to the possibility. It never came to my mind to think, gee let's go by train. It was almost ten years before I had last travelled by train. Now that*

I have the card I have used it already seven or eight times within the last six months. So it has absolutely been an incentive.

Mobility card user

I find it easy to use. You can easily book the journey via the internet. It is easily accessible, uncomplicated and easy to use. So this is really ideal. Also for trajectories which do not come first to mind I make the choice for the mobility card, if only to experience how it goes. I am enthusiastic about the mobility card.

Mobility card user

Nevertheless, on closer inspection the distinction between the two user groups is not as large as it would seem. While the mobility card users are positively surprised about travelling by public transport and the ease of use of the mobility card, they only make sporadic use of it. The users only utilize the mobility card in specific circumstances, mostly for a specific business trip.

This is a specific destination which I am enthusiastic about, also because my expectations were so low. So, these low expectations were exceeded. However, it is not the case that I suddenly have become a public transport adherent, certainly not.

Mobility card user

There were a number of reasons given why the car remained the favourite transport mode. For the commuting practice a simple reason for car use was that the office location was not easily accessible by public transport. For many commuters the travel time by public transport was simply a lot longer than by car. For the practice of business travel other concerns played a major role. As explained above, customer contact and quality of service are key characteristics of the company. Reliable travel time, flexibility and representativeness were mentioned as the most important aspects in business travel. The car was deemed the most suitable for that. A further consequence of making the business trips by car is that the commuting trip automatically is also undertaken by car.

One of the core themes within Arval is: an agreement is an agreement. It is completely unacceptable to arrive too late at an appointment. With the car you always have some options to travel via an alternative route, with the train that is a lot more difficult.

Mobility card user

I think that uncertainty is one of the most important things for travelling by public transport: does the train or bus leave on schedule? That for me is very important. And if you miss them, what are the alternatives.

Non-user

Another reason is that a lease company automatically attracts a lot of car-minded employees. Amongst the mobility card users and non-users there were a lot of car-enthusiasts who simply never considered public transport for their trips. In addition, there was little knowledge and almost no use of door-to-door services provided by the mobility card such as the public transport bike and park and ride facilities. Especially these door-to-door services can make public transport more attractive.

Though there were also some employees who commuted to work by public transport, the company can be characterized as car-oriented (not surprising for a lease car company!). The orientation on the car was strengthened by the high-level of ownership of a lease car and a free tank card, also perquisites for the employees who were interviewed. Especially the free tank card can have an adverse effect on the consideration of alternative travel modes in non-work related mobility practices (social and leisure travel). Expenses made for commuting trips and business travel trips are covered by the employer, whether they are made by public transport, bike or car. However, the perquisite of a tank card means that expenses are also covered for other mobility practices undertaken by car. Therefore for most trips there is little to no incentive to consider alternatives to car use.

> *The lease car driver has absolutely no idea what the price of litre fuel costs. Or what his car costs. He fills up his car, he flops his plastic card on the counter and enters his pin code, and he is ready to go.*
>
> Interview Bakker (2008)

Lessons learned

This case study can best be described as an example of stability in and continuity of mobility practices despite the introduction of a company level mobility management program. Notwithstanding the efforts and best intentions of the board of directives the results with regard to a modal shift are relatively meagre and the mobility practices are continued predominantly along the lines of the old routines. One could say that the positive trigger moment, the introduction of the mobility card, was not enough to disrupt the habitual behaviour of the lease car drivers. In addition, the absence of a fatal moment, such as road construction works or an office relocation, decreased the need to experiment with the provided alternatives. Also, free and basically unrestricted fuel because of a tank card further reduced the incentive to weigh alternatives.

A positive impact that the company mobility management plan had is that the users of the mobility card got more acquainted with other transport modes and were positively surprised of its use. Yet, all in all the differences between the users and non-users with regard to the frequency of use of public transport and/or multimodality were quite small. Though it was never the intention of the board of directives to make employees public transport adherents, a more significant behavioural change was hoped for. Therefore the changes in the material dimension of mobility practices, in this case the increase in the mobility portfolio of the employees, did not lead to significant alterations in the practice. Contributing factors that provided continuity in the practice can be found in the material dimension (office hard to reach by public transport), temporal dimension (early closing time of the office), and experience dimension (car-minded organization, arriving on time and representatively is considered crucial and best to be undertaken by car). Finally, an interesting aspect about this case of mobility management is that the possibilities for telecommuting were rather limited. Only the higher staff members were given the technological means (such as a log in token) to work from home, an opportunity which was mostly used to work extra hours in the evening. While various employees expressed a desire to be able to work from home or work flexible hours this was not encouraged (a policy derived from the French mother company).

In contrast, in the next case study we will turn towards a company were telecommuting and working at flexible work hours was indeed actively stimulated as part of the company mobility management strategy.

6.3.3 Company level mobility management: an integral approach

The final case study of mobility management describes an example of an integrated company mobility management plan. The focus is on the Rabobank, the largest bank in the Netherlands, which implemented mobility management measures not as solitary measures but in consideration of the social contexts of commuting and business travel. The Rabobank was one of the first companies to recognize that mobility is highly related to the way work, and to a lesser extent social life itself, is organised. Therefore, the attempt to make changes in the practices of commuting and business travel went side by side with the implementation of a novel approach to working.

As a corporation Rabobank also participated in the Taskforce Mobility Management which was initiated in 2007. This taskforce was called to life by the Dutch employer's branches (VNO NCW and MKB) by request of the Dutch government. In total fifty companies who were considered to be frontrunners in mobility management participated in the taskforce and operated as ambassadors.

From 2005 onwards Rabobank became increasingly interested in the question how the mobility practices of its employees could best be organised. The immediate causes for implementing mobility management were concrete mobility issues such as a shortage of parking places and increasing mobility costs due to a planned relocation of four business offices into one new central office in Utrecht's city centre. Furthermore, Rabobank anticipated to expectations of increasing accessibility issues in the near future which would lead to productivity loss of its employees (De Jager, 2008). Finally, sustainable mobility was also an important element of the corporate social responsibility program of the bank.

The specific mobility management approach adopted was influenced by the participation in the first scientific experiment with rush hour avoidance of which the Rabobank was one of the initiators. The outcome of the experiment was that positive (financial) incentives, making travel alternatives more attractive, in combination with the opportunities for flexible travel choices can be sufficient impetus for travellers to make significant changes in their mobility patterns (*ibid.*). In addition, Rabobank had experienced how their customers had voluntarily adopted internet banking on a large scale, thereby playing a crucial role in the radical change in the practice of private banking. Alterations in consumer behaviour were therefore considered feasible if the right approach was taken. Based on the perspective of mobility management, De Jager (2007) makes a clear distinction between a traditional top down and innovative bottom up approach in influencing mobility patterns. The traditional top down approach could be characterized as a supply driven, government initiated, and modal shift oriented approach which was unpopular by employees and largely ineffective in reaching the goals. In contrast, the novel approach adopted is demand driven, multimodal, voluntary based, and flexible.

In addition, the vision on sustainable mobility is based on a new balance between work life, mobility, and private life at home. Thus, the mobility of employees is seen as intricately linked with the way employees work. In the program 'Rabobank Unplugged' a novel way of working was introduced that corresponds with the changing mobility practices. Based on the principle of 'time and place independent working', flexible working hours and work locations were introduced. Employees also received more individual responsibility over their work processes. Correspondingly, changes in the way of working required a new way of supervision from the employer. The new management style adopted is more based on trust than on control, more based on coaching than on steering, and more based on mutual agreements than on rules (De Jager, 2008). According to Rabobank: 'employees determine themselves where, when and how they come to the best result ... This provides employees with more space to make their own choices, which makes it easier for them to reach their full potential, makes the organization more flexible, and makes the working methods and service more tailored to the desires of the customers (Rabobank, 2011, p. 30, author's translation). A consequence of this new management style is that employee productivity is measured by delivered performance instead of on work hours spent at the office (from input to output based assessment). Clearly, Rabobank aimed to apply the same basic principles to the organisation of work as it does to mobility: demand driven, flexible and based on personal responsibility.

Next to facilitating time and space independent working, other mobility management measures were taken. Similar to Arval, described in the second case study, Rabobank started to experiment with the introduction of multimodal mobility passes for 5,500 lease car owners. In the first half year approximately 20% of the lease car drivers made use of the card while 80% did not. Under the users of the mobility pass the amount of car kilometres decreased with 8%. In contrast, under the non-users car kilometres increased with 6%. In 2007 the 20% lease car owners who made use of the mobility card had travelled over 2.8 million kilometres by train (De Jager, 2008; Rabobank, 2008). As the pilot with the mobility card was considered a success the card was made available to all employees who owned a lease car.

In 2007 Rabobank started a second mobility management pilot which introduced the notion of individual mobility budgets. The philosophy of individual mobility budgets is that each employee is given a yearly financial budget which can be spent on a variety of mobility options. Depending on the individual needs and preferences the employee can make a choice for ownership of a lease car, public transport pass, taxi use, purchasing a bicycle, etc. All these travel expenses made by the employee are booked of the personal mobility budget. If the budget is not fully used at the end of the year, the remainder of the budget is given as a bonus to the employee. The aim of these personal mobility budgets is to increase flexibility of everyday modal choice in commuting and business travel. Lease car owners are therefore the likely target group as the general lease car agreements make other forms of transport less attractive (as discussed in the Arval case study).

To stimulate public transport use even further, in 2007 each employee was given the opportunity to order five free train tickets for the trajectory corresponding with their daily commute. In total

1,800 employees had ordered 9,000 return tickets (Rabobank, 2008). Finally, a firm parking policy was implemented for the main office in Utrecht; employees living within ten kilometres of the head office were no longer allowed to park there.

Though it is difficult to claim a causal relation with the mobility management measures, Rabobank employees increasingly chose to make use of public transport modes in the years following 2006 (Table 6.1). In 2008, the employees had travelled 4.4 million train kilometres. This number increased with another 23% to 5.4 million train kilometres in 2009 (Rabobank, 2011). Thus, from the introduction of the mobility management measures in the second half of 2006 to the end of 2010 train use, measured in kilometres travelled, has doubled. In contrast, the mileage for lease cars remained more or less stable.

In addition to an increase in public transport usage, 7,000 on a total of 44,000 Dutch employees were considered flex-workers in the year 2008 (De Jager, 2008). Interesting in this respect is that Rabobank supports employees and middle management of the 'unplugged working method' by specially designed workshops (e.g. on healthy work-life balance, on cooperation, and on leadership in a virtual work environment).

Lessons learned

The pro-active mobility management approach adopted by Rabobank provides a variety of important learning points, especially in comparison with the Arval case study. Though both cases can be labelled under the heading of company level mobility management and in both cases the managerial support was clear and visible, the effect and outcome is quite different. The Rabobank case shows that an approach which focuses not only on the supply of mobility innovations but also on the social contexts of commuting and work can be successful in facilitating structural changes in commuting practices. Though the number of train kilometres is still relatively small in comparison with the number of car kilometres (Table 6.1), the case does show a clear increase in train kilometres due to the introduction of the mobility card. In addition, the number of car kilometres of the Rabobank has stabilized while in the Netherlands as a whole commuting distances have continued to increase.

The pursued approach consisted of a mix of measures: the introduction of flexible working hours and e-working, strict parking policy, mobility cards for lease car owners, mobility budgets, and an ABC-energy label policy. This integral approach of mobility management is likely to be more successful as the various components are mutually reinforcing. While Arval introduced

Table 6.1. Changes in mobility behaviour of Rabobank employees (Rabobank, 2011).

	2007	**2008**	**2009**	**2010**
Train use (in million km)	2.8	4.4	5.4	5.5
Lease car use (in million km)	Unknown	266	273	264
Share of energy efficient lease cars (ABC energy label)	71%	73%	79%	87%

the mobility card in a car-dependent location, Rabobank purposely decided to build its new head office next to the largest and most centrally located train station in the Netherlands. From a consumer perspective the financial incentives (mobility budget, cost for parking places) also stimulate non-automotive transport. All in all, because the context in which mobility decisions are taken have been permanently altered, for many employees this has led to different outcomes in mobility choices.

Especially interesting about this approach of mobility management is that changes in mobility were directly related to alterations in the way of working. Changes in mobility practices involve much more than technological changes, organisational and cultural changes are just as important to facilitate innovations in mobility practices. For example, 'time- and space-independent working' is a precondition for teleworking and travelling outside rush hours. This required important changes in the company organization and management style (such as the focus on output-orientation and trust in employees). By emphasizing personal responsibility and flexibility in choices, both in the ways of working as in the ways of mobility practices, the company clearly opted for a demand driven approach in which it placed its trust in its employees and provided them the opportunities to make their own choices. Though this approach to working and travelling is especially promising for the highly mobile professional, this case study has shown that with the right approach structural alterations in the practices of commuting are possible.

6.3.4 Synthesis

In this paragraph three case studies have been presented in which an attempt was made to orchestrate a shift in the mobility practices of commuters. The outcomes of these case studies help us to better understand stability and change in the practices of mobility. The routinization of commuting activities is one of the key mechanisms that explain the relative stability in commuting. Therefore, in most circumstances a trigger moment is required in order to shake up the taken for granted character of day to day activities and bring human agents into a state of discursive consciousness. Thus, most long-term changes in consumer behaviour require that a change in the contextual situation occurs. Mobility management builds on these principles. The idea behind mobility management is that routines in the organisation of mobility, not only from the perspective of employees but also from the perspective of employers, are opened up. The case studies provided various examples of both negative and positive trigger moments that provide such a context change. It is interesting to notice that these fatal moments do not only apply to individual consumers (in the form of moving or road constructions) but can apply to a whole company as well. In the case of the Rabobank, the relocation to the city centre of Utrecht formed a strong incentive to reconsider the organisation of mobility on the level of the company.

Based on the case studies we can conclude that mobility management may lead to innovation in mobility practices through various means. First, mobility management may facilitate alterations in mobility practices by making already existing means of transport more easily accessible for example through the provision of multimodal travel information and mobility cards. Second, mobility management may facilitate modal change by making new modes of mobility available for example through vanpooling, car-sharing and bike-sharing. Third, mobility management measures may alter the decision context by breaking down (financial) regulations that keep the routine locked in place, for example by implementing personal mobility budgets. In addition to making travel alternatives more attractive, the case studies have shown that practitioners are proactively 'confronted' with socio-technical innovations and may start to experiment with these innovations. Thus, the mobility management measures can have the effect that new ways of doing things are introduced in mobility practices.

Theoretically, the cases show that alterations in practices imply that a shift in one or more dimensions of the mobility practices occurs. That is, socio-technical innovations need to be integrated in the specific mobility practices which require adjustments from practitioners in their daily life routines. For these practitioners some dimensions in the practice remain stable, while other elements are completely or partially changed. The specific nature of these alternations clearly is dependent on the specific socio-technical innovation which is implemented. Participants of rush hour avoidance predominantly made alterations in the temporal dimension of mobility practices (travel time shift), or the spatial dimension of mobility practices (travel route shift). On the other hand, the South pass and the mobility cards had a stronger effect on the modal choices of the practitioners. Often the innovations are shown to have an effect on various dimensions at the same time. Practitioners who experimented with telecommuting or public transport had different time-spatial patterns, used different transport modes and often changed their opinion about travel modes altogether. However, clearly innovation in mobility practices does not always take shape, even if the mobility portfolio of practitioners is broadened.

The various cases also showed how the various practices are connected with each other, both through the actions of human agents as through institutions arrangements. Clearly, from the viewpoint of the practitioners, the mobility practices have relations because practitioners make use of the same motility, the potential for mobility, for different practices. This applies most directly to the personal mobility portfolio. The Arval case shows that the free tank card for many lease car owners not only has a strong effect on the modal choice in commuting and business travel but also on the other mobility practices as non-automotive transport has become less attractive. Simultaneously, the mobility card had the effect that several card users became more accustomed with public transport, not only during commuting and business travel but also during leisure travel.

Furthermore, mobility practices and work practices are intricately connected via various institutional arrangements such as corporate management style, human resource management, and national rules and regulations such as the occupational health and safety act, collective labour agreements, and fiscal arrangements. All these institutional arrangements have an influence on how innovation in mobility practices takes shape. For example, the reluctance to allow telecommuting on a corporate scale, as shown in the second case study, inhibits the possibilities for this travel alternative. It is no coincidence that smart working and smart travelling are increasingly seen as two sides of the same coin. That the previously mentioned Taskforce Mobility Management has recently changed into the 'platform smart working and smart travelling' is a clear indication of the close-knit relation between working and travelling.

6.4 Dynamics of mobility practices and the transition perspective

An important question that remains to be answered is how these case studies can be placed into perspective of transition theory and transition management. Though the described cases provide clear learning points for influencing travel behaviour it is not yet clear how this relates to alterations on the level of regimes. Pro-active mobility management strategies using positive and negative trigger moments may lead to alterations in the individual conduct of mobility practices or even lead to alterations on the level of whole corporations, but how do processes of de- and re-routinization work on the level of collective routines? In this concluding paragraph we will attempt to formulate an answer to this question.

Undoubtedly, the social-technical regime of the car is dominant in the domain of mobility. Currently, the private internal combustion engine car is the benchmark for all transport alternatives with regard to speed, flexibility, comfort, convenience, and cost calculation. Indeed, the words 'transport alternatives' literally indicates what is considered as ordinary and what is considered as out of the ordinary or unconventional. Based on an innovative car-sharing project Canzler (2008, p. 113) concludes that it is extremely difficult to find a functional equivalent to the privately owned car. In an attempt to increase the collective use of private vehicles in this project, participants received a financial incentive for those time periods in which the private car was rented out to other participants in the project. Despite the financial incentive, this flexible return option was used only to a very limited degree. The collective scheduling of time and space for cars was considered unattractive as it reduced the optional spaces for being mobile (*ibid.*). As a result of this pilot, Canzler concludes that improved transport information and financial incentives have little effect on the modal choice.

Canzler is correct in stating that it is hard to find a functional equivalent to the privately owned car. However, as Sahlins (1976, p. 169) has emphasized: 'utility is not a quality of the object, but a significance of the objective qualities'. There is a difference between the functional properties of a certain object, in this case the private internal combustion engine car, and the relevance and meaning that is attached to this function. For instance, that the more limited driving range and maximum speed of an electric vehicle is considered a strong barrier for the electrification of the car regime is not (only) a matter of functional differences, but also a matter of how much significance is attributed to these functions. In this respect, it is worthwhile to look into the conclusions drawn by Hoogma *et al.* (2002) on the basis of various transport experiments. These conclusions illustrate how learning processes are vital for understanding how innovations may become integrated in mobility practices. Hoogma *et al.* (2002) describe a pilot with electric vehicles in La Rochelle in the early 1990s which was set up to learn more about user preferences. The authors explain that via a complex learning process users gradually developed a new relationship with these electric vehicles. Over time users learned that the electric vehicle was certainly a real car, however, with its own distinctive features which were different from an internal combustion engine car. Furthermore, the advantages of driving an electric vehicle (low noise levels, ease of driving, cleanliness, low costs) ultimately led to a very positive assessment. Also, during this process users learned to modify their mobility routines, which were originally based on the characteristics of a regular car, in ways that fitted with the properties of the electric vehicle (*ibid.*, pp. 89-93). This means that the users adapted their mobility practices so that the electric vehicle was used for short intra-urban trips, and a regular car was used for longer trips. Thus, a shift from internal combustion engine cars to electric vehicles means that not only the material dimension of mobility practices is altered, but also the symbolic dimension ('what constitutes a car?') and the spatio-temporal dimension of mobility practices ('for which time interval and spatial distances can I use the vehicle?'). In certain circumstances this might even lead to changes in the locations visited by the users of electric vehicles. Based on interviews with electric vehicle owners, Kurani & Turrentine (2002) indicate that 36% of the EV owners made some adjustments in the planning of trips, while 11% of the owners actually changed the location of common activities, for example by shopping closer to home or planning shopping trips to stores with recharging facilities.

As the case studies of mobility management in this chapter have shown, changes in the conduct of mobility practices also apply when human agents shift from car usage to public transport services. In theory, a significant number of social activities could be undertaken by means of non-automotive transport modes (Dijst *et al.*, 2002; Katteler & Roosen, 1989)[80]. However, the shift to non-automotive transport modes also implies a shift in the spatio-temporal organization of mobility practices: 'for the time to be acceptable, the order in which activities are carried out and/or the location choice may have to be altered' (Dijst *et al.*, 2002, p. 425). It is clear that the high degree of flexibility in activity schedules by car drivers can never be attained if only public transport modes are used. Therefore, some authors claim that society is becoming increasingly car dependent because many people see no or little alternative for the car to maintain their mobility patterns (see Jeekel, 2011). However, it is especially this necessity of rescheduling time-spatially dispersed activities which is ignored by car users. As Kaufmann makes clear: 'exclusive car drivers omit in their evaluation the fact that if they were users of public transport their schedules would have different spatio-temporal characteristics' (Kaufman, 2000, p. 14).

Illustrative is another experiment described by Hoogma *et al.* (2002) which focuses on car-sharing in combination with the use public transport modes. In contrast to the car-sharing experiment conducted by Canzler (2008), in this experiment car users gradually began to reduce the number of trips made by the car, limiting their use primarily to those cases in which public transport was not suitable. Interestingly enough, though the number of car kilometres decreased substantially, the participants did not feel constrained. What is more important than the relative success of one experiment in comparison to another is the fact that, apparently, for these participants access to cars via the car-sharing system was sufficient enough to maintain the minimum level of motility[81]. Another interesting aspect which is a result of the learning process is that the shift in the modes of mobility also altered the way users experienced these modes, resulting in a re-evaluation of car driving (which was experienced as more and more stressful) and an increase in the appreciation of public transport modes (not having to drive, being able to read or sleep). Thus, also in this case the change in the material dimension of mobility practices also had an implication on the symbolic dimension of mobility practices, and in the spatio-temporal

[80] Based on a pilot study in a Dutch neighbourhood Dijst *et al.* (2002) indicate that travellers have more time-spatial opportunities to use non-automotive transport means than is generally expected. Simulations on activity patterns revealed that 47% of the car users in the pilot project were in a position to shift to environmentally friendly alternatives. While this pilot was based on simulations, the research by Katteler & Roosen (1989) on the substitutability of car use was based on diaries of actual trips made. This study showed that respondents did not exclude or heavily objected to an alternative choice for 60% of the car trips.

[81] The uncertainty expressed by the project participants, with regard to costs and sufficient access to a car, seem to be similar to those expressed in the German car-sharing program. However, one reason for the relative success of this transition experiment according to Hoogma *et al.* (2002) was the frequent and open forms of communication between pilot members and pilot managers (see pp. 153-158).

dimensions of mobility practices. An example of the latter is that some participants even altered their shopping and leisure-time practices by searching out opportunities closer to home (*ibid.*).

The abovementioned conclusions indicate that learning processes are of importance in the management of regime shifts. Learning processes are required in order for practitioners to know how to apply various socio-technical innovations to their full advantage[82]. As socio-technical mobility innovations, such as car- and bike-sharing, make new mobility patterns possible they need to be integrated in the whole of time-spatially dispersed activities in social life. Practitioners have to experiment with the alterations this requires, not only in their travel patterns, but also in their other social practices (working, shopping, etc.). Second-order learning takes place when practitioners not only learn how to integrate innovation in mobility practices but also alter the decision criteria upon which decisions are made. In second-order learning existing ideas and beliefs are put to the test, are reflected upon, shaken, and potentially changed. In the abovementioned experiments this happened when the internal combustion engine car or car driving in total were re-evaluated.

In sum, de- and re-routinization requires that the way a human agent conducts his/her mobility practices is placed under close inspection. An important element in processes of de- and re-routinization is the question whether or not changes occur in the framework of ideas upon which decision are made. To a certain extent, a regime shift on the level of (commuting) practices is prone to these same dynamics. The regime encompasses the whole of implicit and explicit rules and associated thinking that guide practical behaviour of people while being reconfirmed by everyday practices (Loeber, 2003). Thus, once these rules and thoughts start to shift the expectations of what is everyday mobility may change. This may involve a change in evaluation of various modes of transport, but also a shift in the specific time-spatial patterns of which the mobility practices are constructed. Next to alterations in the conception of 'the automobile', a shift might also occur in the general (ownership) structure of mobility portfolios. This would lead to a situation in which it becomes the 'normal way of doing' to integrate various means of public and private transport services into tailor-made mobility packages. The alteration would involve a reconfiguration of mobility patterns in which practitioners determine for every trip which mode(s) of mobility would best fit the circumstances. In short, in the everyday conduct of mobility practices human agents would become habitual multi-modal mixers instead of habitual mono-modal car drivers.

Similar to transition management, mobility management aims to build on existing windows of opportunity to facilitate such a regime shift. Indications of a shift taking place in the practice of commuting, both from the private car towards other modes of transport and in the uptake of smart working and smart travelling are for example:

- alterations in the conventions or structural stories on time and space independent working, public transport and (lease) cars;
- alterations in the institutional settings and the rules and resources surrounding commuting and work practices;
- alterations in the socio-technical mobility options made available.

[82] Especially in the early stages of development, learning processes are as much of importance for designers and providers of socio-technical innovations as they are for practitioners. In most cases new products and services are the result of co-evolution between behaviour of consumers/users and producers of innovations.

All of these shifts are currently taking place in the practice of commuting, albeit with varying degrees of intensity. Indicative for these on-going alterations is a headline from mobility management magazine: 'The mobility transition has only just begun. In many companies the work and travel culture is changing, and many people see the turn to smart working and smart travelling as a societal transition'. Another sign of the structural change in the collective routines is the headline of the Telegraaf, the largest and most pro-automobile newspaper in the Netherlands: 'the lease car is out of grace'. When given the choice, employees would no longer automatically opt for a lease car but would be more interested in personal mobility budgets.

So, do these changes indicate towards a societal transition taking place, and which pathway may the transition from the car regime entail? While there are certainly shifts taking place in the domain of mobility, currently there are no hard signs that mobility practices in the near future will be undergoing a transition from the car-regime towards a completely new socio-technical regime dominated by other transport modes. However, it similarly unlikely that the car-regime will remain completely unchanged. If a transition is to occur from single car use towards multimodal mixtures it will not be because of one single socio-technical innovation. Nevertheless, due to the constant landscape pressure (accessibility, global warming, fuel prices, consumer demand for modal flexibility), already niche innovations are adopted by the regime players. The massive uptake of multimodal travel cards by the car leasing branch is a clear example of this adoption.

Chapter 7.
Sustainable mobility behaviour: a matter of general environmental concerns or context-specific mobility portfolios

Love only rarely listens to reason. So it seems here as well. At stake are our needs and preferences – and they have proved astoundingly resistant to clever arguments and cunning mathematics of the critics of the automobile'.

Sachs, W., For the love of the automobile (1992, p. vii).

7.1 Introduction

Throughout this thesis we have elaborated the social practices model to emphasize and stipulate an approach which aims to analyse the involvement of citizen-consumers in on-going transition processes in the domain of everyday mobility. According to Jackson (2005, p. 22-23), conceptual models of consumer behaviour, such as the social practices model, may have two distinctive purposes. First, the conceptual models may function as 'heuristic devices' for exploring and identifying those factors which are of importance in consumer behaviour. The identification of these general factors are of use to policymakers who attempt to influence consumers to conduct more environmentally benign consumer behaviour. Holistic and structurally complex models of consumer behaviour are often listed as belonging to this category. Second, models of consumer behaviour may provide a conceptual and theoretical framework for conducting detailed empirical research on specific types of (environmentally relevant) consumer behaviour (*ibid.*). This second purpose is primarily empirical, aiming particularly to test the strength of various relationships in consumer behaviour and the influence of intervention methods.

It is clear that so far the social practices model has been applied in line with the first purpose, namely to increase the conceptual understanding of environmentally-relevant consumer behaviour in relation to (changing) systems of provision in the mobility domain. While the previous chapters showed that the social practices model is a useful approach in analysing transition processes in the practices of new car-purchasing and commuting, it did not systematically examine in a quantitative way the specific relationship between the components of the model. Neither did it thoroughly investigate the (individual variation in) environmental concerns and portfolios of the consumers participating in these practices. This chapter can be read as an attempt to meet one of the dominant challenges of the social practices approach, which is to develop a method which allows for analysing the individual variety in the conduct of social practices in the domain of mobility while avoiding the traps of the micro-approaches which continue to dominate the sustainable consumption debate. This challenge is met by investigating the dispositions and possibilities of individual human agents towards environmentally relevant social change in direct relation with the levels of available sustainable alternatives on offer and the specific modes of provision by which these innovations are made available.

The aim of this chapter is threefold. The primary goal of this chapter is to empirically apply the social practices model, as developed in the CONTRAST research program, in the domain of everyday mobility behaviour. This application refers to an empirical test by translating the various components of the social practices model in order to analyse the role of context-specific dynamics in the domain of everyday mobility behaviour in relation to individual values and concerns. Though the various social practice theories offer a sophisticated understanding of how knowledgeable and capable human agents conduct their everyday routines while being enabled and constrained by the social and technological systems of provisions, they have been applied only to a very limited extent to environmentally relevant behaviour (Hargreaves, 2008). Furthermore, to our knowledge no quantitative analysis has been conducted which has applied social practice theory to the domain of everyday mobility. With the help of two large-scale consumer-surveys conducted in 2007 and 2008, in this chapter it is explored if and how the contextual approach of the social practices model may contribute to studies of environmentally relevant consumer behaviour and behaviour change in the domain of everyday mobility.

The second aim of this chapter is to investigate how consumers assess and evaluate the various environmentally friendly alternatives on offer in the domain of everyday mobility. Instead of one narrow and pre-defined vision of sustainable mobility, these alternatives refer to the wide range of available options to be mobile in a more sustainable manner. Furthermore, various strategies aiming to increase the availability and usability of these environmentally friendly alternatives are explored to investigate if, and under which conditions, the respondents can be interested in changing their mobility behaviour in concurrence with their preferred sustainable mobility options. The intention is to determine the role of contextual factors in mobility behaviour by examining (consumer's perception of) the quality and quantity of the available sustainable products and services on offer in combination with the possible strategies of producers and suppliers in the modes of provision.

The third and final aim is to determine the role of contextual factors from the individual and life world perspective by integrating the notion of mobility portfolios into research on pro-environmental behaviour and behaviour change. Thus, while the second aim is directed to the provision side, with the third aim the focus of attention is on the individual and life-world perspective. Based on the many studies that have analysed the relationship between environmental concern and environmentally significant behaviour there is common agreement that there is only a very weak relationship between general environmental concerns and specific environmental behaviour. For instance, Bamberg (2003, p. 30) argues that situation invariant general environmental concerns cannot influence specific behaviours directly[83]. Instead of focusing (only) on general and domain-specific environmental concerns, context is introduced by exploring the notion of mobility portfolios (see Chapter 5). In this chapter we will investigate whether or not we can distinguish greener or less green mobility portfolios. Furthermore, it is investigated if these green mobility portfolios are related to general environmental concerns. Finally, it is analysed whether

[83] Instead Bamberg (2003) proposes that situation-specific cognitions, referring to the consequences of the behaviour, are direct determinants of specific behaviours while general environmental concerns remain important indirect determinants of behaviour.

or not individuals with green mobility portfolios are prone to conduct mobility practices in a more sustainable way.

In Paragraph 7.2 we will first describe the preliminary research on sustainable consumer behaviour conducted in the CONTRAST research program. Next, the details of the quantitative survey are laid out in the methodology paragraph. The following two paragraphs focus on the general dispositional and the conjunctural specific environmental concerns of the citizen-consumers. The main body of the chapter focuses on the citizen-consumer assessment of the various sustainable mobility alternatives. Paragraph 7.6 consists of the general consumer assessment while Paragraph 7.7 investigates individual differences in consumer assessment based on mobility portfolios and environmental concerns.

7.2 Preliminary research on citizen-consumer behaviour in context

In 2004 a preliminary research was conducted by Telos and Motivaction which aimed to broaden consumer behaviour research by including everyday routines of social practices in various consumption domains next to variables such as environmental values, awareness and knowledge[84]. The research was conducted to search for both individual and contextual variables that explain sustainable consumer behaviour. Based on the literature the following conceptual model (Figure 7.1) served as a guideline for the empirical research.

On the basis of value statements the respondents were divided into three sustainability segments: high, medium, low. There was a clear relation with the other elements (parts B-D) of the sustainable awareness indicators. Respondents in the 'high sustainability'-category had more knowledge of sustainability issues and of sustainable alternatives (part B), they were more willing to pay extra for a sustainable product (part C), and they felt more personally responsibility for achieving a sustainable society (part D).

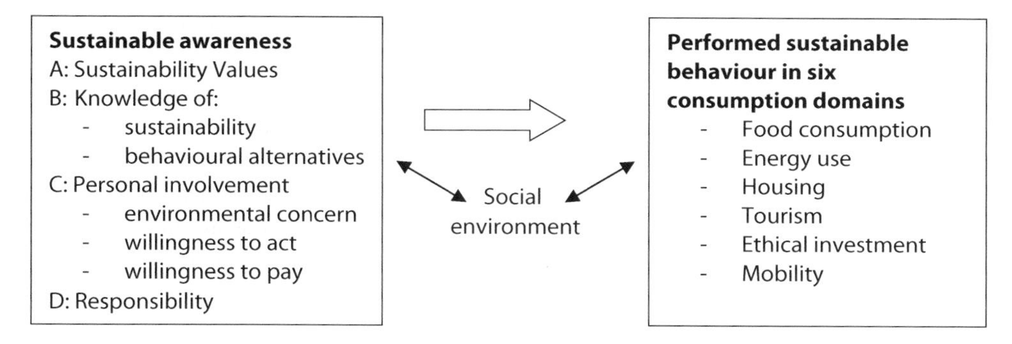

Figure 7.1. Societal valuation of sustainable development (RIVM, 2004).

[84] The research conducted by Telos and can be seen as one of the predecessors of the Contrast research programme.

For each of the six consumption domains a small number of questions regarding performed routines were asked to measure the performed sustainable consumption behaviour. Based on a regression analysis it was concluded that the four elements (A-D) in the above described conceptual model were able to explain 56% of the variance in portrayed consumer behaviour (RIVM, 2004). However, because only a small selection of the questions were used to measure the relation between sustainable awareness and performed sustainable behaviour the coefficient is only indicative.

A more important conclusion from this research is that the extent to which consumers perform sustainable behaviour varies significantly per consumption domain (RIVM, 2004). In the report a division is made between the differentiation (spread of the performed behaviour within a consumption domain) and the sustainability level of the domain as a whole (see Table 7.1). For the domains of everyday mobility and tourism this means that, in 2004, on the one hand the performed behaviour is relatively unsustainable when compared to the other domains, and on the other hand the differentiation between the respondents is low. The authors indicate that the minimal differentiation within these domains is probably related to the quantity of products on offer: when there is ample choice it becomes easier to act sustainable and the performed behaviour is more likely to differ (*ibid.*) (see also Paragraph 3.4.2).

However, the differences between the various consumption domains are not that large that people portray highly sustainable behaviour in one domain and highly unsustainable behaviour in the next. Citizen-consumers who attach great importance to sustainable development behave more environmentally friendly than those who find sustainable development of no or less importance.

In addition to the research in 2004, a preliminary consumer research was conducted by the Contrast research group in 2007. This research was part of the 'Sustainability Monitor 2007', a yearly investigation of the motives and barriers of Dutch consumers for sustainable consumption behaviour[85]. While the Sustainability Monitor provided worthwhile insight into (the motives of) sustainable consumer behaviour in various consumption domains, the most interesting aspect of the survey was related to abovementioned RIVM study. With regard to the transition phases of the various consumption domains, Figures 7.2 and 7.3 largely confirm the expectations expressed by the RIVM study of 2004.

Table 7.1. Differentiation and level of sustainability per consumption domain (adapted from RIVM, 2004, p. 106).

		Stage of the transition process in the consumption domain	
		-	+
Differentiation within the domain	-	tourism, mobility	energy
	+	food consumption, investment	housing

[85] The sustainability monitor is conducted on a yearly basis by the Internetwork on Sustainability (see www.insnet.nl). This specific internet survey was conducted amongst 1,359 respondents (92% response rate).

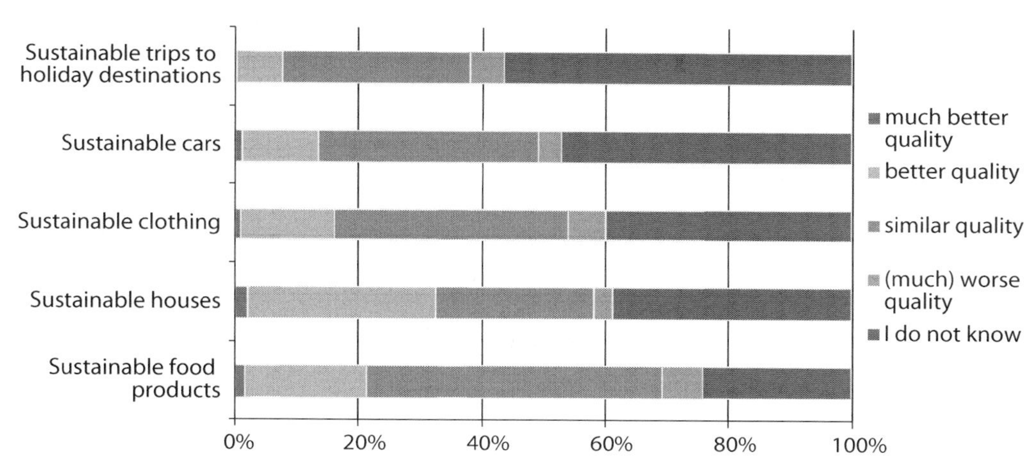

Figure 7.2. Consumer perception of the quantity of products and services on offer in 2007 (n=1,265, Insnet, 2007).

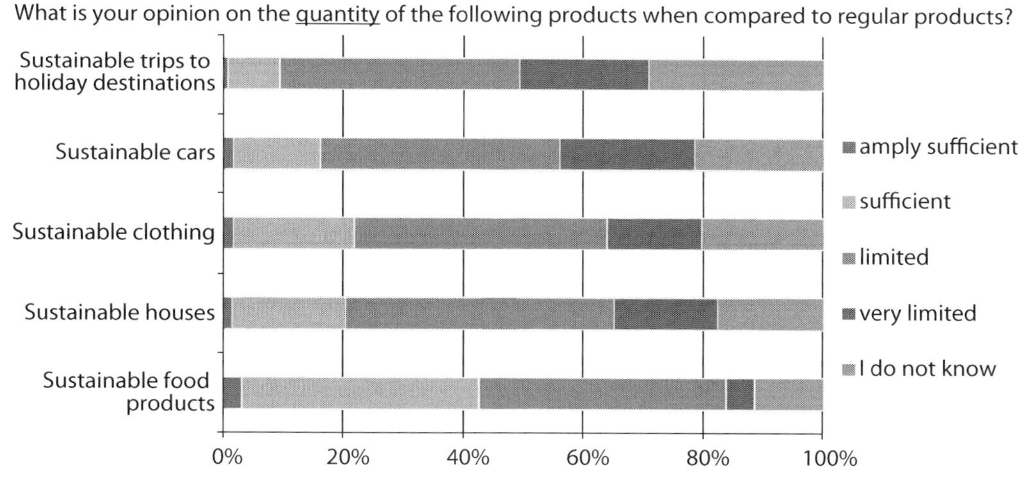

Figure 7.3. Consumer perception of the quality of products and services on offer in 2007 (n=1,265, Insnet, 2007).

There are two aspects of these figures which are worth noting. First, the perceived quantity and quality of the sustainable products and services on offer differs substantially per consumption domain. The quantity of sustainable products on offer, when compared to regular products, is considered (very) sufficient by only 10% of the respondents for holiday trips and by 15% for sustainable cars. In comparison, in the domain of food consumption the quantity is considered sufficient by over 40% of the respondents of the survey. Also, generally speaking the demand on offer, when compared to regular products, is considered limited or very limited by the respondents for all the domains. Second, a large proportion of the respondents found it difficult to assess the quantity and especially the quality of the sustainable products and services on offer. This is shown by the large proportion of respondents who answered 'I do not know'. It is likely that this percentage can be seen as an indication of familiarity with sustainable products and services in a certain consumption domain. Not surprisingly, in those consumption domains with less perceived quantity of sustainable products, the familiarity with regard to the quality of sustainable products was low as well (indicated by the higher proportion of respondents answering 'I do not know').

In sum, the previously expressed expectation by the RIVM that the tourism and mobility domain are still in the first transition stage(s) is strengthened by the outcome of the Sustainability Monitor 2007 (at least for the categories holiday trips and sustainable cars). The quantity of products, when compared to regular products, is considered highly insufficient, and the respondents are unfamiliar with the specific quality.

7.3 Methodology

In order to meet the aims stated in Paragraph 7.1 a quantitative survey among Dutch citizens was conducted. This survey investigated consumer concerns and sustainable consumption behaviour in the five following consumption domains: tourism mobility, everyday mobility, clothing, housing and food consumption[86]. However, in this chapter we will primarily focus on the results within the domain of everyday mobility and we will not elaborate on the specific results within the four other domains.

The research was performed by Motivaction, the same market research bureau who conducted the preliminary research in 2004. Motivaction's online marketing panel, containing over a 100,000 panel members, was used to bring in the respondents.

The field research was conducted between 11th of July and 4th of August 2008. The survey started with twenty questions about general consumer attitudes and concerns in relation to sustainable development (general dispositional lifestyle element). The following segments contained questions with regard to each of the five consumption domains (conjunctural specific lifestyle element). By and large, the domain-specific segments contained questions on domain-specific consumer concerns, an identification of portfolios, an assessment of the supply on offer (sustainable alternatives), and an assessment of provider strategies. For the domain of everyday mobility five sustainable alternatives were investigated (elaborated in Paragraph 7.6). The citizen-consumer assessment focused on issues of usability, reliability and availability of the innovations discussed.

Because of the length of the questionnaire it was decided to send out the questionnaire in three blocks with a time interval of one week. Block A consisted of the general concerns and everyday mobility; Block B consisted of clothing and tourism mobility; Block C consisted of food consumption and housing. Since the overall aim of the Contrast-survey is to analyse and compare consumer behaviour in different domains it was essential that the same respondents filled in all the three blocks. To spread the number of drop outs evenly among the blocks, a third

[86] The questions used, layout and form of the survey was the result of a combined effort by the whole project-group: four PhD researchers and supervisors, the Netherlands Environmental Assessment Agency, and LEI.

of the respondents received the blocks in the order of ABC, another third in the order of BCA, and the final third in the order of CAB. In total 2,906 respondents had filled in one or more parts of the questionnaire; Block A was filled in by 2,242 respondents, Block B by 2,302 respondents, and Block C by 2,288 respondents. Of the total number of respondents 1,594 had filled in all the three questionnaires blocks (drop-out rate is therefore 45%).

An important methodological point is that the survey is highly dependent on the self-reported behaviour and behavioural intentions of the respondents. Instead of an 'impact-oriented analysis' which measures actual consumer behaviour, this chapter must predominantly be seen as an example of an 'intent-oriented analysis' with all the uncertainties that come with it (see also Gatersleben *et al.*, 2002; Poortinga *et al.*, 2004).

7.4 General dispositional environmental concerns

In Chapter 3 we have elaborated extensively on the concept of lifestyle, also in the way that it is defined and used by Giddens (1991). To reiterate, the lifestyle is the sum of all the social practices in which the individual participates, together with the storytelling that goes along with it (*ibid.*). The lifestyle component is on the one hand an individual, unique, and identity providing aspect of a human agent, while on the other hand lifestyle experiences are always shared experiences based on narratives surrounding consumption domains and specific social practices. To make an empirical distinction between these two elements of the lifestyle in the project we make use of the terms 'general dispositional lifestyle' and 'conjunctural specific lifestyle' as introduced by Rob Stones (2005). The general dispositional lifestyle element contains knowledge, experiences and beliefs that rise above a specific context or action and exercises influence in a whole range of contexts. In the conjunctural specific element of the lifestyle, on the other hand, the knowledge, experiences and beliefs are primarily related to a specific consumption domain or social practice. In Figure 7.4, which visualises the social practices model without the systems of provision, the conjunctural specific lifestyle segment is illustrated for the domain of everyday mobility.

Many of the models on pro-environmental consumer behaviour focus on the general dispositional part of lifestyles. General value orientations such as Rokeach values and altruism-scales are typical for the general dispositional lifestyle. The same applies for general environmental concern, knowledge of sustainability, and overall perspectives on sustainable production and consumption. Though one should not overemphasize the rigidity of the distinction between the two lifestyle segments – both segments are of course part of the same individual – the distinction is useful to understand why current research has indicated that situation invariant general environmental concerns show a relatively weak relation with practice specific behaviour.

In the survey the general dispositional element was measured with 20 statements (divided into four segments) regarding the respondent's position in various environmental issues. These four segments aimed to represent four societal discourses of sustainable consumption and production (discussed in Spaargaren *et al.*, 2007) which guide the thoughts and actions of citizen-consumers in the area of sustainable consumption. These societal discourses are based on three underlying principles. First, citizen-consumers may acknowledge the necessity to deal with environmental challenges with which our society is faced, or they may feel sceptical or even opposed to the environmental goals set by the scientific and political community. Second, citizen-consumers differ

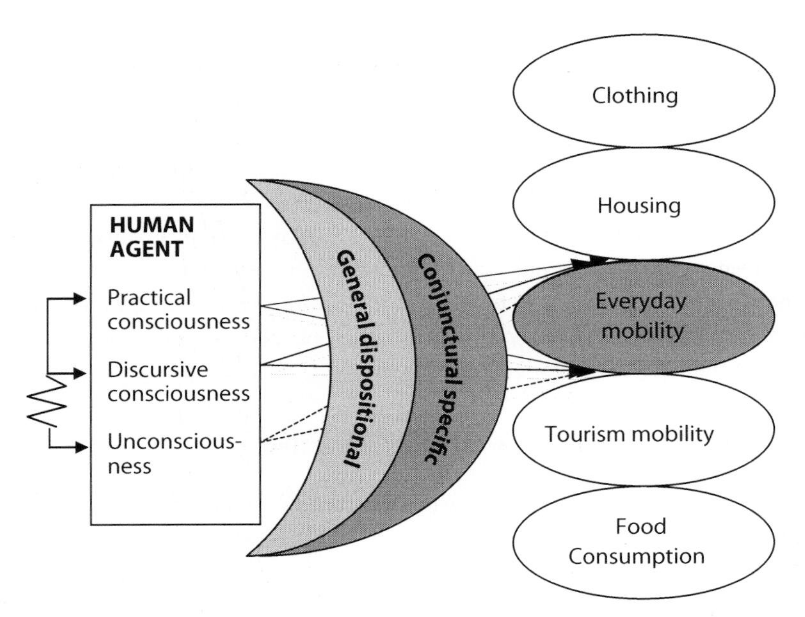

Figure 7.4. General dispositional and conjunctural specific elements of the lifestyle (adapted from Spaargaren et al., 2007).

in the degree to which they themselves feel responsible for reducing the environmental impact of consumption and production patterns. Citizen-consumers may become actively involved in greening their consumption patterns or they may feel that environmental issues are primarily the responsibility of policy-makers and companies. Third, citizen-consumers may differ in their perspective to the degree of change which is desired or deemed necessary. While some citizen-consumers may consider a radical break with current institutions and practices a necessary precondition for sustainable development, other citizen-consumers may opt for a more gradual environmental change within the current institutional frameworks (*ibid.*).

Based on the variation in the abovementioned basic principles we identified four different positions as a tool to measure general environmental concern. The first segment represents those citizen-consumers who are sceptical to the environmental problems as portrayed in the main societal debate. People who score high in this segment either doubt the existence of human-induced global environmental challenges such as climate change, or they consider other problems more important. The viewpoint represented in the second segment is based on a strong confidence in the dynamics and self-correcting capabilities of society. Global environmental challenges are acknowledged as problematic but citizen-consumers have confidence in technological innovation driven by the market and national government. Furthermore, the responsibility for the greening of consumption and production chains is ascribed to these same parties as environmental challenges are not seen as phenomena which they have any significant impact on. Citizen-consumers represented by the third segment have the viewpoint that, next to active environmental policies, they themselves are at least partially responsible in reducing the environmental impact of consumption and production chains. Finally, the fourth segment represents the viewpoint of citizen-consumers who feel that the gradual changes which are currently being pursued are not sufficient to meet existing environmental challenges. In order to attain a sustainable society more radical measures are needed; the consequent large-scale impact on the organisation of everyday life (such as a general decline in overall consumption level) is accepted by these citizen-consumers (see Spaargaren *et al.*, 2007).

Figure 7.5 shows the respondents' reactions on the twenty questions. This gives an overview of the overall response to the four segments and of the consistency in the answers. Looking at the first segment ('sceptical to necessity of environmental change'), generally speaking the respondents acknowledged the existence of environmental problems. For example, over 70% of the respondents (strongly) agree that environmental problems are something to worry about. However, only a quarter of the respondents let their lives be directly influenced by environmental problems.

The second segment reveals a very interesting aspect of the survey. While it was expected that citizen-consumers would lay responsibility for environmental change either in the hands of corporations/governments or in their own hands, this is not what the answers point out. Over 50% (strongly) agrees with the statement that the business community should take care of existing environmental problems. Also, citizen-consumers have a relatively strong confidence in cleaner technologies. On the other hand only 18% of the respondents (strongly) agree with the statement that they, as a consumer, can do little to reduce the climate change problem. Similarly, only 16% of the respondents have the viewpoint that they do not need to solve environmental problems. Thus, citizen-consumers feel that dealing with environmental challenges is a shared responsibility.

In the third segment the abovementioned aspect of personal responsibility is further examined. The answers to these questions do not reveal stark differences, both with regard to the response between the questions as within one question. On the aspect of co-responsibility the citizen-consumers are evenly spread out.

With regard to the final segment, which focuses on the necessity for radical change, there are two points to make. First, on a general level the majority of the respondents agree that radical change is necessary to deal with (global) environmental problems. Only 16% of the citizen-consumers (strongly) disagree with the statement that we need to consume less, and even a limit on a person's total amount of climate change emissions is rejected by no more than 12% of the respondents. Second, with regard to how these environmental problems should be solved there is much less consensus.

Because the four segments are a theoretical construct based on existing societal debates it is uncertain whether or not they form a consistent scale of the general dispositional part of the lifestyle. To test if these segments are consistent, a principal component analysis was conducted. This data reduction technique is often used by social science researchers in the development and evaluation of scales (Pallant, 2007). Data reduction techniques attempt to reduce a large number of variables into a smaller amount by grouping interrelated variables. The principal component analysis more or less confirms the descriptive analysis portrayed above (Table 7.2). The first component clearly shows that those citizen-consumers who are sceptical to environmental issues respond to the statements in a consistent manner. The third segment (co-responsibility) is clearly present in the second component, that is, citizen-consumers who are environmentally committed and consider themselves personally responsible. The existence of the other two segments could not be confirmed with the principal component analysis. That does not mean that the underlying

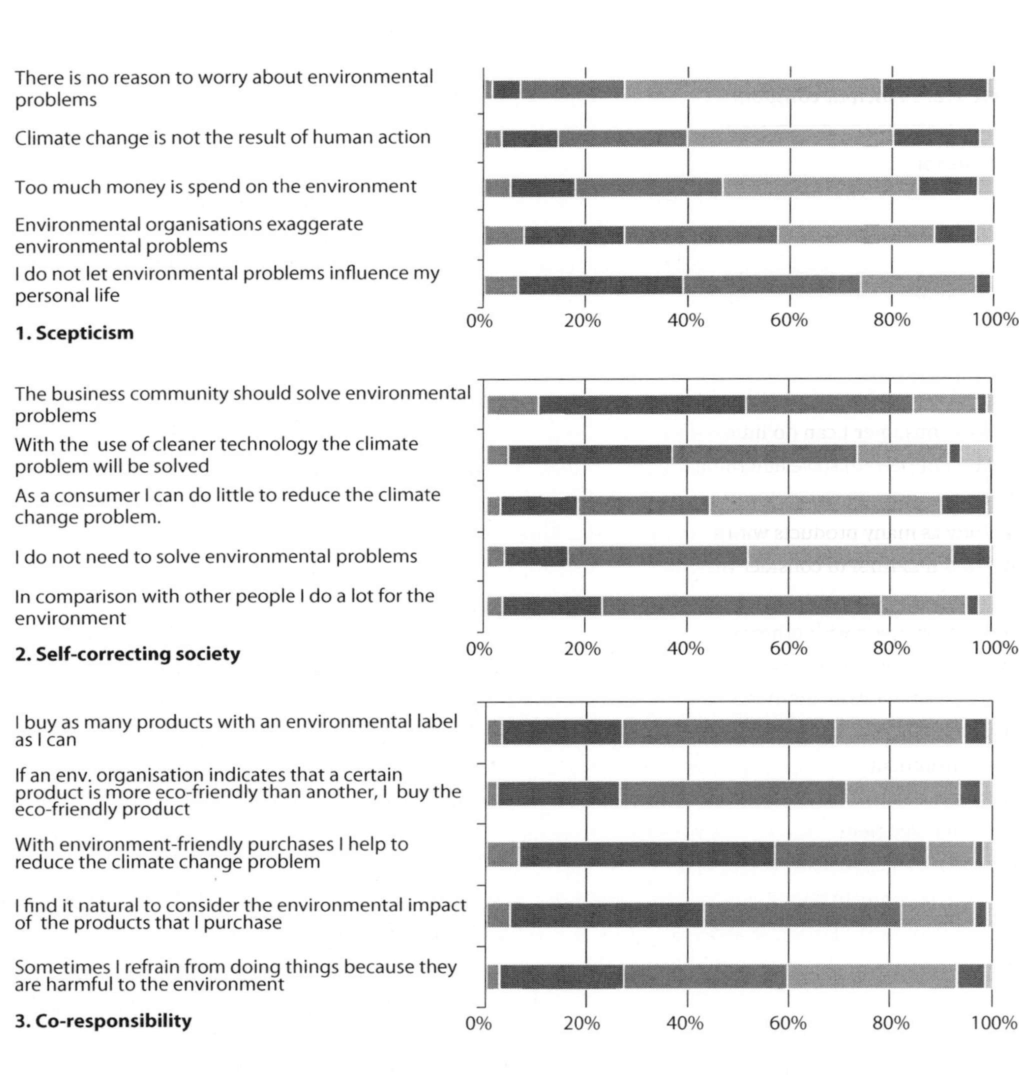

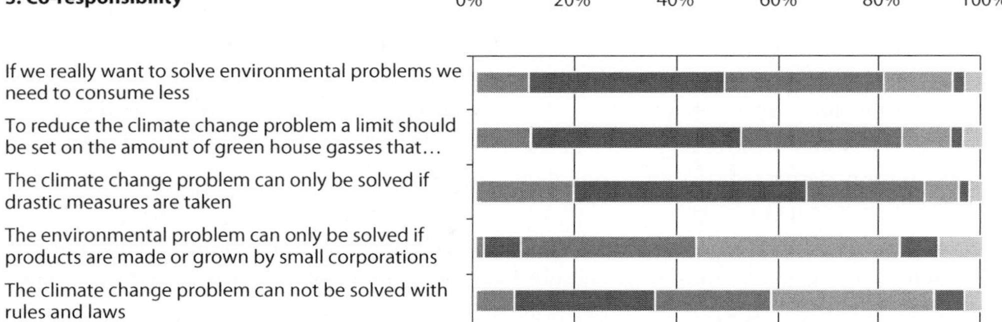

Figure 7.5. Statements representing general dispositional environmental concerns (n=2,242, Cronbach's alpha's 1=0.66; 2=0.60; 3=0.74; 4=0.66).

Table 7.2. Principal component analysis of the general dispositional environmental concerns[1].

Statements	Components[2]			
	1	2	3	4
3 Too much money is spend on the environment	0.739			
1 There is no reason to worry about environmental problems	0.730			
2 Climate change is not the result of human action	0.711			
4 Environmental organisations exaggerate environmental problems	0.671			
8 As a consumer I can do little to reduce the climate change problem	0.656			
9 I do not need to solve environmental problems	0.654			
20 The climate change problem cannot be solved with rules and laws	0.608			
11 I buy as many products with an environmental label as I can		0.788		
14 I find it natural to consider the environmental impact of the products that I purchase		0.736		
10 In comparison with other people I do a lot for the environment		0.674		
12 If an environmental organisation indicates that a certain product is more eco-friendly than another, I buy the eco-friendly product		0.651		
15 Sometimes I refrain from doing things because they are harmful to the environment		0.634		
13 With environment-friendly purchases I help to reduce the climate change problem		0.564		
5 I do not let environmental problems influence my personal life	0.463	-0.481		
19 The environmental problem can only be solved if products are made or grown by small corporations		0.404		
16 If we really want to solve environmental problems we need to consume less				
6 The business community should solve environmental problems			0.719	
18 The climate change problem can only be solved if drastic measures are taken	-0.461		0.566	
17 To reduce the climate change problem a limit should be set on the amount of greenhouse gasses that people may emit			0.522	
7 With the use of cleaner technology the climate problem will be solved				0.875

[1] Prior to performing principal component analysis, the suitability of data for factor analysis was assessed (methods based on Pallant, 2007). The KMO value was 0.92, exceeding the recommended value of 0.6 and Bartlett's Test of Sphericity reached statistical significance ($P<0.000$), thus supporting the suitability of the factor analysis of the correlation matrix. The PCA revealed four components with eigenvalues exceeding 1, each component subsequently explaining 21.4%, 18.4%, 8.7% and 5.6% of the variance.

[2] Rotation converged in 6 iterations. For ease of interpretation only loadings above 0.4 have been displayed. Extraction method: principal component analysis. Rotation method: Varimax with Kaiser normalization.

storylines are not used or recognized by citizen-consumers, however, from the survey they could not be verified as separate components of the general dispositional lifestyle.

7.5 Conjunctural specific environmental concerns

In the previous paragraph the general environmental debates surrounding sustainable consumption and production were analysed and discussed. One of the suppositions of the social practices approach is that these debates may vary from one domain to the next. Therefore in this section the environmental debate surrounding sustainable mobility is investigated in more detail. Comparable to the general dispositional lifestyle segment, the four segments of the sustainable mobility debate were investigated with four groups of questions. This will give us insight into how citizen-consumers perceive (environmental) problems in the domain of mobility. Do citizen-consumers acknowledge the existence of persistent problems related to automobile-based lifestyles? What role for themselves do they see?

In Figure 7.6 the responses to the twelve statements measuring the conjunctural specific environmental concerns are presented. The response to the first segment ('environmental scepticism') reveals a consistent pattern: only a small part of the population is sceptical to environmental issues in the domain of mobility and most citizen-consumers acknowledge that environmental problems are important[87]. Only 13% of the respondents (totally) agree with the statement that the contribution of transport to environmental problems is negligible. Furthermore, 69% of the respondents find it important that traffic and transport are environmentally friendly.

The second segment ('environmental issues must be solved by government regulations and technological fixes') does not show a clear pattern. The far majority (totally) agrees with the statement that there should be stricter European laws on the polluting emissions of new cars. On the other hand, most citizen-consumers do not think that a technological fix such as particle filters will quickly solve local air pollution. The response to the final statement of this segment is also very interesting. A hefty 72% of the respondents feel that one cannot expect car drivers to travel with public transport modes as long as public transport fails to be good alternative to the car!

The third segment focuses on whether or not the respondents see an active role for themselves in the greening of the mobility domain. The responsibility of citizen-consumers in the domain of mobility is higher than one might expect considering the preliminary research described in Paragraph 7.2. The RIVM report, based on a consumer survey conducted in 2004, considered the domain of mobility to be in the initial phases of sustainable development (see Table 7.1). The outcomes in Figure 7.6 seem to indicate that in 2008 citizen-consumers are reasonably willing to actively become part of transition processes to sustainable development in the domain of mobility[88]. Almost half of the respondents (totally) agree with the statement that car drivers should alter their behaviour as they are primarily responsible for transport-related environmental problems. Moreover, the respondents not only feel co-responsible, they are also positive of the market effects of green consumer choices: almost 80% beliefs that car producers will increase the

[87] The average response in the first segment, based on the five-point Likertscale, is only 2.39 (with 3 being neutral).

[88] The average response in the third segment, based on the five-point Likertscale, is 3.63 (with 3 being neutral).

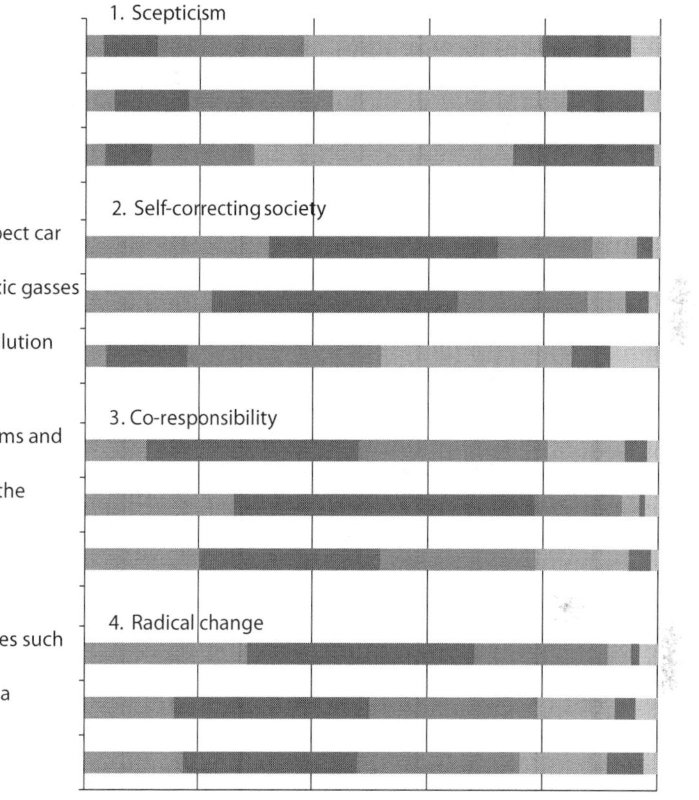
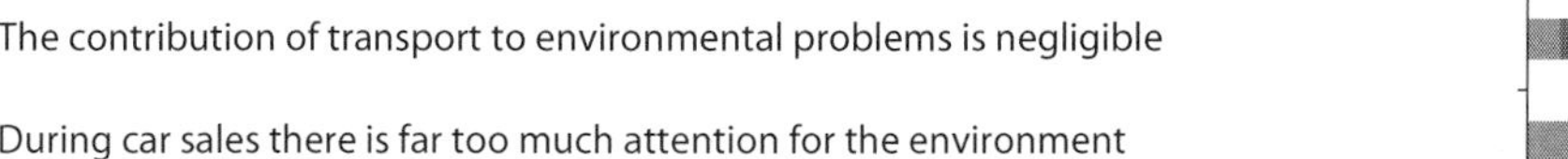

Figure 7.6. Statements representing conjunctural specific environmental concerns (n=2,242, Cronbach's alpha's 1=0.59; 2=0.42; 3=0.61; 4=0.52).

number of fuel-efficient cars on the market when more consumers start buying these cars. The question whether the relative position of the domain of everyday mobility has shifted in relation to the other consumption domains is addressed shortly hereafter.

Finally, the fourth segment represents the storyline that radical changes in the current system of mobility are necessary to cope with transport-related environmental problems. A surprisingly large proportion of the respondents agree with these firm statements which would imply a major shift in the current ways of doing things. Most illustrative of this is that exactly 50% of the respondents (totally) agree with the statement that the car as a mode of transport has reached its limits and that we should swiftly strive for a completely different and environmentally friendly transport system! In contrast, only 17% (totally) disagrees with this statement. The rapid shift to alternative fuels such as hydrogen and biofuels is supported by an even larger proportion, namely 68%. A latent market for sustainable fuels seems to be present.

An initial conclusion is that environmental change in the domain of mobility is supported by a surprisingly large part of the Dutch citizen-consumers. Clearly the domain of mobility has surpassed the phase of scepticism. Environmental issues are recognized and acknowledged as problematic and almost a majority of respondents feels at least partly co-responsible for dealing with these issues. More importantly, the current system of petrol-based automobility is seen as unfavourable and a systemic change to other modes of transport or fuels is supported by most.

To test whether or not the abovementioned four segments are supported by the empirical data as a coherent scale, a principal component analysis has been conducted (Table 7.3). The principal component analysis shows coherence in the second component which contains the statements of the first segment (scepticism). The other three segments are primarily found in the first component and to a lesser degree in the third component. This outcome implies either that three of the four segments are not present as mutually exclusive storylines in the Dutch sustainable mobility debate or that the statements were not illustrative of these story-lines. This, however, does not alter in any way the significance of the citizen-consumer's response to the individual statements portrayed in Figure 7.6. It does mean that we need to search for other ways of grouping citizen-consumers in the domain of everyday mobility. This will be carried out in the next paragraphs where we will discuss a typology of travelling and go into the mobility portfolios (on the basis of citizen-consumer's experience with innovations).

7.5.1 Comparison with the other consumption domains

Next to describing environmental concerns in the domain of everyday mobility, it is also interesting to relate these outcomes to the other consumption domains investigated in the Contrast research programme. Because the previous sections revealed that the four proposed segments do not come out as mutually exclusive storylines it is difficult to compare these five consumption domains. An exception to this is the storyline 'environmental scepticism' (or 'environmental change is not a priority') which was shown to be an existing component both in the general dispositional and in the conjunctural specific environmental concerns. Therefore for each of the five consumption domains

Table 7.3. Principal component analysis of the conjunctural specific environmental concerns[1].

	Statements	Components[2]		
		1	2	3
12	Cars that heavily pollute the air should be banned from driving into the city	0.661		
9	For short trips (less than 6 km.) people should predominantly use the bicycle	0.657		
7	Car drivers themselves are responsible for transport-related environmental problems and therefore should change their own behaviour	0.655		
11	The car as a mode of transport has reached its limits so we should swiftly strive for a completely different and environmentally friendly transport system	0.613		
5	There should be stricter environmental regulations which prescribe how much toxic gasses a new car may maximally emit	0.526		
6	If particle filters become obligatory for freight traffic, diesel cars and busses, air pollution will soon be over	0.446		
2	During car sales there is far too much attention for the environment		0.743	
1	The contribution of transport to environmental problems is negligible		0.718	
3	I do not find it important that traffic and transport are environmentally friendly		0.662	
10	Instead of driving on petrol or diesel we should rapidly switch to radical alternatives such as hydrogen or biofuels			0.665
8	If more consumers would buy energy-efficient cars, then car producers would increase the number of efficient cars available on the market			0.632
4	As long as public transport fails to be a good alternative to the car, you cannot expect car drivers to travel with public transport modes	-0.460		0.631

[1] Prior to performing principal component analysis, the suitability of data for factor analysis was assessed (methods based on Pallant, 2007). The KMO value was 0.84, exceeding the recommended value of 0.6 and Bartlett's Test of Sphericity reached statistical significance ($P<0.000$), thus supporting the suitability of the factor analysis of the correlation matrix. The PCA revealed three components with eigenvalues exceeding 1, each component subsequently explaining 21.5%, 16.6% and 12.3% of the variance.

[2] Rotation converged in 5 iterations. For ease of interpretation only loadings above 0.4 have been displayed. Extraction method: principal component analysis. Rotation method: Varimax with Kaiser normalization.

the three statements representing the first segment have been added together (Figure 7.7)[89]. The degree to which the respondents (totally) agree with the statements of the first segment can be seen as a measure of the environmental concern in that domain; the more respondents that agree with these statements the more environmental change is perceived to be unnecessary.

[89] In order to make a reliable comparison of the consumption domains only those respondents were selected who had filled in all of the first segment statements. After removing the missing cases 1,158 respondents remained who had filled in every statement.

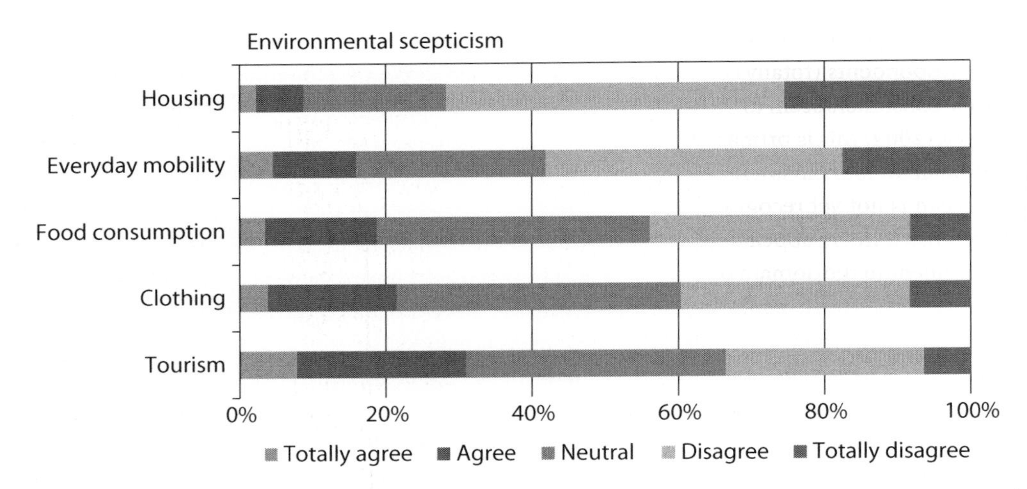

Figure 7.7. Domain specific response to first segment statements on environmental concerns (n=1,158).

Figure 7.7 confirms the expectation that citizen-consumers are not consistent in their environmental concern. A consistent environmentally friendly lifestyle, which is enacted in every consumption domain, is not evidently recognizable (see also Steg, 1999).

Not surprisingly, the consumption domain housing is the 'leading' domain with a far majority of the respondents rejecting the statements that environmental change is not a priority; 71% of the respondents (totally) disagree with these statements. This was to be expected as the first generation environmental innovations in housing (double glass windows, energy saving lights, energy efficient boilers) have become well established in most households[90]. Furthermore, currently much of the attention for climate change is focused on housing and home maintenance.

Everyday mobility is located at a surprisingly high position in the comparison of consumption domains; 58% of the respondents (totally) disagree with the notion that no environmental change is necessary. In the domain of everyday mobility environmental concerns are (at least to a certain extent) considered important and apparently they have also become more important when compared to 2004. This is in line with the conclusions of Chapter 4 in which the increasing existence and use of environmental information and monitarisation in the practice of car purchasing pointed towards a realisation of an ecological rationality and as the first steps of a process of ecological modernisation. There is a (growing) concern and a sense of urgency that a change in the current system of automobility is necessary, a sense of urgency which is larger than in most other consumption domains. The processes described in chapter four, such as the growing media attention for sustainable mobility (especially with regard to biofuels, hybrid cars and electric cars), the vast number of taxation schemes for energy-efficient cars implemented by the Dutch government, and the increase in green car commercials, seem to have made a lasting impression on the Dutch citizen-consumers.

---

[90] To indicate, in the previously mentioned 'sustainability monitor 2007' 91% of the respondents had double glass windows, and 64% had energy efficient boilers installed.

In rather sharp contrast to these developments lies the domain of tourism mobility; only 33% of the respondents (totally) disagree with the statements pointing to environmental scepticism. Citizen-consumers seem to be least concerned about this consumption domain. As is described by Verbeek (2009) this is primarily due to the fact that the debate about sustainable development in the domain of tourism is a relatively recent phenomenon. Sustainable consumption in the domain of tourism is not yet recognized as a possible and meaningful alternative by citizen-consumers, nor are there many appealing sustainable travelling options on offer. The debate of sustainable development in the domain of tourism is dominated by the idea of 'The environment? Not during my holiday' (Verbeek, 2009).

7.6 Consumer assessment of sustainable mobility innovations

So far it was shown how contextuality plays a significant role in the sense that each consumption domain is positioned differently in the debate of sustainable consumption and production. In this paragraph the second aim of this chapter is investigated, namely to determine the role of contextual factors in everyday mobility behaviour by examining (citizen-consumer's perception of) the quality of sustainable products and services on offer. This citizen-consumer assessment was conducted by asking respondents to indicate the attractiveness of a number of sustainable transport alternatives. In addition to assessing the general attractiveness of these sustainable mobility options, the response to potential improvements in the quality of the systems of provision was investigated. The respondents were asked to indicate whether or not these potential provider strategies would convince them to perform the sustainable mobility behaviour more frequently in the future. These answers can be seen as an indication of the receptiveness of citizen-consumers for specific strategies in the domain of everyday mobility. The main aim is to answer the question if and how these provider strategies, as part of the system of provision, can play a role in consumption-production chains.

7.6.1 Attractiveness of sustainable mobility alternatives

The pursuit of sustainable development in the current system of mobility is undertaken by a broad variety of approaches. Brought back to its essence the sustainable mobility alternatives can be positioned in either of the following three paradigms: (1) alternative car and fuel technologies; (2) improved use of non-automotive (public) transport modes; (3) reduced need to travel (see also Holden, 2007; Nykvist and Whitmarsh, 2007). In the first paradigm the system of automobility remains the most probable transport system for the future, therefore (radical) innovations in vehicle and fuel technologies are the most promising niches. The second paradigm is traditionally oriented towards a modal shift from car use to public transport use. Recently, the scope has also broadened so as to include smart ways of travelling where multi-modal users combine different (automotive and non-automotive) transport modes. The third paradigm, reducing the need to travel, focuses on various niches which reduce the physical need to travel, for instance through land-use planning and ICT. Nykvist and Whitmarsh (2007) also relate this last paradigm to a post-industrial lifestyle in which citizen-consumers, based on their values and world views, actively choose for a local and green way of living.

Table 7.4 displays the three paradigms as well as various examples of niche innovations corresponding to these paradigms, and (potential) strategies used by policy-makers and producers in the provision of these socio-technical innovations. The sustainable mobility alternatives, as part of a specific sustainable mobility paradigm, show the various ways in which citizen-consumers can conduct mobility practices more environmentally by embedding environmental socio-technical innovations into their normal ways of doing things. For instance, 'green automobility' may not only encompass the previously discussed practice of car purchasing, but also tanking for fuels, car driving, car maintenance and climate compensation. 'Localism' on the other hand contains the various forms of mobility-reduced lifestyles: local and green living in a car-free neighbourhood; replacing actual mobility with virtual mobility; travelling with slow travel modes. 'Modal flexibility' is specifically related to multi-modal travelling and the innovations which broaden and facilitate citizen-consumers' travel options (such as various forms of mobility management described in Chapter 6).

Only the following five alternatives in Table 7.4 have been presented to the respondents: purchasing a green car, car-free community, downsizing, slow travelling, multi-modal travelling[91]. Because slow-travelling and downsizing are likely to be unfamiliar terms for the respondents they have been operationalized as 'using the bicycle more often' and 'teleworking/working at home'.

Of these five potential sustainable mobility options the purchase of a new car is overall considered as most attractive, though none of the options is seen as entirely unattractive (Figure 7.8). Living in a car-free or car-restricted area is deemed the least appealing alternative. This outcome might result from the respondents' belief that car-free areas are inhibiting to their

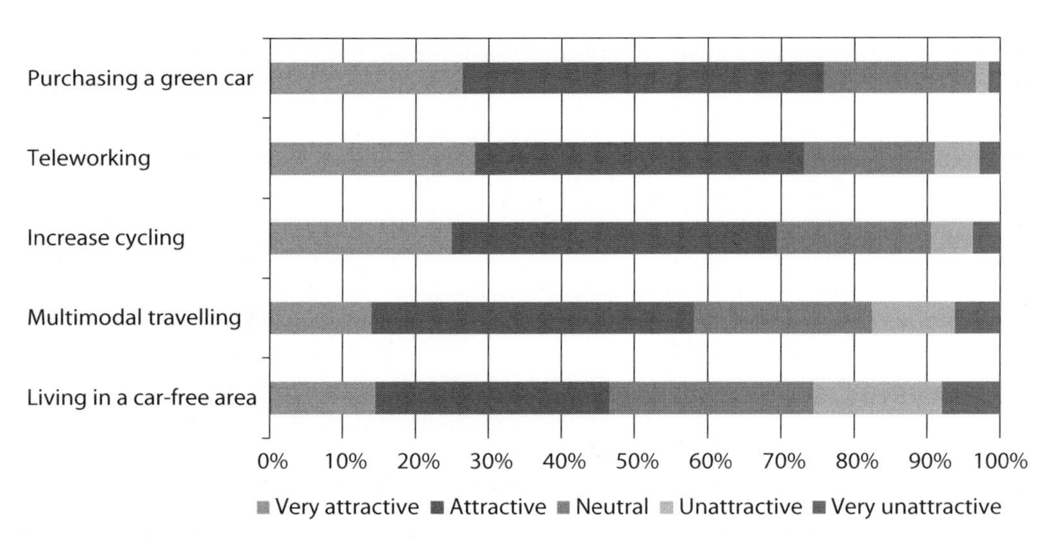

Figure 7.8. Attractiveness of potential sustainable mobility alternatives (n=2,242).

[91] This means that for the paradigm of 'green automobility' only the purchase of a green car is investigated; this option was selected because it was considered most favourable in the focus group research discussed in Chapter 4.

Table 7.4. Sustainable mobility alternatives, their related innovations and potential provider strategies (the underlines mark the sustainable alternatives and provider strategies which were part of the questionnaire).

Sustainable mobility alternatives	Examples of (niche) innovations	Examples of (green) provider strategies
1. Greening of automobility		
Purchase green car	• Energy efficiency label, Eco Top 10, etc. • (Radical) change in vehicle technology: HEV, EV, low emission ICE	• Environmental information tools + ecological monitarisation • Extended warranty on novel technologies • Increase accessibility to renewable fuels
Tank green fuels	• Shell V-Power, BP Ultimate • Biofuels, CNG, EV chargers	• Provision of cleaner petrol and diesel at regular fuel pomp • Increase accessibility to renewable fuels
Green driving style	• In-car information systems, Greenscan • 'Het Nieuwe Rijden', Eco driving	• Monitoring with feedback on driving performance • Integrating eco driving in driving lessons and examination
Green maintenance	• 'Band op Spanning' (tyre pressure monitoring service for fleet for fleet owners) • Green car washing • Reuse of car parts	• Information about timely car maintenance (tyre pressure, fluids) • Ecolabel for car-wash companies • Green car insurance
Climate compensation	• Greenseat • Greenlease, greenplan, etc.	• Climate compensation service at fuel pomp • Climate compensation as part of sustainable leasing program
2. Localism		
Car-free community	• Car-free or car-restricted neighbourhoods	• Promotion of product service systems: car sharing, bike-sharing, etc. • Parking policies
Downsizing	• Teleworking, teleconferencing, teleshopping • werkdichterbijhuis.nl; wisselwerknederland.nl	• Provision of ICT and workplace at home • New style of managing / Internal corporate communication • Organised mobility reduction: work closer to home
Slow travelling	• Public bicycle sharing program • Bicycle lanes (intracity); bicycle highways (intercity) • New vehicles: e-bikes; cargo bikes; segways	• Bicycle route planner • Creation of 'landscape experience routes' • Provision of product service systems • Mileage allowance for cycling commuters
3. Modal flexibility		
Multi-modal travelling	• Mobilitymixx, NS Business card • Personal mobility budget • Product service systems: vanpooling, car-sharing, bike-sharing • Transferium; Park and Ride • 9292ov.nl	• Organised provision of mobility services • Access to mobility services facilitated by employers • Organised provision of mobility services • Infrastructural connections of transport systems • Integrated travel information (car and public transport)

freedom of movement or that they want to be able to have a closer watch over their vehicle (in the introductory text it was mentioned that many car-free areas have collective parking lots at the border of the neighbourhood). In the next sections these five sustainable mobility alternatives, the reasons why respondents are attracted to these alternatives, and various provider strategies are investigated (in the order of most attractive to least attractive).

7.6.2 Purchasing a clean and energy-efficient car

The first sustainable mobility alternative discussed here is a familiar one, considering the chapter on car-purchasing. In this survey the respondents were asked why they considered the purchase of a clean and energy-efficient car an attractive option of sustainable mobility. The respondents had to choose one of the pre-given reasons, or give an open-end alternative answer. Figure 7.8 shows that 76% of the respondents considers purchasing a green car an appealing option. Surprisingly the benefit of ecological gains is most often given as the reason why this alternative is interesting; 50% of the respondents mention this as the main reason (not visualized in figure). Saving fuel cost is mentioned by 36% of the respondents as the main argument. These clearly are the two most leading arguments. The third reason, given by only 8% as the main argument, is the ability to benefit from environmental subsidies. Summarized, ecological reasons (50%) and economical reasons (44%) are more or less equally important considerations. It is hard to interpret whether this outcome is predominantly the result of the genuine pro-environmental attitudes of the respondents or the result of a strong social desirability bias. It could also be that these considerations are seen as mutually reinforcing, as one of the statements of the open-end answers indicates:

> *This is a good solution for every stakeholder. The environment and my wallet are both availed.* (male, 47)[92]

As mentioned in the beginning of this paragraph, next to assessing the attractiveness of sustainable mobility alternatives, the perceived quality of the systems of provision was investigated. Here the overall citizen-consumer assessment is discussed (in terms of usability, reliability and availability). Figure 7.9 represents the respondents' assessment of the provision of clean and fuel-efficient cars on three aspects: the provision of environmental information, the technical reliability, and the availability. With regard to the assessment of environmental information provision, Figure 7.9 shows that 38% of the respondents consider energy efficiency of cars not clearly indicated.

> *I would rather buy a second-hand car because of the cost, unfortunately it is almost impossible to take environment considerations into account.* (female, 32)

The technical reliability of novel automotive technologies, here questioned in the form of hybrid and bio-fuel technology, is basically trusted by the respondents. Only 17% of the citizen-consumers see this as an obstacle to purchase a clean and energy efficient car. Much more problematic is the current availability of filling stations for cars fuelled by alternative fuels (here questioned in the

[92] All the quotes from the survey are translated by the author.

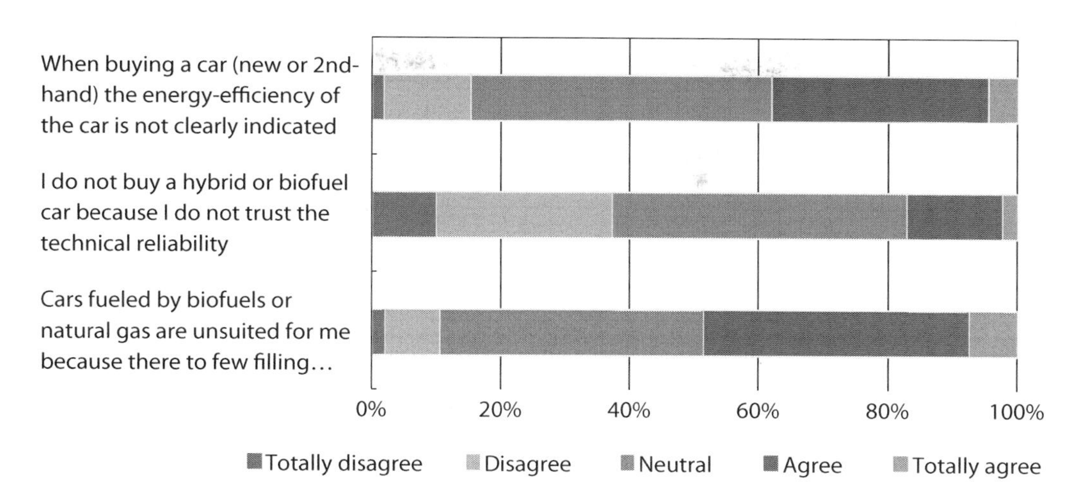

Figure 7.9. Assessment of the system of provision of green car purchasing (n=857).

form of bio-fuels and natural gas). 49% of the respondents agree there is a lack of a widespread distribution of filling stations making it an inhibiting factor for buying a green car.

In direct response to the assessment of the system of provision, provider strategies were presented to the respondents. These strategies specifically target the existing barriers for sustainable mobility behaviour. The respondents were asked to indicate whether or not these potential provider strategies would convince them to take the sustainable mobility alternative more into consideration or perform the sustainable mobility behaviour more frequently in the future. Figure 7.10 shows that the proposed provider strategies increase the self-reported receptiveness of the respondents for the purchase of a green car. 50% of the respondents mention that they will take the energy efficiency of second-hand cars sooner into consideration if an energy label would be available.

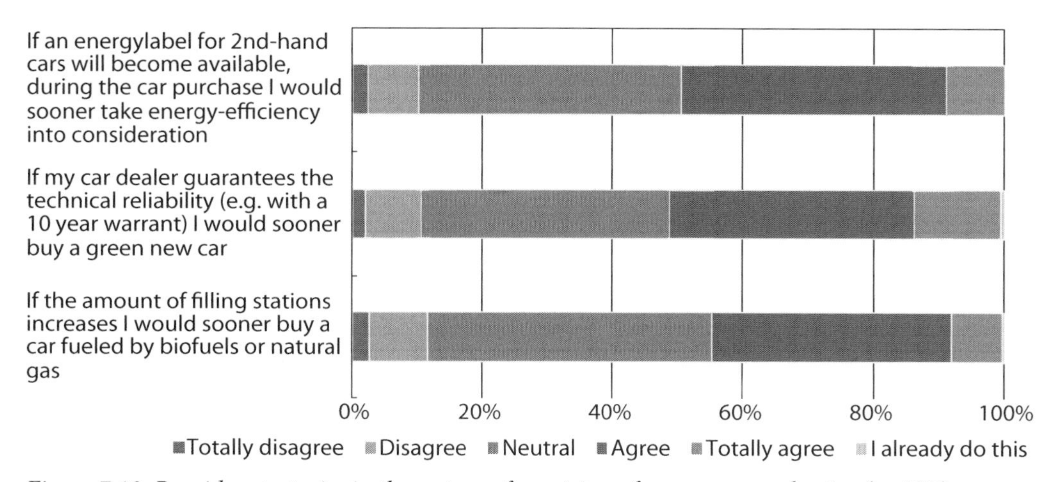

Figure 7.10. Provider strategies in the system of provision of green car purchasing (n=857).

Though such a label does not increase the availability of green cars it will increase the availability of environment-specific information of these cars.

Furthermore, 51% will sooner consider buying a car with novel vehicle technologies (such as hybrid or biofuel cars) if the technical reliability of the car is guaranteed by the dealer, for instance with a ten year warrant. This strategy focuses on the reliability of the innovation; the risk of purchasing novel and in some cases unproven technologies is reduced for citizen-consumers when the dealer (on behalf of the car manufacturer) provides a long-lasting guarantee. Finally, 44% of the respondents would sooner buy a car fuelled by biofuels or natural gas if the amount of fuelling stations increases. Increasing the availability of filling stations is therefore likely to have a strong positive effect on the willingness of consumers to take these specific green cars into consideration.

In sum, the outcome of this survey shows that the potential for sustainable vehicles is enormous. The in-depth analysis of the car-purchasing practice showed that in recent years more fuel efficient cars are sold already. It is safe to conclude that in addition many more citizen-consumers are at least willing to consider purchasing a sustainable vehicle.

7.6.3 Virtual mobility through teleworking

A completely different approach of sustainable mobility is the reduction of mobility by 'downsizing' the amount of kilometres travelled, for example by teleworking or telecommuting. This sustainable mobility alternative is perceived almost as attractive as the purchase of a clean and energy-efficient car (73%, see Figure 7.8). There is a variety of reasons why the respondents find this alternative attractive. First of all is that it provides them a better balance between work life and private life (32%). Another important benefit of teleworking is that it reduces the feeling of hurriedness in everyday life (23%). Third is saving fuel costs (21%). The final argument, mentioned by 15%, is the ability to avoid traffic jams. These arguments are illustrated by the following quotes:

> *Teleworking allows me to spend more time with my children and also the children do not need to go to an expensive baby-sitter.* (female, 37)

> *Because I do not get stuck in a traffic jam I can work during my travel time which also makes my boss happy.* (female, 35)

While many of the citizen-consumers are very positive about teleworking, the possibilities to do so are not always present. Only 26% of the respondents have a job which allows them to work from home for at least one day a week (Figure 7.11). The need to be physically present at work therefore poses a strong barrier for telecommuting, even if it is only for one day a week. The feeling that teleworking can be disadvantageous for one's career is not shared by the majority of the respondents. This is a somewhat surprising outcome, since a previous survey on teleworking in the Netherlands, conducted by Walrave en De Bie (2005), showed that the majority of non-teleworkers felt that teleworking could be disadvantageous for one's career. Fear for the negative effects of teleworking for one's professional career may be less stringent now than in 2005. However, over 40% has given a neutral response so it could also be that they are simply not sure about this

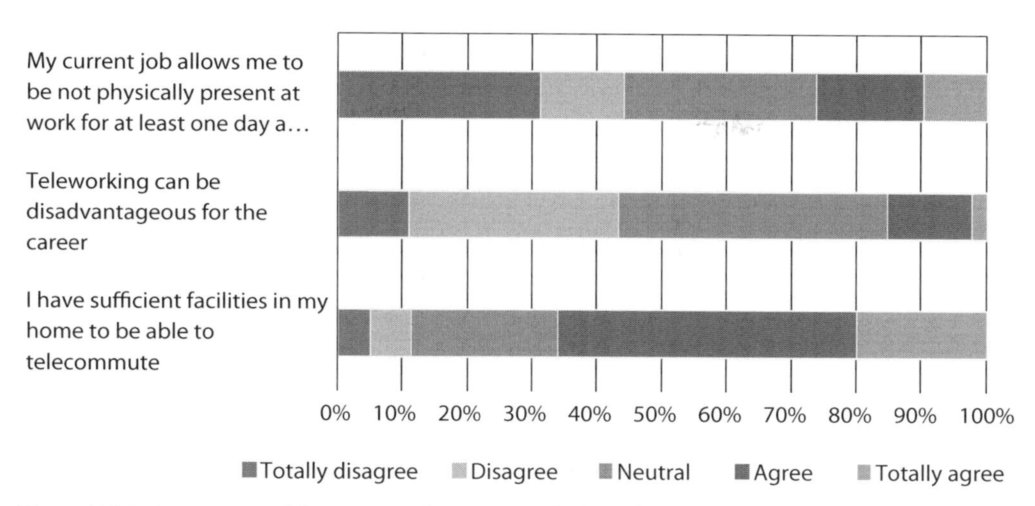

Figure 7.11. Assessment of the system of provision of teleworking (n=850).

issue for instance because of lack of experience. Finally, the far majority of the citizen-consumers have the necessary facilities at home to be able to telecommute (65%).

In Figure 7.12 the respondent's reactions on the potential provider strategies in the system of provision of teleworking are displayed. Approximately 5% of the respondents already are regular teleworkers. The first strategy to facilitate teleworking, improved time planning by the employer will make it easier for 30% of the respondents to telework. Hence, for the majority of the respondents this strategy does little to solve the issue of needing to be physically co-present at work.

Open support for teleworking by the employer, for instance expressed through workshop sessions on teleworking and internal communication pointing out the possibility of teleworking,

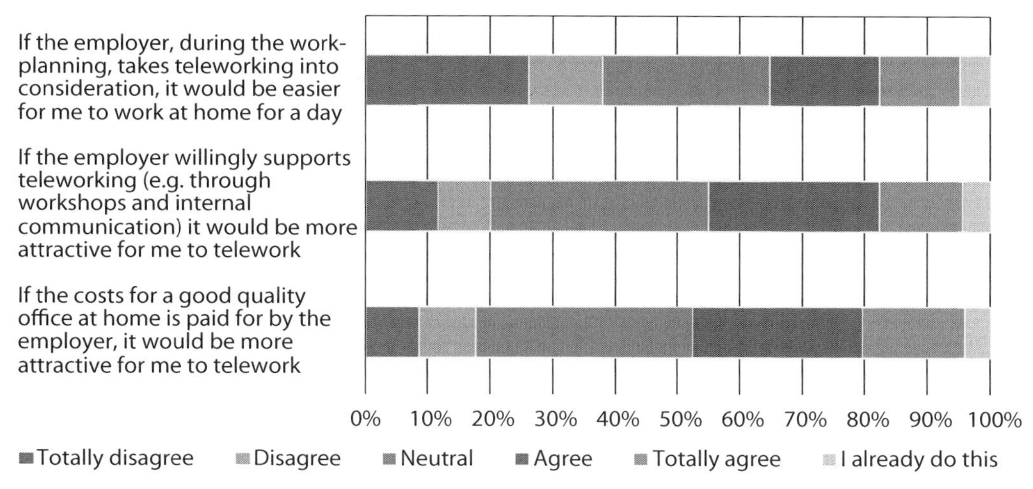

Figure 7.12. Provider strategies in the system of provision of teleworking (n=850).

makes teleworking more attractive for 41% of the respondents. Finally, for 44% of the respondents telework becomes more attractive if the employer covers the costs for a good quality home office. This is a surprisingly high outcome as only one-tenth of the respondents had previously indicated that they lack the facilities to telework (Figure 7.11).

In sum, for many travellers teleworking is an attractive option. Social acceleration has led to an increase in the pace of life causing many people to consider time a scarce good, and to feel under constant time pressure and stress (see Rosa, 2003). As teleworking eliminates travel time, thereby reducing the scarcity of time, it can accommodate a de-acceleration or slow-down in the pace of everyday life. Nevertheless, the need to be physically present at work remains a barrier for telecommuting. Though one may assume that in the near future, with increasing employment in the services industry and growing possibilities of ICT, it is likely that time and place independent working will increase.

7.6.4 Increase in cycling

Another form of localism is slow travelling. Slow travelling, well-known in the domain of tourism, is a form of travelling which emphasizes the experience dimension of travelling. Instead of travelling as quickly as possible, based on the premise that all travel time is wasted time, in slow travelling the journey itself is seen as an integral part of the holiday (see also Verbeek, 2009). In the tourism domain the experience dimension often exists of a connection made with local communities or natural areas, and of course the travel experience itself. Clearly, slow travelling in everyday mobility differs substantially from tourism mobility where time is much more in abundance. Slow travelling can refer to many travel modes such as travelling on foot, by bicycle and various forms of public transport. For sake of clarity and to simplify, in the survey this sustainable mobility alternative has been limited to (an increase in) cycling.

70% of the respondents find cycling (very) attractive. This is perhaps not surprising considering that there are two times more bicycles in the Netherlands than there are cars[93]. The far majority of the Dutch citizens is at least an occasional cyclist. Nevertheless, even in the Netherlands a third of the travellers use the car for trips between 1 and 2.5 kilometres, and the car is already the dominant mode of transport for trips between 2.5 and 5 kilometres. An increase in cycling remains a promising sustainable mobility alternative for these short trips.

By far the most important reason for respondents to choose this alternative is that cycling is beneficial for one's health and physical shape (46%). Another 24% considers economic benefits (saving fuel or not having to own a car) the most important reason. Ecological considerations are the motivation for 18% of the respondents. Finally, the ability to avoid traffic jams is given by only 3% as the dominant argument. This is probably because the bicycle is primarily a mode of transport for short trips. The open-end answers revealed that many respondents prefer cycling because it is more enjoyable and more fun than other modes of transport.

[93] In 2009 there were 18 million bicycles and 7.7 million cars present in the Netherlands (www.bovag.nl).

A known barrier for increase in bicycle-use is the difficulty of cycling when being outside of the home-town (for example as the last chain in a multi-modal trip)[94]. Many of the respondents (47%) agree that it is not easy to use a bicycle for short trips when being out of the home-town (Figure 7.13). While most travellers have indeed a bicycle in the place of residence, in other locations this is not the case which makes cycling trips much more complicated. Another barrier is the possibility of cycling while conveying children or goods. Especially for grocery shopping a commonly used argument not to use the bicycle is that it is difficult to transport goods (see AVV, 2006a). Though many respondent regard it difficult to convey children or goods with a bicycle (45%), there is also a significant proportion of the respondents who (strongly) disagrees (31%). It is possible that this last group has already coped with this issue, for instance through a higher frequency of visits and a lower quantity of groceries per visit.

As a potential provider strategy to improve the accessibility of cycling for short distance trips, the well-known self-service bicycle system in Paris (Velib) was introduced and explained to the respondents. In Paris thousands of bicycles, spread out over 750 stations, have been made available for citizens and visitors; these bicycles can be rented via pass for a small fee (the first half hour is even totally free).

Figure 7.14 shows that 47% of the respondents would sooner make use of the bicycle for short-distance trips (in foreign cities) if such a self-service system would be accessible. Increasing the accessibility of bicycles to consumers in out-of-home-town situations therefore seems a promising option to promote the use of cycling. The second strategy, increasing the accessibility of small cargo bikes to consumers, can count on much less enthusiasm. For only 27% of the respondents the availability of communal cargo bikes would sooner make them do the grocery shopping, or bring the children to school, by bicycle. Thus, even though these cargo bikes aim to increase the

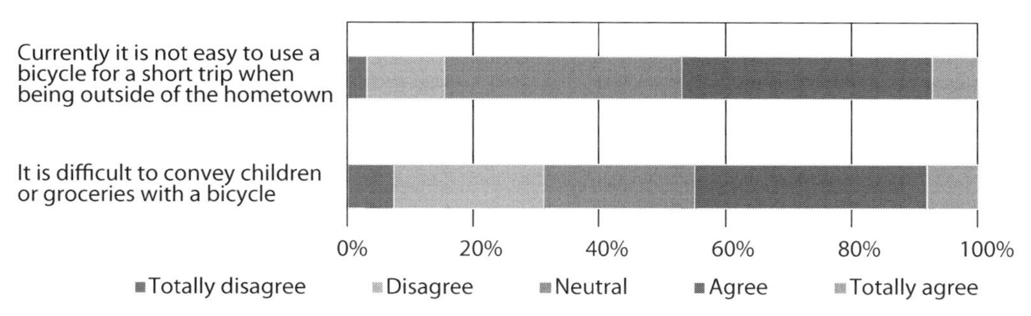

Figure 7.13. Assessment of the system of provision of cycling (n=926).

[94] Research on multimodal (public transport) trips makes clear that the last part of the trip is often the weakest link in the multi-modal chain. That is, the choice of available transport modes is more limited. For example, after train use only 12% of the travellers use the bicycle to reach their destination while in contrast 39% of the travellers use the bicycle to reach the train station of origin. This percentage is much higher because most travellers do have a bike in their home-town but not in other locations. There are three possibilities of cycling after train use: embarking a (folding) bike on to the train, having a second bicycle at the destination station, renting a bicycle (Fietsersbond, 2003). Though even less common, the bicycle could also be used in multi-modal trips involving the car, for instance as a form of P+R.

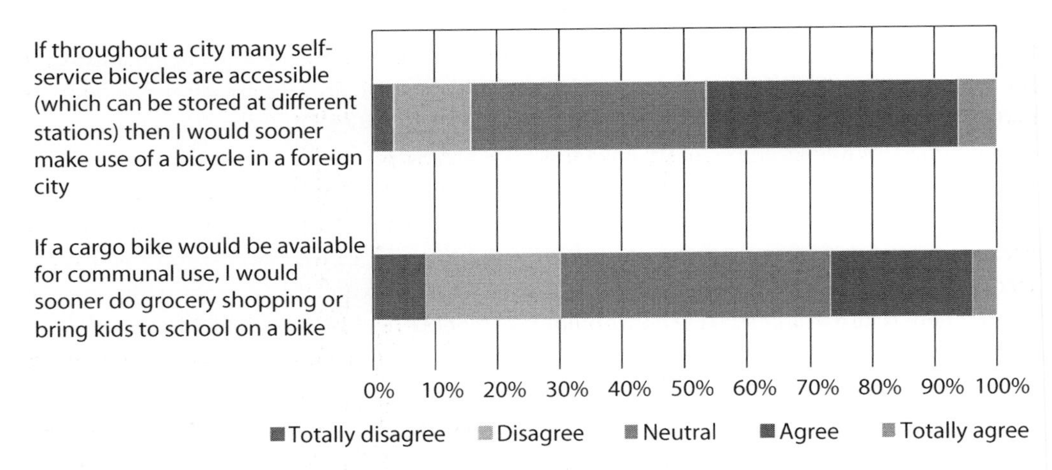

Figure 7.14. Provider strategies in the system of provision of cycling (n=926).

usability of regular bikes (so as to include the possibility of conveying goods and children) it is generally not perceived as a positive contribution.

That Dutch citizens are amongst the world's most fanatic cyclers makes slow travelling a logical alternative. Though public transport bicycle systems (such as the OV-fiets and Velib) remain relatively unknown, these systems can play an important role in multi-modal trips, especially by strengthening the last part of trip.

7.6.5 Multi-modal travelling

After several decades of increasing mono-modality, heading into the direction of car use, there is a resurging interest in the possibilities of modal flexibility and multi-modal travelling. The number of citizen-consumers with a broad choice in the modes of transport is on the increase. Travellers have upgraded their mobility portfolio and options because of the complexity of car travelling due to parking problems and congestion, and the many socio-technical innovations which facilitate multi-modal travelling.

In transport literature a so-called 'choice traveller' is someone who has the skills, positive appropriation and access to make use of either an automobile or public transport modes. It is estimated that between 1991 and 2002 the percentage of these choice travellers has increased from 25% to 45% (Stienstra, 2002). Modal flexibility expresses itself not only in the possibility to choose either a car or a public transport mode for a certain trip, in certain cases travel modes can be combined, for instance via a Park and Ride facility or the use of car-sharing. To facilitate modal flexibility and multi-modal travelling various innovations such as mobility cards, product-to-service systems and integrated travel information have been developed to make the organisation of smart travelling easier.

As with most sustainable mobility alternatives, the majority of the respondents find the option of modal flexibility and multi-modal travelling (very) attractive. The most important reasons to choose for this mobility option were environmental reasons (31%), costs saving because of the avoidance of parking fees and fuel costs (28%), and the avoidance of congestion (19%). Less important arguments were that multi-modal travelling leads one faster to one's destination (8%) and that train travelling provides the possibility of working during travel time (5%).

Figure 7.15 gives an indication of the respondent's assessment of the system of provision of modal flexibility and multi-modal travelling. Though the majority of the respondents are positive about multi-modal travelling only a quarter can combine car use with public transport use for some of the trips they make. This small percentage is perhaps not surprising when the second statement is taken into consideration: only 15% of the respondents receive multi-modal information (such as the possibilities of travelling via P+R facilities) during their travel information search. This means that existing forms of integrated travel information do not reach a large part of the travellers; either this information is not sufficiently made available for travellers or it is not sufficiently accessed by them. Finally, a surprisingly large portion of the respondents expresses a desire to travel more by train; the distance from the train station to the destination is seen by 44% as an inhibiting factor to do so.

In comparison with the provider strategies discussed in the previous sections, the provider strategies of multi-modal travelling are met with moderate enthusiasm (Figure 7.16). 21% of the respondents indicate that he/she would more often shift to public transport modes during a car trip if there was a higher amount of Park and Ride facilities close to the highway. Next, 34% (totally) agrees that it would be easier to combine car use with public transport modes if journey planners would clearly indicate the possibility to travel via a P+R facility. Almost similarly, 31% of the consumers state that a better organisation of the last part of the multi-modal chain would lead to them to travel more often by train. The suggested provider strategies therefore only partially contribute to a better consumer assessment of the system of provision and a better integration of various travel modes.

The survey shows that the possibilities for multi-modal travelling are perceived as much more limited as the other three alternatives discussed so far. It is easy to understand that modal

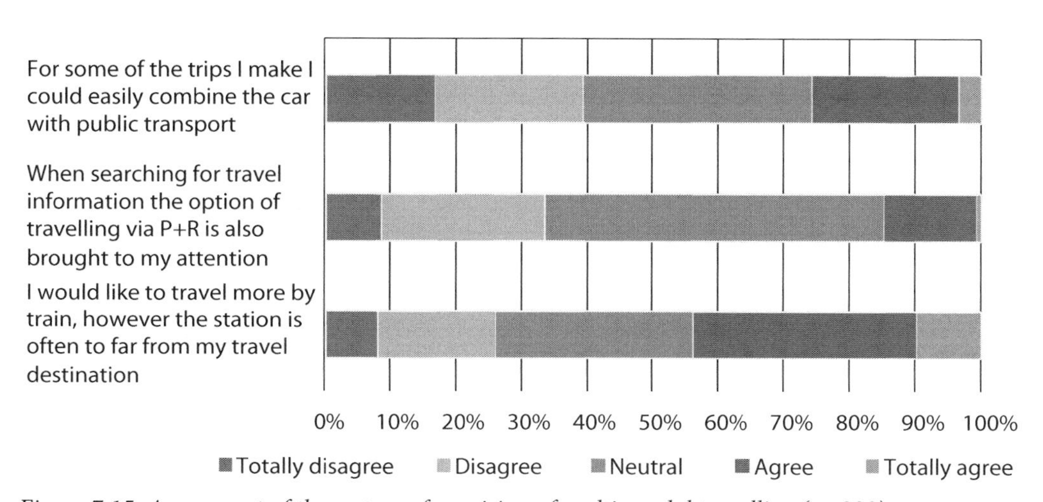

Figure 7.15. Assessment of the system of provision of multi-modal travelling (n=933).

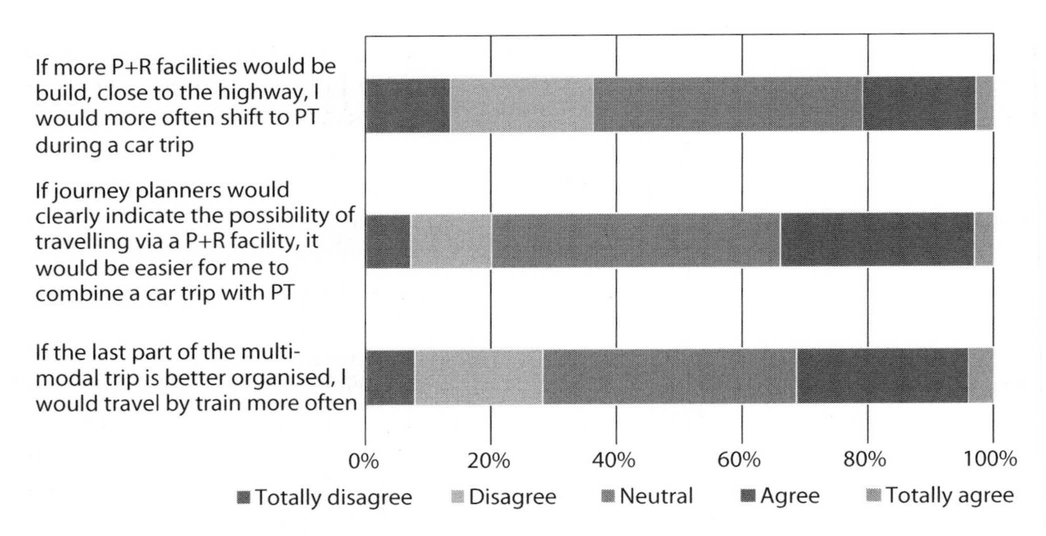

Figure 7.16. Provider strategies in the system of provision of multi-modal travelling (n=933).

flexibility requires a more radical shift in the mobility patterns of citizen-consumers than shifts in car use or cycling. A more in-depth analysis of this alternative will be presented in Paragraph 7.7.

7.6.6 Car-free communities

The final sustainable mobility alternative, living in a car-free community, is an extraordinary form of 'localism' because it is about sustainable urban neighbourhoods and therefore bears close relationships with the domain of housing. In many ways car-free communities can be seen as the most radical sustainable alternative; on the one hand these residential areas often have substantial legal, physical and/or financial restrictions on car ownership[95] and on the other hand car-free neighbourhoods have a distinctive community orientation. The typical car-free neighbourhood has a stronger than average social cohesion and environmental orientation[96]. In short, most car-free neighbourhoods not only have an impact on one's mobility patterns and transportation modes; living in such a settlement often attracts citizens with a common interest and community-oriented lifestyle. Considering that for many respondents a car-free environment is a radical change from their current everyday situation it is perhaps not surprising that it is the least attractive sustainable mobility alternative (47%). The most important reason for wanting to live in a car-free or car-restricted environment is that it provides more living space and green

[95] Reduced car-ownership in these neighbourhoods may be accomplished via means such as physical restriction to car access or car parking, or fiscal measures such as parking permits, allowing the use of a single parking place, which in rare occasions exceed 10,000 euros (see Scheurer, 2001).

[96] To indicate, Ornetzeder *et al.* (2008) in a research on car-free housing clearly show that inhabitants of a car-free community, when compared to a reference group, have more social contacts, identify more with the settlement and have higher levels of environmental concern.

areas (55%). The remaining motives are less negative influence of air pollution (16%), a safer traffic condition in the neighbourhood (12%), and less hindrance of traffic noise (10%). In addition, in the open-ended survey response many respondents mentioned that a car-free or car-restricted area provides a better environment for their children to grow up in.

Not surprisingly considering the low amount of car-free or car-restricted areas (the relatively widespread and typically Dutch 'Home Zone areas' not included), only 5% of the respondents indicate that during the search for a new home they come across a property within a car-free area (Figure 7.17). The supply of houses within car-free areas can therefore be considered very low. The statement that the number of parking places in one's living area may be reduced if this is beneficial for the liveability, leads to a very mixed response; the group of proponents and opponents and the neutral group are almost equal in size.

Almost a third of the consumers (strongly) agrees with the statement that, when searching for a new house, they would sooner take the car-restricted characteristics of a neighbourhood into account if these would be more clearly indicated by the provider (Figure 7.18). This gives an

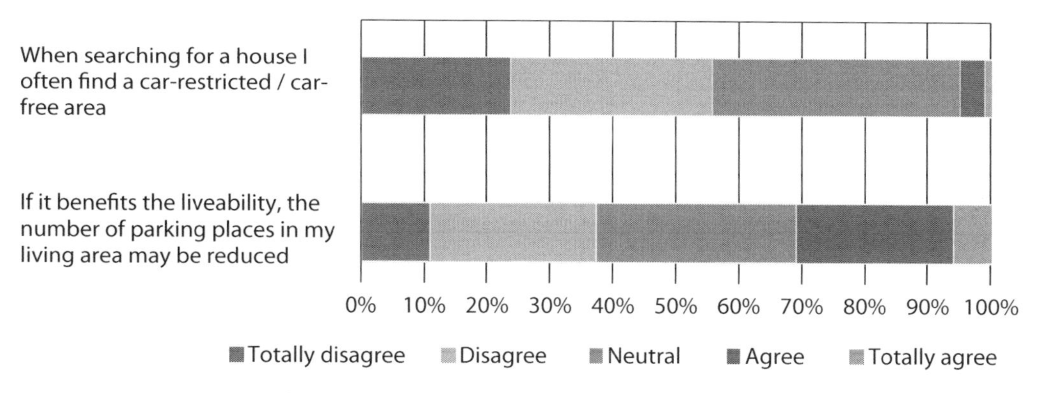

Figure 7.17. Assessment of the system of provision of car-free communities (n=918).

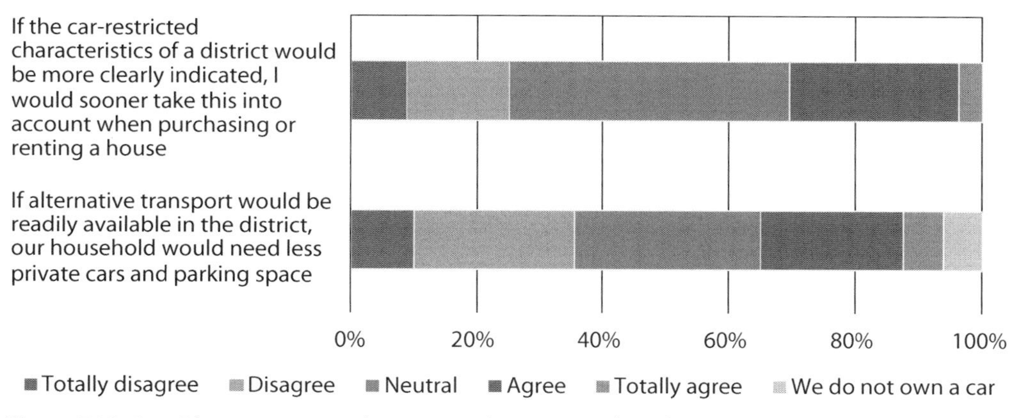

Figure 7.18. Provider strategies in the system of provision of car-free communities (n=918).

indication of the willingness of the respondents to take these characteristics into consideration. The last provider strategy, the provision of alternative transport modes in the district (such as car-sharing facilities, high-quality public transport) is for 29% of the respondents a reason that they would need less private cars and parking space.

7.6.7 Resumé

In this paragraph the general consumer response to five sustainable mobility alternatives has been investigated. Generally speaking citizen-consumers are very positive about the attractiveness of the mobility alternatives. The receptiveness of socio-technical innovations in the domain of mobility is therefore, at least latently, present in the Dutch population. Of the alternatives investigated 'purchasing a green car' is considered more attractive than modal flexibility and the three forms of localism. Green automobility to many consumers may seem as less of a dramatic change to their everyday mobility practices than multi-modal travelling or living in a car-free community.

The evaluation of consumers also to a large extent depends on the specific system of provision and green provider strategies. With the assessment of the provider strategies it is shown how green provider strategies may target existing barriers for the integration of socio-technical innovations in existing mobility practices. Furthermore, by asking similar types of questions (for each of the five sustainable mobility alternatives) which focused on usability, accessibility and trust issues, a comparison between alternatives can be made. Table 7.5 gives a summary of the system of provision and the green provider strategies as assessed by the total consumer population. The table can be read in two ways: it shows for each of the alternatives whether the respondents answered positively, neutrally or negatively to the posted questions. On the other hand, the table shows which type of

Table 7.5. Consumer assessment of the system of provision and potential provider strategies.[1]

	Accessibility		**Usability**		**Trust**	
	System of provision	**Provider strategy**	**System of provision**	**Provider strategy**	**System of provision**	**Provider strategy**
Purchasing a green car	0	+	-	+	0	+
Teleworking	+	+	-	-	+	+
Cycling	-	0	-	+	.	.
Multimodal travelling	0	0	-	-	.	.
Living in a car-free area	-	0	-	0	.	.

[1] + = positive assessment; 0 = neutral assessment; - = negative assessment; . = not measured. Rating in either positive, negative or neutral was designated on the basis of the largest answer category of the consumers for each statement asked. If the largest answer category was neutral, the system of provision or the provider strategy receives a '0', if the largest answer category was a positive evaluation a '+' is added, etc. Trust issues were only examined in two of the five alternatives to limit the number of survey questions.

barrier (accessibility, usability or trust) generally is the most relevant. Because this table negates which questions lay behind the answers no hard claims can be made on the basis of it, however, the table does show some interesting aspects.

First, more than trust issues and accessibility to socio-technical innovations, the usability of these innovations in everyday life is the foremost barrier. For each of the five sustainable mobility alternatives the usability is seen as a barrier by the citizen-consumers. To indicate, while the far majority of the respondents regard teleworking attractive, and considering that trust-issues and ICT-connections are no major obstacles, the self-reported negative usability seems to inhibit that teleworking is conducted on a large-scale basis.

Second, consumers expect that most green provider strategies will have a positive impact on their mobility behaviour in the sense that they are more likely, or will more often, conduct their practice in a more sustainable manner. Third, though circumstantially, it provides an answer to the question why purchasing a green is car is appreciated the most by the respondents and living in a car-free neighbourhood is appreciated the least. Purchasing a green car scores highest on issues of accessibility, usability and trust while living in a car-free area scores the lowest.

7.7 Individual differences in consumer assessment: the role of mobility portfolios and environmental concerns

In this paragraph we will investigate individual differences in mobility portfolios of the respondents as the third and final contextual approach to sustainable mobility behaviour[97]. In Chapter 5, it was explained that the notion of motility, a human agent's potential to be mobile, encompasses and integrates the various elements which are important to be able to participate in mobility practices. Motility consists of access, skills and cognitive appropriation. Flamm & Kaufmann (2006) used the term mobility portfolios to indicate all the mobility products and services available to individuals or households. This mobility portfolio includes the privately owned vehicles, and access to mobility services provided by mobility providers through public transport passes, and membership(s) of new mobility services such as car-sharing, bike-sharing.

The main goal of this paragraph is to investigate whether or not citizen-consumers with a 'greener' mobility portfolio are indeed also more receptive to sustainable mobility alternatives. The rationale behind this hypothesis is that citizen-consumers with a larger mobility potential – i.e. more access to mobility products and services, and with a positive evaluation of these products and services – are more likely to integrate the sustainable mobility alternatives into their mobility practices.

Next to variation in mobility portfolios the individual differences in the environmental concerns will be investigated. The reason for this analysis is that many models of consumer behaviour have a strong focus on environmental concerns to explain environmentally relevant behaviour. It is therefore interesting to compare differences in environmental concern in relation to the assessment

[97] The term 'contextual' is somewhat debatable for this paragraph because the main focus is on the individual differences of citizen-consumers. We have chosen to use the label contextual because this paragraph focuses on practice-specific contexts in the form of mobility portfolios and their relation with the system of provision.

of sustainable alternatives. The analysis of individual differences is applied only to the sustainable alternative multi-modal travelling[98].

7.7.1 Mobility portfolio in the consumer survey

The starting point of the mobility portfolio is a traveller typology based on the various transport modes. Instead of trying to decipher which type of traveller each respondent was, the respondents were simply asked to indicate in which description of traveller types they recognized themselves the most (Table 7.6). It is interesting to see that a relatively large portion of the respondents places him- or herself in the category of local traveller (pointing out the importance of short-distance travelling) and the categories of choice traveller (pointing out that at least one fifth of the respondents has direct access to, and experience with, both public transport and car use).

When relating the attractiveness of the sustainable alternatives with each traveller type, interesting differences emerge (Figure 7.19). In the first place, some traveller types are in general much more positive about sustainable alternatives than others. More specific, the 'car traveller' and 'other type' are much less positive than the 'choice traveller, mainly PT' and 'local traveller'. On average the five sustainable alternatives are evaluated as (very) attractive by 57% of car travellers. For choice travellers with an emphasis on public transport and local travellers, this is 77% and 75% respectively. Second, most traveller types show wide variation in the attractiveness of the sustainable alternatives. For example, car travellers clearly prefer teleworking and the purchase of

Table 7.6. Segmentation of respondents' traveller types (n=2,207).

	Description of traveller types	Share
Car traveller	I always travel by car, for short distances I sometimes use a bicycle	37%
PT traveller	I always use public transport, sometimes in combination with a bicycle. I do not own a private car.	12%
Choice traveller, mainly car	I regularly use the car, public transport and bicycle. However, the car I use the most.	15%
Choice traveller, mainly PT	I regularly use the car, public transport and bicycle. However, public transport I use the most.	6%
Local traveller	I almost never travel over long distances. I cover short distances by walking or cycling. For long distances I use public transport or ride along with someone else	20%
Other	None of the above	10%

98 The survey questions which measure mobility portfolio focused predominantly on multi-modal travelling.

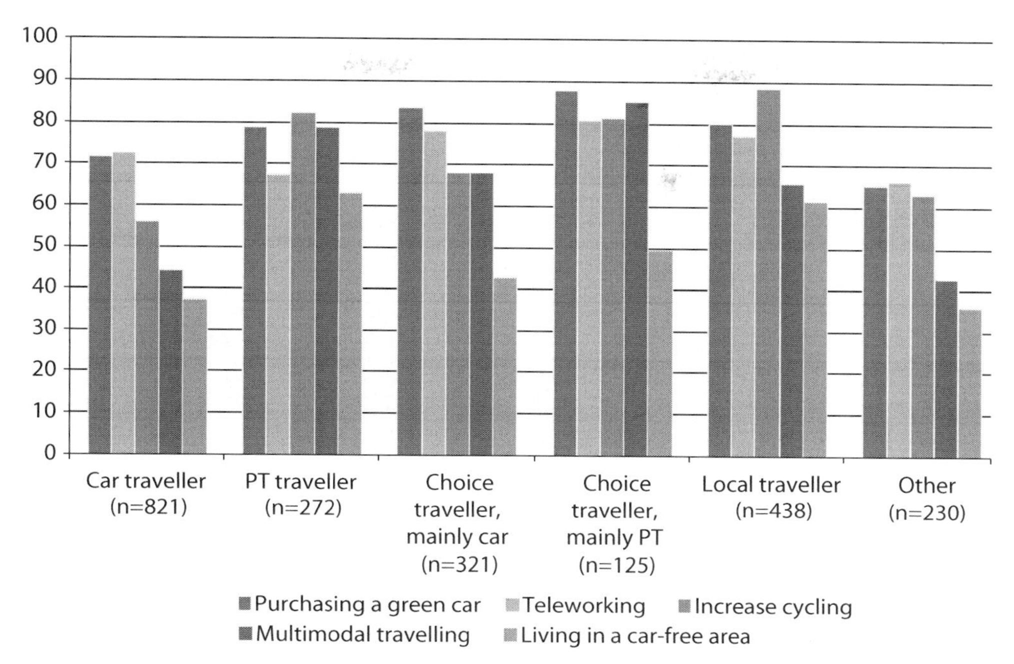

Figure 7.19. Attractiveness of sustainable alternatives per traveller type (n=2207, traveller type 'other' not visualized).

a green car, and are least attracted to multimodal travelling and car-free neighbourhoods. Local travellers on the other hand have the highest preference for cycling.

In line with the concept of motility, the respondents' experience with socio-technical innovations in the mobility domain was inquired (Table 7.7). The experience with these innovations was based on two aspects: whether one has used the innovation or not, and whether the experience of the user was positive or negative. As Flamm & Kaufmann (2006) make clear, past experience is one of the preconditions to easily use a means of transportation. The same applies to the innovations

Table 7.7. Respondents' experience with socio-technical innovations (n=2,242).

	Bike-sharing	**Car-sharing**	**Park and Ride**	**PT route planners**	**Mobility cards**
1. User, positive experience	2.9%	1.2%	17.3%	47.7%	1.2%
2. User, neither positive nor negative experience	0.3%	0.8%	6.2%	11.6%	0.3%
3. User, negative experience	0.4%	0.3%	1.7%	3.6%	0.3%
4. Non-user, familiar with innovation	35.6%	50.0%	57.6%	16.8%	9.6%
5. Non-user, unfamiliar with innovation	60.9%	47.8%	17.2%	20.3%	88.6%

in Table 7.7. Simply because of former utilization, users, when compared with non-users, are more likely to have the necessary skills for using the presented innovation. Positive or negative experiences conceptualise the aspect cognitive appropriation which refers to the subjective element of motility, the (functional and symbolic) judgement of transport modes, based on individual and collective representations (see Chapter 5).

Based on the abovementioned aspects of experience, Table 7.7 shows five categories of respondents for each of the five social-technical innovations which facilitate modal flexibility. The five innovations were presented to the respondents with a short introductory text and image. Table 7.7 reveals that the majority of the respondents have not used these socio-technical innovations. For instance, while many respondents are familiar with the concept of car- and bike-sharing, only a few percent has ever used it. Public transport route planners are by far the most used innovation by the respondents. The least used and least familiar are the mobility cards, those multi-modal travel cards which allow bookings of train-tickets, taxi's, and car-sharing via the internet. The reason for this outcome is likely to be that mobility cards are predominantly used for business travel.

In short, with the exception of PT route planners the percentage of actual users of the presented innovations is low. However, the experiences of the users are in general positive or at least neutral. The respondents are most unfamiliar with mobility cards and to a lesser degree with bike-sharing.

7.7.2 Mobility portfolios and the assessment of sustainable alternatives

The previous section focused on the mobility portfolio of the respondents in general. In this section we will go into the relationship between individual differences in mobility portfolios and the assessment of the sustainable alternatives. For that purpose the respondents are divided into three groups based on their experience with the abovementioned socio-technical innovations. Each respondent is placed into the category of low, medium or high mobility portfolio.

To designate each respondent to one of the three categories, points were assigned to the level of experience with the innovations. For example, users who had positive experiences with bike-sharing receive more points than non-users who are not even aware of the existence of bike-sharing. Based on Table 7.7, user category 1 (user, positive experience) was awarded twenty points, user categories 2 (user, neutral experience) and 4 (non-user, familiar with innovation) were awarded five points, and user categories 3 (user, negative experience) and 5 (non-user, unfamiliar with innovation) were awarded zero points. Thus, each respondent received 0, 10 or 20 points for each of the five innovations. All respondents are designated a score ranging between zero and hundred (Figure 7.20). The low portfolio group consists of respondents with 20 points or less (33% in size), the medium portfolio group ranges between 30 and 40 (41% in size), and the high portfolio group consists of 50 points or more (26% in size)[99].

Though the notion of mobility portfolios is of course much more sophisticated and complex than the numerical representation of Figure 7.20, this way of conceptualisation allows us to make

[99] The border lines between these three categories were also based on a statistical necessity. The spread in the group with high portfolio is unevenly large. However, if the high portfolio group consisted only of respondents with 60 points or more the group would become too small for statistical techniques to compare groups (such as ANOVA).

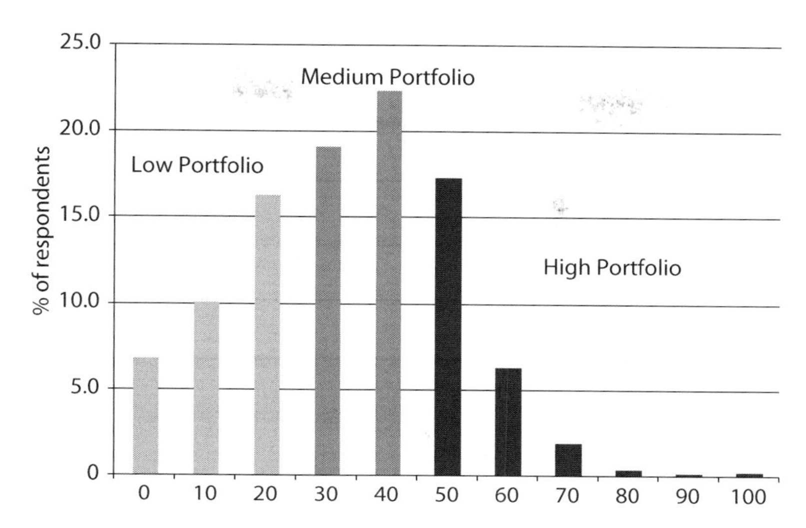

Figure 7.20. Distribution of respondents (numerical) mobility portfolio (n=2,242).

quantitative comparisons between different groups of citizen-consumers. The most important question is of course whether or not citizen-consumers with a high mobility portfolio respond differently to the statements than citizen-consumer with a medium or low mobility portfolio. Part of the answer is depicted in Figure 7.21 which, for each of the three portfolio groups, shows the assessment of the system of provision and the provider strategies of multi-modal travelling. With the exception of two statements, the figure reveals a clear hierarchy in which citizen-consumers with a high mobility portfolio respond most positively and citizen-consumer with a low portfolio respond the least positively to the nine statements. The lower outcome for the seventh statement ('I would like to travel ...') can also be explained since the statement refers both to a desire to travel more by public transport and to an experienced barrier to do so. It is quite possible that the respondents with a high mobility portfolio feel fewer barriers to travel by train more often.

To test whether the differences portrayed in Figure 7.21 are significant, a one-way between-groups ANOVA was conducted (see Textbox 7.1 for detailed analysis). With the exception of the abovementioned seventh statement, the high mobility portfolio group scored significantly higher ($P<0.05$) than the medium and the low mobility portfolio group on each of the statements. The implication of these differences is reflected in two ways. First, it shows that citizen-consumers with a high mobility portfolio are already more adept at multi-modal travelling. They mention that there are more trips for which they are able to combine public transport with car use (statement 1); also they have above average access to travel information with Park and Ride facilities (statement 4). Second, citizen-consumers with a high mobility portfolio have more intensions for multi-modal travelling if the described provider strategies are implemented. They are therefore much more likely to be receptive to the provider strategies of multi-modal travelling.

In addition to the analysis of variance, it is interesting to compare the three mobility portfolios with the abovementioned traveller typology. Which type of travellers have an above average mobility portfolio? When the three portfolio groups are compared one can see that the high

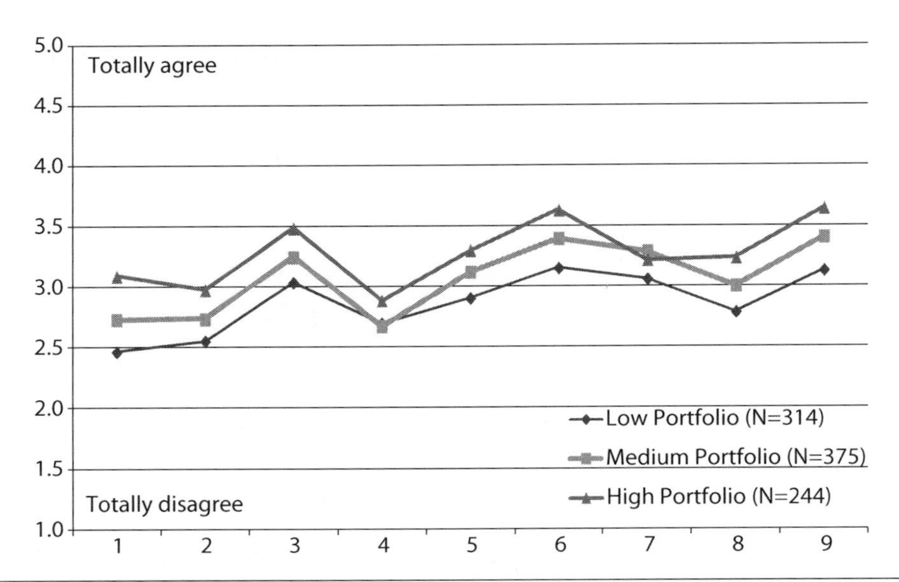

1. For some of the trips I make I could easily combine the car with public transport
2. If more P+R facilities are build, close to the highway, I will more often shift to PT during a car trip
3. If more P+R facilities are build, close to the highway, other travellers will more often shift to PT during a car trip
4. When searching for travel information the option of travelling via P+R is also brought to my attention
5. If journey planners clearly indicate the possibility of travelling via P+R facility, it will become easier for me to combin a car trip with PT
6. If journey planners clearly indicate the possibility of travelling via P+R facility, it will be easier for other travellers to combine a car trip with PT
7. I would like to travel more often by train, however the station is often to far from my travel destination
8. If the last part of the multi-modal trip is better organised, I will travel by train more often
9. If the last part of the multi-modal trip is better organised, other travellers will travel by train more often

Figure 7.21. Comparison between portfolio groups on assessment of multi-modal travelling.

portfolio group consists much less of sole car users and consists much more of choice travellers and public transport users (Figure 7.22). The graph also shows a very consistent multistage distribution of the three groups (low, medium, high) over the different traveller types. An interesting outcome is that the high portfolio group cannot directly be affiliated with one specific traveller type.

7.7.3 Environmental concerns and the assessment of sustainable alternatives

In this section a more commonly used analysis is conducted by investigating the relation between citizen-consumer's environmental concerns and the consumer's assessment of sustainable alternatives. To analyse the environmental concerns, use is made of the first-segment statements

Tekstbox 7.1. Results from the one-way between-groups ANOVA with post-hoc tests.

ANOVA is an one-way analysis of variance which compares the variance (variability in scores) between the three different groups (independent variable) with the variability within each of the groups. With the use of ANOVA one can test whether there are significant differences in the mean scores on the dependent variable between the three groups (Pallant, 2007, p. 243). In this case it means that with the use of one-way ANOVA it is analysed whether or not the three groups (low, medium, and high portfolio) vary significantly with respect to the assessment of the sustainable alternative multi-modal travelling.
There was a statistically significant difference at the $P<0.05$ level in the assessment scores for the three portfolio groups: statement 1, $F(2, 930) = 23.8$, $P=0.00$; statement 2, $F(2, 930) = 12.7$, $P=0.00$; statement 3, $F(2, 930) = 18.6$, $P=0.00$; statement 4, $F(2, 930) = 5.48$, $P=0.00$; statement 5, $F(2, 930) = 13.5$, $P=0.00$; statement 6, $F(2, 930) = 26.2$, $P=0.00$; statement 7, $F(2, 930) = 3.9$, $P=0.01$; statement 8, $F(2, 930) = 15.1$, $P=0.00$; statement 9, $F(2, 930) = 27.8$, $P=0.00$.
The one-way between-group ANOVA only indicates that there is a significant difference in the mean scores of the seven statements between the three portfolio groups. The post-hoc test shows exactly where these differences between the three groups come from. Post-hoc comparisons using the Tukey HSD test indicated that all the mean scores for the three portfolio groups where significant different at the $P<0.05$ with the exception of statements 4 and 7. For statement 4 the post-hoc test indicated that the mean score (M) for the low portfolio group (M=2.70, SD=0.738) was not statistically significantly different from the medium portfolio group (M=2.67, SD=0.834). The high portfolio group (M=2.89, SD=0.938) was indeed significantly different from both other portfolio groups. For statement 7 the post-hoc test indicated that the mean score for the high portfolio group (M=3.22, SD=1.13) was not statistically significantly different from both the medium portfolio group (M=3.29, SD=1.12) and the low portfolio group (M=3.06, SD=1.01).

('environmental scepticism') of the general dispositional and conjunctural specific lifestyle which were presented earlier in this chapter. Three groups of citizen-consumers were created to vary in environmental concerns. Thus, for the general dispositional and the conjunctural specific lifestyle element one group with 'low' environmental concerns, 'medium' environmental concerns and 'high' environmental concerns was created[100]. The three groups (low, medium, high environmental concerns) were related to the assessment of the system of provision and the provider strategies of multi-modal travelling.

Generally speaking the differentiation over the nine statements showed the same pattern as the mobility portfolio (Table 7.8; graphs not included). The general picture is that the group consisting of environmentally sceptical citizen-consumers responded more negatively to the nine statements than the other two citizen-consumer groups (medium or high environmental concerns). This

[100] To clarify, the group with low environmental concerns consist of those citizen-consumers who scored relatively high on environmental scepticism, etcetera. Group size for the general dispositional environmental concerns: Low, n=672; Medium, n=901; High, n=504. Group size for the conjunctural specific environmental concerns: Low, n=549; Medium, n=1055; High, n=595.

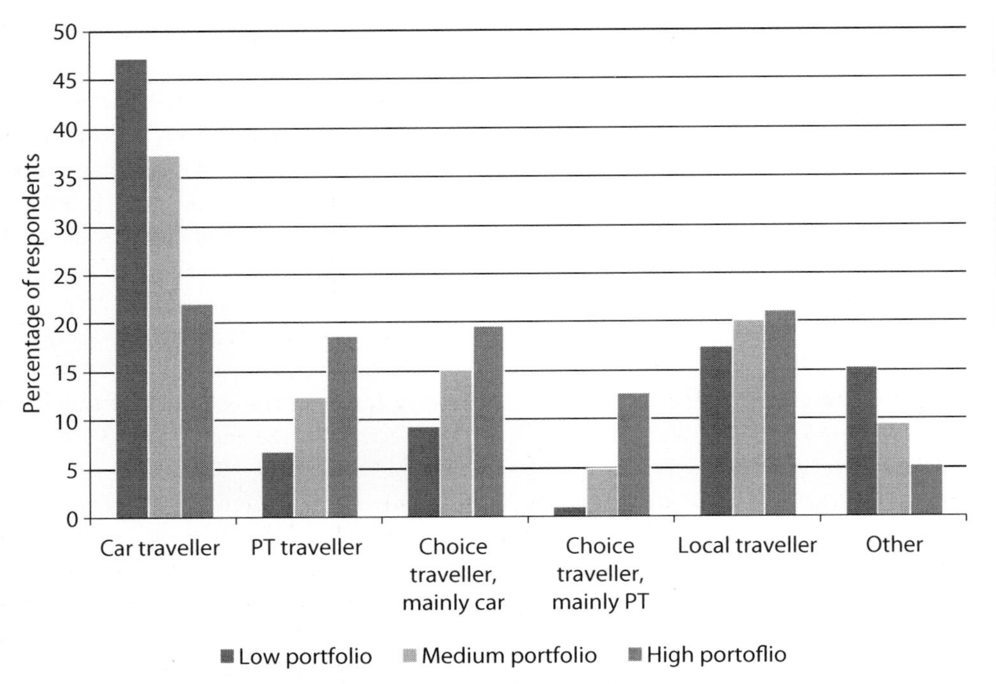

Figure 7.22. Traveller typology split up for the mobility portfolios.

Table 7.8. Differentiation in environmental concerns in relation to assessment of multimodal travelling.

	Mean scores on the nine statements on multimodal travelling								
General dispositional environmental concern	Stat 1	Stat 2	Stat 3	Stat 4	Stat 5	Stat 6	Stat 7	Stat 8	Stat 9
High (n=269)	2.87	2.97	3.50	2.69	3.28	3.59	3.28	3.20	3.62
Medium (n=396)	2.77	2.70	3.17	2.79	3.11	3.35	3.21	2.98	3.30
Low (n=205)	2.52	2.54	3.06	2.74	2.85	3.17	3.11	2.79	3.25
Conjunctural specific environmental concern	Stat 1	Stat 2	Stat 3	Stat 4	Stat 5	Stat 6	Stat 7	Stat 8	Stat 9
High (n=236)	2.85	2.72	3.37	2.58	3.19	3.56	3.19	3.06	3.52
Medium (n=433)	2.77	2.76	3.27	2.74	3.12	3.39	3.24	3.01	3.40
Low (n=250)	2.57	2.71	3.08	2.88	2.98	3.18	3.12	2.91	3.21

holds both for the general dispositional environmental concerns as for the conjunctural specific environmental concerns.

However, after conducting a one-way between-groups analysis of variance these differences were shown to be negligible and in most cases not statistically significant. This lack of variance was to be expected for statements four and seven (based on the previous analysis of mobility portfolio), but for many of the other statements this lack of significant differences is applicable as well (Textbox 7.2). This outcome means that citizen-consumers with a high environmental concern did not have significantly more experience with socio-technical innovations in multimodal travelling, nor do they appear more receptive for the presented provider strategies! The outcome

Textbox 7.2. Results from the one-way between-groups ANOVA with post-hoc tests.

General dispositional environmental concerns:

A one-way between-groups analysis of variance was conducted to investigate the impact of general dispositional environmental concerns on citizen-consumer assessment of multimodal travelling. There was a statistically significant difference at the $P<0.05$ level in the assessment scores for the three portfolio groups: statement 1, $F(2, 867) = 6.2$, $P=0.00$; statement 2, $F(2, 867) = 11.4$, $P=0.00$; statement 3, $F(2, 867) = 17.6$, $P=0.00$; statement 5, $F(2, 867) = 12.9$, $P=0.00$; statement 6, $F(2, 867) = 18.3$, $P=0.00$; statement 8, $F(2, 867) = 11.0$, $P=0.00$; statement 9, $F(2, 867) = 16.4$, $P=0.00$. However, statement 4 and statement 7 showed no significant group differences.

Though the ANOVA showed many statistically significant relations, further analysis by way of post-hoc comparisons using the Tukey HSD test point to another direction. The mean scores for the three groups (low, medium, high general dispositional environmental concern) were only significantly different at the $P<0.05$ for statement 6 and 9. For the other seven statements at least one of the groups did not differ significantly from one of the other two groups. As was already shown by the ANOVA, statement 4 and statement 7 showed no significant group differences at all.

Conjunctural specific environmental concerns:

A one-way between-groups analysis of variance was conducted to investigate the impact of conjunctural specific environmental concerns on citizen-consumer assessment of multimodal travelling.

There was a statistically significant difference at the $P<0.05$ level in the assessment scores for the three portfolio groups: statement 1, $F(2, 916) = 4.3$, $P=0.01$; statement 3, $F(2, 916) = 6.6$, $P=0.00$; statement 4, $F(2, 916) = 8.3$, $P=0.00$; statement 5, $F(2, 916) = 3.4$, $P=0.03$; statement 6, $F(2, 916) = 14.5$, $P=0.00$; statement 9, $F(2, 916) = 8.4$, $P=0.00$. However, statement 2, statement 7 and statement 8 showed no significant group differences.

Post-hoc comparisons using the Tukey HSD test indicated that the mean score for three groups only differed significantly (at $P<0.05$) from each other for statement 6. For the other statements at least one of the groups did not differ significantly from one of the other two groups.

implies that the differentiation in mobility portfolios better explain variation in receptiveness for provider strategies than the environmental concerns.

Also interesting is that the correlation between the mobility portfolio and the environmental concerns is relatively weak. For mobility portfolio and general dispositional environmental concerns the Pearson correlation coefficient is 0.188, for mobility portfolio and conjunctural specific environmental concern this is 0.192. In contrast, the Pearson correlation coefficient between the two environmental concerns is very high, indicating a strong relation (r=0.580).

This means that environmental concerns and mobility portfolio are relatively separate components; a green mobility portfolio therefore shows little relation with high environmental concerns.

Nevertheless, this does not directly imply that there is absolutely no relation between environmental concerns and the sustainable alternatives. A correlation analysis was used to determine the strength of the relation between the various individual and lifestyle variables and the assessment of multimodal travelling. None of the components (neither mobility portfolio nor environmental concerns) showed more than a weak correlation[101]. From previous research on sustainable consumption this was only to be expected.

Finally, Figure 7.23, shows the attractiveness of multimodal travelling in relation to the mobility portfolios and environmental concerns. For all the three components a consistent multistage

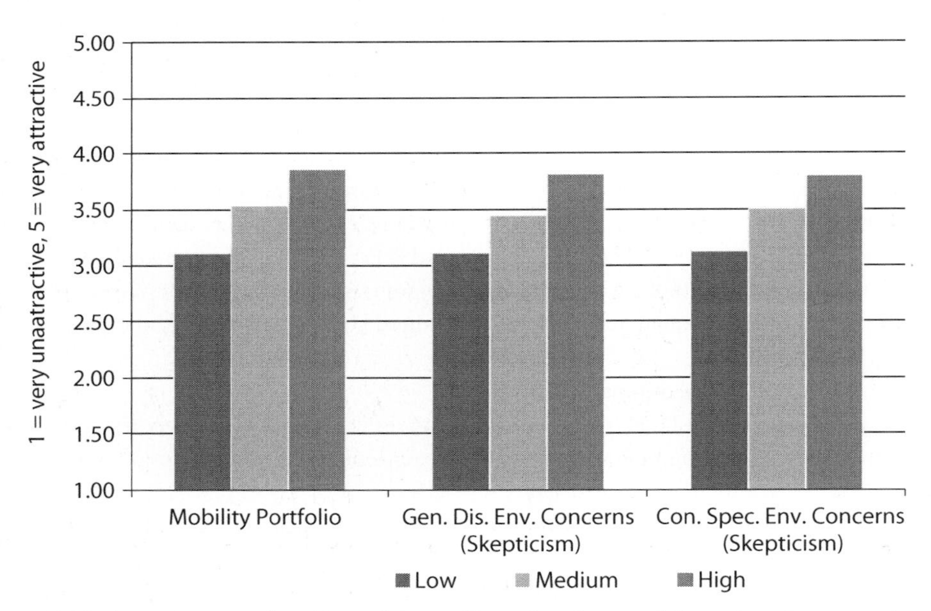

Figure 7.23. Attractiveness of multi-modal travelling related to portfolio and environmental concerns (n=2,232).

[101] The value of the correlation coefficient (r) can range from -1.00 to 1.00. A correlation of 0 indicates no relation at all; a correlation of 1.0 (or -1.0) indicates a perfect positive (or negative) relation. r=0.10 to 0.29 is considered small, r=0.30 to 0.49 is considered medium, r =0.50 to 1.0 is considered large. All correlations were smaller than 0.30.

variation is visible with the 'high' citizen-consumer group being more attracted to multimodal travelling than the 'medium' and 'low' groups. Thus, citizen-consumers with a high mobility portfolio are more attracted to multimodal travelling, the same applies for citizen-consumers with high environmental concerns. The significance of this outcome is discussed in the concluding paragraph.

7.8 Conclusions

The main goal of this chapter was to analyse (individual variety in) the conduct of social practices in direct relation to the context in which the behaviour takes place. Therefore the various components of the social practices model have been made applicable for a large-scale consumer survey. This chapter explored the role of contextual factors in sustainable consumption in the domain of mobility via three routes. First, 'inter-domain', through comparing environmental concerns of citizen-consumers over different consumption domains. Second, 'intra-domain', through comparing the quantity and quality of the products and services and potential improvements in the system of provision. Third, through individual differences based on mobility portfolios and environmental concerns.

In all the three routes the role of context was shown to be a significant factor of influence on citizen-consumer behaviour. First, it was shown how contextuality plays a significant role in the sense that each consumption domain is positioned differently in the debate of sustainable consumption and production. While in the domain of housing, and to a lesser extent the domain of everyday mobility, the necessity for changing the current regime towards a more sustainable one is recognized and supported by a majority of the citizen-consumers, in the domain of tourism this is far less the case. The inter-domain comparison therefore showed that citizen-consumers have a strong feeling that a change in the current system of mobility is necessary. This is an interesting outcome considering the fact that in previous studies the domain of everyday mobility was amongst the most stable regimes (resistant to change) with little indication of a change towards sustainability. This survey supports the outcome presented in the chapter on new car purchasing that in a relatively short time-period in the domain of mobility possibilities for sustainable change have emerged.

Second, the intra-domain comparison showed that especially purchasing a green car is considered a very attractive sustainable option by citizen-consumers. From a consumer point of view sustainable car mobility may be seen as the least disruptive alternative to current mobility patterns. The intra-domain comparison of the quality and quantity of the products and services on offer showed the types of barriers why citizen-consumers refrain from adopting sustainable mobility alternatives. Furthermore, the presented provider-strategies were shown to have a positive effect on the willingness of citizen-consumers to take the behavioural alternatives into consideration.

Third, it is shown that citizen-consumers with a high mobility portfolio are already more adept at multi-modal travelling than citizen-consumers with a low mobility portfolio. In addition, citizen-consumers with a high mobility portfolio have more intentions to increase multi-modal travelling if the described provider strategies are implemented. The mobility portfolio is therefore an important factor in the receptiveness of citizen-consumers to multi-modal travelling. While citizen-consumers with high environmental concerns also show a higher than average attraction to multimodal travelling (as portrayed in Figure 7.23), this does not translate into a higher receptiveness to provider strategies. Nor are citizen-consumers with high environmental concerns more adept at multi-modal travelling.

What do these outcomes tell us? The contextual approach adopted in this chapter is a relatively novel one. The focus adopted differs from that of the dominant literature in social psychology and transport studies which often try to trace back the variables for car use and public transport use to general individual characteristics (e.g. socioeconomic groups, value patterns, etc.) or attributes of transport modes (Diana & Mokhtarian, 2009). Theoretically, the three-step exploration is therefore more important and interesting than the question which sustainable mobility option citizen-consumers prefer over the other, or to illustrate the specific barriers and possibilities for the five sustainable alternatives presented. Partly this chapter is a confirmation of earlier conducted research which emphasized that a policy solely based on increasing citizen-consumer's environmental knowledge and awareness will be ineffective.

However, even though micro-based analyses of sustainable consumption is criticized, it is important to note that we do not claim to label the widely used social and cognitive psychology approaches to environmentally relevant consumer behaviour redundant. That would do great injustice to the many relevant studies which have been conducted in the field of sustainable consumption. Undoubtedly environmental concerns do play a role, one way or the other, in environment-relevant behaviour. Nevertheless, we are indeed in search of different explanatory variables which will likely lead to a different type of suggestions for consumer-oriented environmental policy. Environmental consciousness predominantly plays a role in consumption behaviour in which the behavioural alternative requires little effort. This is clearly not the case for alternatives to car driving.

To point out the relevance, this chapter supports the active provision of socio-technical innovations to citizen-consumers such as integrated mobility cards, product-to-service systems, etc. The example of multimodal travelling clearly showed that the high mobility portfolio groups are above average found in the 'choice traveller' groups while the solely car drivers are underrepresented. As access to and experience with mobility innovations has a positive relation with multimodal travelling, the initiatives of mobility management must be seen as a positive development. Nevertheless, not every sole car users will be 'persuaded' to become a modal merger, nor is that necessary. The evaluation of the five sustainable mobility alternatives by citizen-consumers indicates that for solely car travellers the purchase of a green car and teleworking might be more easily attainable.

Chapter 8. Conclusions

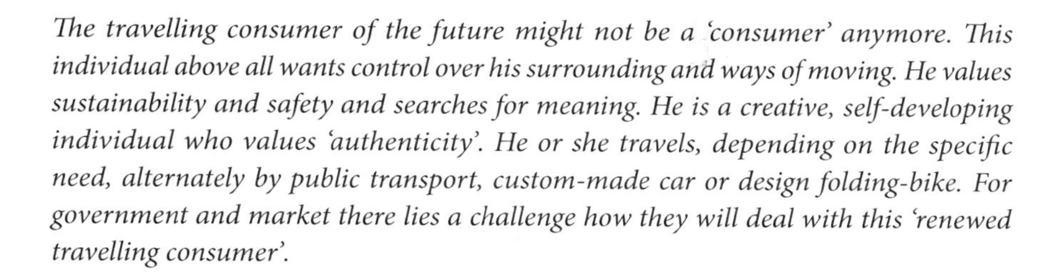

The travelling consumer of the future might not be a 'consumer' anymore. This individual above all wants control over his surrounding and ways of moving. He values sustainability and safety and searches for meaning. He is a creative, self-developing individual who values 'authenticity'. He or she travels, depending on the specific need, alternately by public transport, custom-made car or design folding-bike. For government and market there lies a challenge how they will deal with this 'renewed travelling consumer'.

Ministry of Transport, special edition on the Traveller of the Future (2005).

8.1 Introduction

The possibility to be mobile forms an essential part of society and economy. More than other consumption domains everyday mobility is inherently intertwined with the fabric of social life. Everyday mobility connects a wide array of time-spatially dispersed social practices that make up our daily life. Mobility forms the cement without which other social practices would only be loose bricks. The freedom to invest mobility capital for work, social life and recreation is therefore seen by society as an unconditional right.

While it is virtually impossible not to participate in mobility practices, the problem with the currently dominant socio-technical system of the car is that it is associated with many negative societal consequences. Furthermore, characteristic of these mobility-related problems is that they are multi-facetted (congestion, climate change, air pollution, and safety), that many actors are involved and that due to the large infrastructural investments one can speak of an inert system. On the other hand we can witness that actors such as citizen-consumers, producers and policy makers acknowledge that the current system of mobility has reached its (ecological) boundaries, or even exceeded them. With the term sustainable mobility various actors have aimed to develop systems of mobility that are able to provide for the societal need for mobility while simultaneously taking the (ecological) boundaries into account. Structural changes in the form of a societal or socio-technical transition are required to deal with the ecological, social and economic issues related to the current socio-technical regime of the car. Nevertheless, there is still a wide diversity of visions on the form and pathway of such a sustainable mobility system. Consensus primarily consists on the urgency and persistency of problems while even the problems themselves have a tendency to differ on the (political) agenda's.

The transition (management) perspective offers promising ways to analyse and facilitate various potential system innovations within the domain of everyday mobility. However, despite various pilot projects, policy measures and scientific research the perspective on the role of citizen-consumers in sustainable mobility transitions is still relatively underdeveloped. In this thesis the challenge to facilitate a socio-technical transition in the domain of everyday mobility has been investigated from a citizen-consumer perspective. Investigating and emphasizing the role of human agents in mobility transitions has particular relevance for mobility policies and scientific debate. In the theoretical and empirical chapters an attempt has been made to contribute to the growing body of scientific knowledge of (sustainable) transition processes in the field of mobility. The specific contribution to this body of literature consists of an elaborate attempt to analyse the role of human agency in mobility transitions. With the use of the social practices model, the role of human agents and their ways of thinking, saying and doing with respect to more sustainable mobilities have been put centre stage. In this concluding chapter the fruits of this investigation will be presented by answering the research questions which were posted in the introduction. The focus will be primarily on the question which new insights for research and policy have been gained by the conceptualisation of the social practices model for everyday mobility. The chapter will also address the strategic consequences of this thesis for research and policy on everyday mobility. Based on the current discourse on the role of citizen-consumers in strategies to sustainable mobility, we will elaborate on the sustainable development strategies which should be further pursued in order to better incorporate citizen-consumers in their strategies towards sustainable mobility transitions.

8.2 A practice based approach to study socio-technical transitions

One of the main propositions in this thesis is that citizen-consumer engagement is an essential pre-requisite for sustainable transitions. Only when a connection can be made with the concerns and everyday practices of citizen-consumers, and when the inclusion of citizen-consumers is no longer approached as an obstacle to change but rather as a source for it, transitions to sustainable development can be realized. From a theoretical point of view, in order to make this connection the problem that needed to be addressed is the question how to fill the gap between theories on individual everyday action of citizen-consumers with theories on (socio-technical) transitions. Since the role of agency in transitions has not been well developed so far, we aimed to explore the theme of agency in transitions and discuss the added value of this theoretical approach in comparison with other – more individualist and voluntaristic – approaches to changing mobility behaviour. By combining scientific insights from transition studies, sociology of consumption and (social) psychology a research approach has been described which aims to bridge this divide between actor-oriented approaches and system-oriented approaches to sustainable consumption in general, and everyday mobility specifically.

Especially practice theories contain characteristics which enable them to study the role of agency in sustainable transitions precisely because they connect human agency with structural properties of socio-technical systems. Drawing upon the work of scholars in the field of contemporary theories of practice it was shown how co-evolution of practices, innovations and systems provides meaningful insight into the ways expectations and technologies may change. The social practices approach was presented as a potentially promising and fruitful practice-based framework to investigate the active involvement of citizen-consumers in transition processes to sustainable development. Throughout this thesis the social practices approach has been applied to study stability and change in everyday mobility practices. From this practice perspective we have investigated both the persistence of mobility related environmental problems and the robustness of unsustainable mobility practices as well as the leverage points for green transformations in the field of mobility.

The three components of the social practices model (systems of provision, practices, individual and lifestyle) provided an intricate way to investigate these dynamics. A complex and core part of this thesis was to define and specify what social practices of mobility actually are as to date virtually no practice-specific mobility research has been conducted. An important element of this thesis therefore focused on the question which types of mobility practices we can discern and what their major characteristics are. We described mobility as the result of participation of human agents in social occasions and social activities which take place at specific moments in time at specific sites. We therefore distinguished the main mobility practices on these social bases of travelling, namely: commuting, business travel, home-school travel, visiting family/friends, shopping and leisure travel. These everyday journeys allow travellers to connect the most important social activities and encounters which construct daily life. In addition to defining these mobility practices, we investigated which basic analytical characteristics are shared by these mobility practices. It was argued that mobility practices can be described as the interplay between four different constitutive dimensions: the temporal, the spatial, the material and the symbolic dimension.

The individual and lifestyle part of the social practices approach was conceptualised to understand individual differences in mobility practices. As practices are internally differentiated, and with individuals differing in levels of understandings, skills and attributions of meaning, there are multiple ways of doing things in practices. We used the notions of 'motility', a human agent's potential to be mobile, and 'mobility portfolios' (Flamm & Kaufmann, 2006) to describe the ways that human agents get access to and appropriate the capacity for mobility.

With three empirical case studies the working of the social practices approach in the domain of everyday mobility has been investigated. The fruits of this conceptual model has been illustrated, both qualitatively and quantitatively with these case studies. The first case study focused on environmental monitoring and monitarisation in the practice of new car-purchasing. It centred on the question how environmental information is provided in the automotive production-consumption chain, at which access points (consumption junctions) and how citizen-consumers integrate this environmental information into their car purchasing various decision. The second case study concerned mobility management in commuting practices. It centred on the question how, for the practice of commuting, existing mobility routines may be (deliberately) altered and new routines may be formed by using positive and negative trigger moment. In the third and final case study the interaction between the modes of access and the modes of provision in the domain of everyday mobility was investigated. On the one hand it focused on citizen-consumer assessment of potential provider strategies for sustainable mobility alternatives. On the other hand the aim was to explore, quantitatively, the role of (green) mobility portfolios in explaining individual differences in mobility practices.

The investigation of these three empirical cases is naturally not enough to make hard propositions about the added value of mobility practices. In order to do so one needs to perform an in-depth analysis of multiple policy-relevant mobility practices. However, with this thesis a new research and policy agenda has been opened up which allows to investigate thoroughly the role of end-users in the (co-) construction of sustainable mobility transitions. In the remainder of this chapter these insights will be elaborated, thereby showing why it is worthwhile to investigate transitions in practices.

8.3 Scientific insights on sustainable innovation and everyday mobility

Research question 3: *What can the social practices model, and the conceptualisation for everyday mobility, contribute to the knowledge on transitions and system innovations?*

8.3.1 The role of context in consumer behaviour

An important scientific insight is that each of the empirical case studies has shown the important role of contextual factors in understanding and influencing consumer behaviour in general, and mobility behaviour specifically. The system of provision makes clear that the possibilities for sustainable consumption to a large extent depend on the amount and kind of socio-technical innovations available in a specific domain. This is clearly the case when comparing practices from different consumption domains (such as food consumption, housing and everyday mobility). As Fine & Leopold (2002, p. 5) have emphasized: 'different systems of provision ... are the consequence of distinct relationships between the various material and cultural practices comprising the production, distribution, circulation and consumption of the goods concerned'. Clearly, in the practice of purchasing a car other producers, distributors and consumer preferences play a role than in purchasing a house or food consumption. Not only the actors and locations (consumption junctions) are different for each of these production-consumptions chains, also the historically grown social connotations vary tremendously. An important implication for consumer behaviour is that innovation processes are similarly domain-specific. Depending on the quality and quantity of socio-technical innovations on offer, it is more or less easy for citizen-consumers to integrate innovations in their social practices in the various consumption domains. What we have seen in the consumer survey is that the (perceived) quality and quantity of the sustainable products and services on offer differ vastly for each consumption domain. This contextual variation is also an important explanation why a citizen-consumer may have different environmental profiles for each consumption domain.

Contextual factors, such as the system of provision, have dynamic properties as well, thereby exerting an dynamic influence on social practices. The practice of new car purchasing showed that the automotive market has gone through a significant change in the way information is available for citizen-consumers. The internet has become one of the most dominant sources of information and the role of the salesmen likewise has shifted from leading to guiding, and from salesman to advisor. Also, formerly fixed customer-dealer ties have become more and more fluid. An important outcome of the abovementioned trend is that the information age has disclosed information about the environmental impact of automobiles that was until recently inaccessible to consumers. At the access points where producers and consumers meet, environmental information is displayed in various ways. Especially information about the fuel efficiency of vehicles is a lot easier to find and compare than a few years ago. This case study showed that it is worthwhile to analyse specifically what information is provided, via which structures of provisions and how they are meaningful supported at important sites of the consumption junction.

In the case of mobility management in commuting practices it was shown how most long-term changes in consumer behaviour require that a change in the contextual situation occurs. In each of the three examples of mobility management influential actors in the system of provision of commuting attempted to orchestrate a shift in the mobility practices of a designated group of commuters. With these strategies alterations were made in the socio-technical mobility options available to commuters. More importantly it was shown how commuting is not a solitary practice but is connected intricately with the way work itself is organised. It demonstrated how the various institutional arrangements which surround commuting and work practices are important elements in a mobility strategy.

Finally, in the consumer survey we investigated how the receptiveness of citizen-consumers for socio-technical innovations in the domain of mobility is also dependent on the specific system of provision and provider strategies. With the assessment of provider strategies it was shown how these strategies may target barriers for the integration of socio-technical innovations in mobility practices.

8.3.2 The dynamics of mobility portfolios and practices

A focus on mobility practices may provide an important contribution to the analysis of innovation processes in the domain of mobility. As shown in the theoretical Chapters 3 and 5 and empirical Chapters 4, 6 and 7 this thesis contributed to relevant questions such as: how do citizen-consumers come into contact with mobility innovations? How do socio-technical innovations become integrated into mobility practices? Concepts like mobility practices, motility and mobility portfolios offer new ways to investigate transformation and resistance to change in the domain of everyday mobility.

The concept of mobility portfolios illustrate how the individual and lifestyle component of human agents not only contains general attitudes and values of citizen-consumers but also includes practice-specific knowledge and skills. Certain skills, appropriation and access portfolios are prerequisites for conducting mobility practices in a certain way. The existing mobility portfolio, as we have seen, provides an important clue to the question how flexible travellers are in their practices. These mobility portfolios of groups of citizen-consumers are therefore important indicators for the possibility to conduct mobility practices in a greener or less green manner.

Theoretically interesting is that motility and mobility portfolios provide the means to analyse mobility behaviour between the level of the individual and lifestyle on the one hand and the social practices and collective conventions on the other, an issue addressed in Chapter 3. There the question was postulated how (individual) differences in habits and routines relate to the specific way of using objects in practices and in the formation of new practices. Practices, on a societal level, have a certain cohesion in the sense that they consist of routinized ways of doing things, based on the specific integration of objects, meanings and skills. Simultaneously, it is clear that practitioners show differences in levels of understandings, in access to objects and in levels of competences of handling the objects. By showing how citizen-consumers have different levels of understandings, competences and access to mobility options, and more importantly, how these have an impact on the way citizen-consumers participate in practices, a relevant contribution has been made to theories of practice.

Motility provides not only important clues how to investigate and explain differences in practices. Another contribution of the empirical case studies is that they have shown how motility and mobility portfolios are not static but dynamic aspects of the individual and lifestyle component of citizen-consumers. While citizen-consumers may instigate an active search for environmental socio-technical innovations without an external impetus, often innovations are made accessible to citizen-consumers. By analysing the effects of orchestrated attempts to increase the mobility portfolios of commuters this thesis has provided information on the structuring influence of mobility providers, employers and governmental organisation in the constitution of mobility portfolios. Empirically interesting is that there were varying ways in which the mobility portfolios can be deliberately influenced by the relevant actors in the practice. These types of strategies may lead to alterations in the mobility portfolio by making already existing means of mobility more easily accessible, by making new modes of mobility available, or by altering the decision context that keep the routine locked into place.

However, as the case of new car purchasing has showed how one cannot assume that citizen-consumers are empowered with information solely by increasing the amount of environmental information available, the same applies for motility. The social shaping of access, as described in Chapter 4, focuses on the question if and how consumers can put the acquired environmental information to use. This same principle applies to mobility practices as merely increasing the mobility portfolio of human agents does not automatically lead to alterations in the conduct of mobility. Socio-technical innovations need to be integrated in the specific mobility practices which require adjustments from practitioners in their daily life routines. It is exactly the perspective of mobility practices that helps to understand that each human agent has to make more or less similar alterations when integrating specific socio-technical innovations in their everyday life practices. The acknowledgement that mobility practices consist of multiple dimensions (temporal, spatial, material/technological and symbolical) helps to clarify the complexity of alterations in mobility practices. As we have shown in the case studies, aside from learning how to manually operate socio-technical innovations, practitioners need to learn how to apply socio-technical mobility innovations in their time-spatially dispersed activities. The case study of commuting showed how a modal shift from car use to public transport use implies not only a change in the technological means of transportation but that it also encompasses different time-spatial patterns and symbolic connotations. The same dynamics take place in a shift from 'conventional' cars to electric vehicles, or an increase in telecommuting and teleconferencing. When mobility is perceived from the viewpoint of practices it is easier to understand how all these elements are connected.

8.3.3 Suggestions for further research

With this thesis only the first few steps have been taken for a new research agenda of a practice based approach to everyday mobility. While a new agenda has been opened up, further investigation on the role of motility on mobility practices is recommended to better understand the intricate relationships between these two elements. Also cohort analysis would provide further information on the mobility options that are available to practitioners throughout their lifetime and how these options are put to use in various mobility practices. Furthermore, cohort analysis would show which transformations takes place in the motility of citizen-consumers thereby allowing us to distinguish increases and decreases in the various mobility options, competences and assessments. But most importantly, a more thorough investigation of the various mobility practices would provide information about the potential leverage points of green transformation both from the perspective of the individual and lifestyle component of citizen-consumers as from the perspective of the system of provision. While in this thesis most of the attention has been dedicated to the practices of commuting and business travel, a similar examination of the other mobility practices would provide useful insight into the dynamics taking place in those mobility practices which are likely to have their own distinct characteristics.

Finally, transition research would benefit from a thorough investigation of large-scale shifts in user preferences, practices and social-cultural images. While it is widely acknowledged that culture, symbolism and user practices are of crucial importance in transitions, what does this mean when we think about the current regime and (possible) regime shifts? As there is a clear increase in the environmental socio-technical innovations available, are we already seeing a shift which can be labelled as radical? Furthermore, what is it that makes a regime shift radical or not? Radical change in this respect is not only a matter of a radical alteration in technologies. Maybe radical change is more necessary in the meaning structures associates with current cars. In the theoretical chapter and the empirical chapter on car purchasing we see that citizen-consumers are especially focused on engine power and speed. In these chapters we have seen that people have all kinds of associations with the car and that these historically grown images and associations are held into place via a variety of mechanisms. The electric car is currently seen as one of the biggest competitors of the ICE car. However, many electric cars are trying to build on the existing images of the car while one could question whether the biggest contribution of the electric car lies herein. Can we not already witness a shift in user practices and associates structures of meaning which point towards another direction? In recent years cars have become more compact again, thereby breaking the trend of increasing car size. Furthermore, we see that ICT and in-car technologies have become more important. The question is whether or not for future generations cars will be judged on the power and torque or on which types of communication is available.

8.4 Recommendations for governance

Research question: *By which means can environmental policy better incorporate citizen-consumers as agents of change in strategies towards sustainable mobility?*

Recommendation 1: citizen-consumer engagement is an essential element in the transition to sustainable mobility

Over time mobility policies, both inside and outside of the Netherlands, shift in the way mobility related problems are defined, while the suggested directions for solving the problems vary accordingly. We have illustrated that similar shifts occur in the storylines of policy-makers and in political debate on the role of citizen-consumers to solve mobility related problems. One can see fluctuations between demand and supply oriented approaches in the discourse and policies in the field of mobility. The specific policy measures implemented (such as increasing road network capacity, traffic management, mobility management or communication campaigns) reflect how the role of mobility in society is perceived at that moment in time. Moreover, the adopted policy approaches reflect the (changing) perception of the role of citizen-consumers in everyday mobility practices and in environmental policy in general. Contemporary mobility policies tend to be supply-oriented in character, thereby lacking a clear cut and convincing orientation on citizen-consumers and their everyday life mobility behaviours. The far majority of public spending in the domain of mobility is invested in large-scale infrastructure development with the aim to solve location-specific congestion. Infrastructure construction requires huge public investments while it is uncertain that congestion levels in the future will be as high as currently projected. Furthermore, demand-oriented approaches to facilitate mobility patterns may prove to be just as (cost-)effective considering the huge infrastructure investments.

While in the last years the situation has changed somewhat for the better, citizen-consumers are often seen as actors which are not willing or able to implement structural changes in their everyday life patterns and are therefore frequently considered as barriers to the successful realisation of sustainable transitions. We provided arguments why a citizen-consumer orientation in mobility is not only a fruitful but also an essential ingredient in (policies targeting) transitions towards sustainable mobility. This requires a different approach as conventional supply-oriented frameworks are often not compatible with a citizen-consumer oriented approach and purely voluntaristic policies have proven to be unsuccessful in the past.

In this thesis citizen-consumers are considered as ecologically more or less committed actors which may become involved in the process of the ecological transformation of production-consumption chains. The research question presupposed that it is possible and also important to make better use of citizen-consumers as agents of change in transitions to sustainable mobility. The empirical chapters provide a clear indication that, contrary to most scientific and policy viewpoints, citizen-consumers do actually acknowledge the need for structural changes in the field of mobility. A majority of the Dutch citizen-consumers perceives environmental issues related to everyday mobility as problematic and feels, at least partly, co-responsible for dealing with these issues. Though this outcome may not come as a complete surprise, more interesting is the acknowledgement of citizen-consumers that the petrol-based system of automobility is seen as problematic and that large-scale systemic changes are supported by a majority. Especially when compared to other consumption domains, in the domain of everyday mobility in the last few years groups of end-users have become enmeshed with sustainable mobility initiatives.

While previous studies often stress that the car-based system of mobility is amongst the most stable regimes the empirical chapters in this thesis have shown that the potential for sustainable mobility behaviour in the Netherlands is on the rise. All the three empirical chapters support the outcome that in a relatively short time-period in the domain of mobility possibilities for sustainable change have emerged. Partially this is also the result of deliberate consumer-oriented environmental strategies such as in the practices of new car purchasing and the practice of commuting.

Recommendation 2: the existence of various groups of citizen-consumers implies that there is not one dominant transition path towards sustainable mobility

While the abovementioned acknowledgement of the need for systemic change is an important and, from an environmental point of view, positive outcome, it does not automatically mean that citizen-consumers are actively investigating new ways of doing things because they are environmentally committed to do so! While there is a small group of citizen-consumers which indeed chooses an energy-efficient car or travels by public transport to work primarily because of environmental reasons, in general citizen-consumers are not actively searching for ways to make their mobility patterns more sustainable. To be able to participate in or potentially alter social practices, citizen-consumers need to be invited and challenged with attractive visions and ideas about sustainability, while at the same time being actively supported to get involved in transition processes. This also implies that citizen-consumers are seen in a broader perspective than purely as adopters or rejecters of sustainable innovations, or to solely see them as users or non-users. In addition to targeting citizen-consumers as (non-) adopters of sustainable innovations it is important to connect green provider strategies with the development of attractive story-lines. As environmental aspects are rarely the main reason for the actions of citizen-consumers, environmental policy also needs to develop inspiring sustainability visions that make a connection with people's quality of life. These visions show citizen-consumers attractive sustainable alternatives in the domain of everyday mobility.

Even more important is that environmental policy should recognize the existence of various groups of citizen-consumers which brings with it the important conclusion that there is not one dominant strategy towards facilitating structural changes to sustainable mobility. There are different groups of end-users with different mobility portfolios and different perspectives on sustainability. We have shown that there is not one sustainable strategy that works for all of these groups. Clearly, there is not one story-line of sustainable mobility, but there is a multitude of possibilities by which means the system of mobility might become more sustainable. The greening of automobility, localism and modal flexibility are three distinct sustainable mobility paradigms which show the various ways in which citizen-consumers can embed environmental socio-technical innovations into their normal ways of doing things. Localism, with its focus on mobility-reduced lifestyles (car-free environments, slow travelling and virtual travelling) provide different sustainable visions than green automobility which focuses on the greening of car-dependent lifestyles. The idea behind these visions is that it is much likelier for car-adepts to become entwined with sustainable automobility than with modal shifts to alternative transport modes. By supplying a multitude of sustainable alternatives in the field of everyday mobility it is possible for end-users to meet their needs and requirements in the shaping of social life while simultaneously greening their mobility patterns. The perspective to keep multiple (sustainable mobility) options open fits well with the fundamentals of transition management. One of the key elements of transition management is to develop a shared transition agenda (Loorbach, 2007). Instead of focusing on the most optimal societal blueprint, this agenda consists of a multitude of transition pathways leading towards a basket of shared visions. Precisely by providing a basket of initiatives, each based on different story-lines, it is possible for end-users to become enmeshed with sustainable mobility.

Recommendation 3: mobility policy could benefit by targeting generic mobility portfolios and specific mobility practices

So far we have concluded that there is great potential for citizen-consumers to become enmeshed with sustainable mobility alternatives. The final piece of the puzzle is how policy can develop strategies to assist citizen-consumers in integrating the various environmental socio-technical innovations in their everyday mobility practices.

In this thesis the concept of mobility portfolio was presented as an important element in the ability of citizen-consumers to participate in mobility practices. To illustrate the importance of mobility portfolios for multimodality the consumer survey showed that citizen-consumers with a high mobility portfolio are more adept at multi-modal travelling than citizen-consumers with a low mobility portfolio. Moreover, citizen-consumers with a high mobility portfolio are more receptive to alterations in practices when provider strategies are implemented. The question arises how citizen-consumers end up in these various mobility portfolios. Also important is the question how policy can assist the transformation in mobility portfolios.

Mobility management can be seen as an active form of mobility portfolio development. The three cases of mobility management in the practice of commuting, as well as the consumer survey, support the active provision of socio-technical innovations to citizen-consumers such as integrated mobility cards, the facilitation of telecommuting, and product-to-service systems (bike- and car-sharing). As access to and experience with mobility innovations has a positive relation with multimodal travelling, the initiatives of mobility management must be seen as a positive development. By deliberately broadening up the mobility portfolios of citizen-consumers their potential for mobility increases allowing citizen-consumers to have modal flexibility.

Policy makers from governments, but even more so from service providers and employers, should build on these principles to increase the modal choices of citizen-consumers. In addition to merely making socio-technical innovations available it is important to focus on the learning processes that are needed to alter mobility patterns in the various mobility practices. Whether or not an innovation becomes integrated in a specific social practice also depends on the historically acquired experience, knowledge and skills, as well as the framing by and evaluation of the practitioner. The de- and reroutinization of mobility practices, however, is not (only) an individual endeavour of citizen-consumers but this process may be actively supported by relevant actors such as employers. Mobility management not only supports portfolio development, by making more socio-technical mobility options available, it can also assist in altering the decision-framework and conventions surrounding work and travel. Actors surrounding the mobility practices play an important role in the question whether or not second-order learning takes place and existing ideas and beliefs on mobility are reflected upon, and potentially changed. Institutional arrangements such as corporate management style, human resource management, and fiscal arrangements in combination with the specific company culture are therefore important elements in the practice of commuting.

Although it is a worthwhile policy approach to focus on mobility portfolio development as a generic consumer-oriented strategy, it important to recognize that mobility practices differ from each other. Even though citizen-consumers apply the same mobility portfolio in various mobility practices and these practices are closely connected with each other, in the practice of commuting other regulations, institutions and actors play a role than in the practice of leisure travel and home-school travel. It is therefore not surprising that the ecological modernisation of these mobility practices shows an uneven development. While some practices, such as commuting and business travel, show great potential for multimodal travelling and down-sizing (teleconferencing, telecommuting), other practices such as social travel, leisure travel and home-school travel show increasing levels of car-usage. Policy-makers aiming to green the consumer behaviour in the domain of mobility could benefit from this knowledge by developing practice-specific consumer-oriented strategies. Though this thesis does not provide ready-to-use consumer-oriented strategies for all of these practices it is an important policy tactic to investigate further. Indeed, it is precisely the (challenging) task of mobility governance to orchestrate and facilitate sustainable mobility alternatives for each of the mobility practices.

Recommendation 4: in mobility governance, take citizen-consumer trends into consideration

The final recommendation is that sustainable mobility governance should not only take the current, but also the future traveller into consideration. As the current supply-oriented approach in mobility governance focuses heavily on infrastructure development, insufficient attention is given to current and future consumer trends. The last years much public spending has been invested in increasing highway capacity based on existing congestion bottlenecks and projected increases in mobility. However, instead of a general increase in (auto)mobility, it is much more likely that mobility will develop in more complex patterns with variations depending on aspects such as region (urban-rural continuum), demographics and lifestyles (age-groups, education levels, ethnicity), and mobility practices.

It is clear that for many people, especially in the field of higher education, the travel and work culture is changing. First, there is more fluidity between work life and private life, allowing more citizen-consumers to work from home and to work at different time-periods. Second, we see an increasing demand for and use of meet and greet and co-working facilities which are predominantly located in the vicinity of public transport services. This new work and travel culture poses promising possibilities for non-automobile transport modes in urban areas. The work and travel culture shows that it is increasingly important for commuters and business travels to have flexibility and the possibility of multimodality in their mobility portfolio. To build upon this trend the barriers between the various modal networks need to be diminished as up to now the Netherlands has not been very proficient in the provision of multimodal networks.

Furthermore, due to increasing levels of urbanisation, the amount of 'cosmopolites' with an inner city-orientation is expected to increase in the next decades. Especially important for these newer generations are flexibility, availability of information and access to services. It is therefore likely that there will be a trend towards increasingly large mobility portfolios of urban citizen-consumers as their strategy is to maximize motility and to have multiple options open. In line with this trend it also probable that access to mobility services might become more important than direct ownership of products (most notably the car) which brings with it possibilities for a policy focusing on a modal shift and multimodality.

While it is plausible that non-automotive transport modes have a great potential in urban areas, the opposite is true in rural areas. Population decline, increasing car ownership and privatisation of public transport already have led to reduced service levels of public transport in rural areas. In non-urban environments car-dependency is therefore more likely to increase than to decrease. While a minimum service level of public transport should be maintained, for instance to ensure the continued participation of elderly people in social activities, from a sustainable mobility perspective it makes more sense to focus on the greening of automobility.

References

Aarts, H., Verplanken, B. & Van Knippenberg, A. (1998). Predicting behavior from actions in the past: repeated decision-making or a matter of habit? *Journal of Applied Social Psychology*, 28: 1355-1374.

ADAC (2005). *Study on the effectiveness of Directive 1999/94/EC relating to the availability of consumer information on fuel economy and CO_2 emissions in respect of the marketing of new passenger cars.* Munich, Germany: ADAC.

Adams, J. (2005). Hypermobility: a challenge to governance. In: C. Lyall and J. Tait (eds.), *New modes of governance: Developing an integrated policy approach to science, technology, risk and the environment.* Aldershot, UK: Ashgate.

Ajzen, I & Madden, T. (1986). Predictions of goal-directed behaviour: attitudes, intentions and perceived behavioral control. *Journal of Experimental Social Psychology*, 22: 453-474.

Anable, J. & Gatersleben, B. (2005). All work and no play? The role of instrumental and affective factors in work and leisure journeys by different ravel modes. *Transportation Research Part A*, 39: 163-181.

Anable, J. (2006). 'Complacent car addicts' or 'aspiring environmentalists'? Identifying travel behaviour segments using attitude theory. *Transport Policy*, 12: 65-78.

Appadurai, A. (ed.) (1986). *The social life of things. Commodities in cultural perspective.* Cambridge, UK: Cambridge University Press.

ARN (2013). *Factsheet recyclingpercentage autowrakken.* Amsterdam, the Netherlands: Auto Recycling Nederland.

Avelino, F., Van Bakel, J., Dijk, M, Nijhuis, J, Pel, B. & Verbeek, D. (2007). *An interdisciplinary perspective on Dutch mobility governance.* 4th Dubrovnik conference on sustainable development of energy, water and environment systems, special session on sustainable social-technical transport system. Dubrovnik, Croatia: Conference Proceedings.

AVV (2006a). *Vervoerswijzekeuze op ritten tot 7,5 kilometer. Argumentaties van autobezitters voor de keuze van de auto, cq de fiets bij het maken van een korte rit.* Rotterdam, the Netherlands: Rijkswaterstaat – Adviesdienst Verkeer en Vervoer.

AVV (2006b). *Evaluatie van de mobiliteitsbeïnvloedende maatregelen tijdens het groot onderhoud A4/A10 Zuid.* Rotterdam, the Netherlands: Rijkswaterstaat – Adviesdienst Verkeer en Vervoer.

Bagwell, P.S. (1988). *The transport revolution 1770-1985.* London, UK: Taylor & Francis.

Bamberg, S. (2003). How does environmental concern influence specific environmentally related behaviors? A new answer to an old question. *Journal of Environmental Psychology*, 23: 21-32.

Bamberg,. S. (2006). Is a residential relocation a good opportunity to change people's travel behavor? Results from a theory-driven intervention study. *Environment and Behavior*, 38: 820-840.

Beck, U., Giddens, A. & Lash, S. (1994). *Reflexive modernization. Politics, tradition and aesthetics in the modern social order.* Stanford, USA: Stanford University Press.

Beige, S. & Axhausen, K. (2008). *The ownership of mobility tools during the life course.* Zurich, Switzerland: Swiss Federal Institute of Technology.

Berkhout, F., Smith, A. & Stirling, A. (2003). *Socio-technological regimes and transition contexts.* Sussex, UK: SPRU Electronic Working Paper.

Bliemer, M., Dicke-Ogenia, M. & Ettema, D. (2009). *Rewarding for avoiding the peak period: a synthesis of three studies in the Netherlands.* Noordwijk, the Netherlands: conference proceedings of European Transport Conference.

Bloor, M., Frankland, J., Thomas, M. & Robson, K. (2001). *Focus groups in social research.* London, UK: Sage.

Boardman, B., Banks, N., Kirby, H.R., Keay-Bright, S., Hutton, B.J. & Stradling, S.G. (2000). *Choosing cleaner cars: the role of labels and guides. Final report on vehicle environmental rating schemes.* Edinburgh, UK: Transport Research Institute.

Bonss, W. & Kesselring, S. (2004). Mobility and the cosmopolitan perspective. In: W. Bonss, S. Kesselring & G. Vogl (eds.), *Mobility and the cosmopotilan perspective. A workshop at the Munich reflexive modernisation research centre.* Munich, Germany: SFB 536.

Bourdieu, P. (1984). *Distinction: a social critique of the judgement of taste.* London, UK: Routledge and Kegan Paul.

Bouwman, M.E. (2004). *Ontwikkelingen in verkeersbeleid in West-Europa.* Rotterdam, Nederland: conference proceedings of Colloquium Vervoersplanologisch Speurwerk.

Bristow, M. (2001). *Could advertisements for cars influence our travel choices? A review of car advertisements and the targeting of different audiences.* London, UK: Transport Planning Society.

Brundtland Report (1987). *Our common future. World Commission on Environment and Development.* Oxford, UK: Oxford University Press.

Cairncross, F. (1997). *The death of distance: how the communications revolution will change our lives.* Cambridge, USA: Harvard Business Press.

Campbell, C. (1989). *The romantic ethic and the spirit of modern consumerism.* London, UK: Blackwell.

Canzler, W. (2008). The paradoxical nature of automobility. In: W. Canzler, V. Kaufmann, S. Kesselring (eds.). *Tracing mobilities. Towards a cosmobilitan perspective* (pp. 105-118). Aldershot, UK: Ashgate.

CapGemini (2007). *Cars online 07/08.* CapGemini Automotive. Available at: http://www.capgemini.com/cars-online-archives.

CapGemini (2010). *Cars online 10/11.* CapGemini Automotive. Available at: http://www.capgemini.com/cars-online-archives.

CBS, PBL, Wageningen UR (2013). *Compendium voor de leefomgeving.* Den Haag: Centraal Bureau voor de Statistiek; Den Haag/Bilthoven: Planbureau voor de Leefomgeving en, Wageningen: Wageningen UR.

CE (2003). *To shift or not to shift, that´s the question? The environmental performance of freight and passenger transport modes in the light of policy making.* Delft, the Netherlands: CE.

CE (2008a). *De planet-kant van duurzame mobiliteit. Top-down visievorming ten behoeve van Transumo.* Delft, the Netherlands: CE.

CE (2008b). *STREAM Studie naar transport emissies van alle modaliteiten.* Delft, the Netherlands: CE.

Clean Fuels Foundation (2011). *E85 and blender pumps: a resource guide to ethanol refueling infrastructure.* Available at: http://www.ffv-awareness.org.

Collins, R. (2004). *Interaction ritual chains.* Princeton, USA: Princeton University Press.

CPB, MNP & RPB (2006). *Welvaart en leefomgeving. Een scenariostudie voor Nederland in 2040.* Achtergronddocument. Den Haag: Centraal Planbureau; Bilthoven: Milieu- en Natuurplanbureau; Den Haag: Ruimtelijk Planbureau.

CROW (2007). *Verkeersmaatregelen bij evenementen.* Ede, the Netherlands: CROW.

Dagevos, H., Van Herpen, E. & Kornelis, M. (2005). *Consumptiesamenleving en consumeren in de supermarkt. Duurzame voedselconsumptie in de context van markt en maatschappij.* Wageningen, the Netherlands: Wageningen Academic Publishers.

Davis, S.C., Diegel, S.W. & Boundy, R.G. (2013). *2012 vehicles technologies market report.* Oak Ridge, USA: Oak Ridge Institute for Science and Education.

De Jager (2007). *Duurzame mobiliteit. Innovaties binnen handbereik.* Presentation given on 7th of June, 2008. Available at: http://www.transumofootprint.nl/documentdetail.asp?id=63062.

De Jager (2008). *Duurzame Mobiliteit. Belonen op de Weg. Dynamisch Mobiliteits Management.* Presentation given on 20th of May 2008. Available at: http://www.transumofootprint.nl/documentdetail.asp?id=66516.

De la Bruheze, A.A., De Wit, O. & Oldenziel, R. (2004). *Mediating practices: technology and the rise of European consumer society in the twentieth century.* Review article of the consumer society network with tensions of Europe project. Available at: http://www.tensionsofeurope.eu/www/en/files/get/file5.pdf.

DGET (2008). *European energy and transport. Trends to 2030. Update 2007.* Brussels, Belgium: European Commission, Directorate-General Energy and Transport.

Diana, M. & Mokhtarian, P. L. (2009). Desire to change one's multimodality and its relationship to the use of different transport means. *Transportation Research Part F,* 12: 107-119.

Diekstra, R. & Kroon, M. (2003). Cars and behaviour: psychological barriers to car restraint and sustainable urban transport. In: R. Tolley (ed.), *Sustainable transport. Planning for walking and cycling in urban environments* (pp. 252-264). Cambridge, UK: Woodhead Publishing.

Dijk, M. (2011). Technological frames of car engines. *Technology in Society,* 33: 165-180.

Dijk, M., Nijhuis, J. & Madlener, R. (2012). Consumer attitudes towards alternative vehicles. In: G. Calabrese (ed.), *The greening of the automotive industry* (pp. 286-303). Basingstoke, UK: Palgrave Macmillan.

Dijst, M. (1995). *Het elliptisch leven. Actieruimte als integrale maat voor bereikheid en mobiliteit.* Delft, the Netherlands: TU Delft.

Dijst, M., De Jong, T. & Ritsema van Eck, J. (2002). Opportunities for transport mode change: an exploration of a disaggregated approach. *Environment and Planning B: Planning and Design,* 29: 413-430.

Douglas, M. & Isherwood, B. (1979). *The world of goods. Towards an anthropology of consumption.* London, UK: Routledge.

Dunlap, R. & Van Liere, K. (1978). The new environmental paradigm – a proposed measuring instrument and preliminary results. *Journal of Environmental Education,* 9: 10-19.

ECN (2009). *Het transitiebeleid voor duurzame mobiliteit. Evaluatie en toekomstvisie.* Petten, the Netherlands: ECN.

Ecorys (2006). *Mobiliteit en evenementen. Een overzicht van ervaringen met mobiliteitsmanagement rond grote evenementen.* Amsterdam, the Netherlands: Ecorys.

Ecorys (2011a). *Fiscale stimulering (zeer) zuinige auto's Onderzoek aanpassing zuinigheidsgrenzen.* Rotterdam, the Netherlands: Ecorys.

Ecorys (2011b). *Zicht op zakelijke (auto)mobiliteit.* Rotterdam, the Netherlands: Ecorys.

EEA (2006). *Transport and environment: facing a dilemma. TERM 2005: indicators tracking transport and environment in the European Union.* Brussels, Belgium: European Environmental Agency.

Elzen, B. & Wieczorek, A. (2005). Transitions towards sustainability through system innovation. *Technological Forecasting & Social Change,* 72: 651-661.

Elzen, B., Geels, F.W. & Green, K. (2004). *System innovation and the transition to sustainability. Theory, evidence and policy.* Cheltenham, UK: Edward Elgar.

Engel, J.F., Blackwell, R.D. & Miniard, P. W. (1995). *Consumer behavior.* Forth Worth, USA: Dryden Press.

Fine, B. & Leopold, E. (1993). *The world of consumption.* London, UK: Routlegde.

Fishbein, M & Ajzen, I. (1975). *Belief, attitude, intention and behavior: an introduction to theory and research.* Reading, USA: Addison-Wesley.

Flamm, M. & Kaufmann, V. (2004). *Operationalising the concept of motility: a qualitative exploration.* Paper presented at the ad-hoc session 'mobility and social differentiation' of the 32nd Kongress der Deutscher Gesellschaft für Soziologie. Munich, Germany: conference proceedings.

Frantzeskaki, N. & Loorbach, D. (2010). Towards governing infrasystem transitions. Reinforcing lock-in or facilitating change? *Technological Forecasting and Social Change*, 77: 1292-1301.

Freudendal-Pedersen. M. (2005). Structural stories, mobility and (un)freedom. In: T.U. Thomsen, H. Gudmunsson & L.D. Nielsen (eds.), *Social perspectives on mobility* (pp.29-46). Farnham, UK: Ashgate.

Gartman, D. (2004). Three ages of the automobile. The cultural logics of the car. *Theory, Culture and Society,* 25: 169-195.

Gatersleben, B., Steg, L. & Vlek, C. (2002). Measurement and determinants of environmentally significant consumer behavior. *Environment and Behavior*, 34: 335-362.

Geels, F. W. & Schot, J. (2007). Typology of sociotechnical transition pathways. *Research policy,* 36: 399-417.

Geels, F.W. & Smit, W.A. (2000). Failed technology futures: pitfalls and lessons from a historical survey. *Futures,* 32: 867-885.

Geels, F.W. (2004). From sectoral systems of innovation to socio-technical systems: insights about dynamics and change from sociology and institutional theory. *Research Policy,* 33: 897-920.

Geels, F.W. (2005a). The dynamics of transitions in socio-technical systems: a multi-level analysis of the transition pathway from horse-drawn carriages to automobiles (1860-1930). *Technology Analysis & Strategic Management,* 17: 445-476.

Geels, F.W. (2005b). Processes and patterns in transitions and system innovations: refining the co-evolutionary multi-level perspective. *Technological Forecasting and Social Change*, 72: 681-696.

Geels, F.W. (2005c). *Technological transitions and system innovations. A co-evolutionary and socio-technical analysis.* Cheltenham, UK: Edward Elgar.

Geels, F.W., (2007). Feelings of discontent and the promise of middle range theory for STS. *Science, Technology & Human Values*, 32: 627-651.

Geels, F.W., Kemp, R., Dudley, G. & Lyons, G. (2012*). Automobility in transition? A socio-technical analysis of sustainable transport*. London, UK: Routledge.

Geerlings, H. & Peters, G. (2002). Mobiliteit en duurzaamheid. Een verkenning van de stand van zaken. In: H. Geerlings, W. Hafkamp, & G. Peters (eds.), *Mobiliteit als uitdaging. Een integrale benadering*. Rotterdam, the Netherlands: Uitgeverij 010.

Giddens, A. (1984). *The constitution of society*. Cambridge, UK: Polity Press.

Giddens, A. (1990). *The consequences of modernity*. Stanford, USA: Stanford University Press.

Giddens, A. (1991). *Modernity and self-identity. Self and society in the late modern age*. Cambridge, UK: Polity Press.

Goffman, E. (1963). *Behaviour in public places*. New York, USA: Free Press.

Goldblatt, D.L. (2002). *Personal vs. socio-technical change: informing and involving householdsers for sustainable energy consumption*. Zurich, Switzerland: Swiss Federal Institute of Technology.

Gram-Hanssen, K. (2006). *Consuming technologies – developing routines*. Paper presented at sustainable consumption and society. An international working conference for social scientist. Madison, USA: conference proceedings.

Grin, J. (2012). The politics of transition governance in Dutch agriculture: conceptual understanding and implications for transition management. *International Journal of Sustainable Development*, 15: 72-89.

Grin, J., Rotmans, J. & Schot, J. (2011). On patterns and agency in transition dynamics: some key insights from the KSI Programme. *Environmental Innovation and Societal Transitions*, 1: 76-81.

Grin, J., Rotmans, J. & Schot, J. (eds.) (2010). *Transitions to sustainable development. New directions in the study of long term transformative change*. London, UK: Routledge.

Grin, J., Van de Graaf, H. & Vergragt, P. (2003). Een derde generatie milieubeleid. Een sociologisch en een beleidswetenschappelijk programma. *Beleidswetenschap,* 1: 51-72.

Gronow, J. & Warde, A. (eds.) (2001). *Ordinary consumption.* London, UK: Routledge.

Gudmundsson, H. (2004). Sustainable transport and performance indicators. *Issues in Environmental Science and Technology*, 20: 35-63.

Hägerstrand, T. (1970). What about people in regional science? *Papers of the Regional Science Association*, 21: 7-21.

Hajer, M. A. (1993). Discourse-coalitions and the institutionalisation of practice. The case of acid rain in Britain. In: J. Forester & F. Fischer (eds.), *The argumentative turn in policy and planning* (pp. 43-76). Durham, UK: Duke University Press.

Hajer, M. A. (1995). *The politics of environmental discourse: ecological modernisation and the policy process.* Oxford, UK: Clarendon Press.

Hannam, K., Sheller, M. & Urry, J. (2006). Editorial: mobilities, immobilities and moorings. *Mobilities*, 1: 1-22.

Hannigan, J. (1998). *Fantasy city: pleasure and profit in the postmodern metropolis.* London, UK: Routledge.

Hargreaves, T. (2008). *Making pro-environmental behaviour work: an ethnographic case study of practice, process and power in the workplace.* Norwich, UK: University of East Anglia.

Harms, L. (2003). Mobiel in de tijd. Op weg naar een auto-afhankelijke maatschappij, 1975-2000. Den Haag, the Netherlands: Sociaal en Cultureel Planbureau.

Harms, L. (2006) Op weg in de vrije tijd. Context, kenmerken en dynamiek van vrijetijdsmobiliteit. Den Haag, the Netherlands: Sociaal en Cultureel Planbureau.

Harms, L. (2008). Overwegend onderweg. De leefsituatie en de mobiliteit van Nederlanders. Den Haag, the Netherlands: Sociaal en Cultureel Planbureau.

Heffner, R.R., Kurani, K.S. & Turrentine, T.S. (2005). *Effects of vehicle image in gasoline-hybrid electric vehicles.* Paper presented at the 21st worldwide battery, hybrid, and fuel cell electric vehicle symposium and exhibition, Monaco: conference proceedings.

Heffner, R.R., Turrentine, T.S. & Kurani, K.S. (2006). *A primer on automobile semiotics.* Davis, USA: Institute of Transportation Studies, University of California,.

Hegger, D. (2007). *Greening sanitary systems: an end-user perspective.* Wageningen, the Netherlands: Wageningen University.

Heiskanen, E. & Pantzar, M. (1997). Towards sustainable consumption: two new perspectives. *Journal of Consumer Policy*, 20: 409-442.

Heliview (2008). *Fleetmanagement 2008.* Benchmarkanalyse. Breda, the Netherlands: Heliview Research.

Holden, E. (2007). *Achieving sustainable mobility. Everyday and leisure-time travel in the EU.* Aldershot, UK: Ashgate.

Holloway, L. & Hubbard, P. (2001). *People and place. The extraordinary geographies of everyday life.* Harlow, UK: Pearson Education Limited.

Hoogma, R. & Schot, J. (2001). How innovative are users? A critique of learning-by-doing and -using. In: R. Coombs, K. Green, V. Walsh & A. Richards (eds.), *Technology and the market. Demand, users and innovation* (pp. 216-233). Cheltenham, UK: Edward Elgar.

Hoogma, R., Kemp, R., Schot, J. & Truffer, B. (2002). *Experimenting for sustainable transport. The approach of strategic niche management.* London, UK and New York, USA: Sponn Press.

Ilmonen, K. (2001). Sociology, consumption, and routine. In: J. Gronow & A. Warde. (eds.), *Ordinary consumption* (pp. 9-24). London, UK: Routledge.

Insnet (2007). *Duurzaamheidsmonitor 2007. Deelrapport Universiteit Wageningen.* Available at: http://www.insnet.org/nl/duurzaamheidmonitor/smv07.html.

Jackson, T. (2005). *Motivating sustainable consumption. A review of evidence on consumer behaviour and behavioural change.* Guildford, UK: Centre for environmental strategy, University of Surrey.

Jackson, T. (2006). *Sustainable consumption.* London, UK: Earthscan.

Jeekel, H. (2011). *De auto-afhankelijke samenleving.* Delft, the Netherlands: Eburon Academic Publishers.

Jensen, M. (1999). Passion and heart in transport – a sociological analysis on transport behaviour. *Transport Policy,* 6: 19-33.

Jensen, M. (2003). *Accelerating mobility and the environment.* Paper for the 6th European conference of sociology. Murcia, Spain: conference proceedings.

Jensen, M. (2006). Environment, mobility and the acceleration of time: a sociological analysis of transport flows in modern life. In: G. Spaargaren, A. Mol & F.H. Buttel (eds.), *Governing environmental flows: global challenges to social theory* (pp. 327-350). Cambridge, USA: MIT Press.

Johansson-Stenman, O. & Martinsson, P. (2006). Honestly, why are you driving a BMW? *Journal of Economic Behavior & Organization,* 60: 129-146.

Kahn Ribeiro, S., Kobayashi, S., Beuthe, M., Gasca, J., Greene, D., Lee, D.S., Muromachi, Y., Newton, P.J., Plotkin, S., Sperling, D., Wit, R. & Zhou, P.J. (2007). Transport and its infrastructure. In: B. Metz, O.R. Davidson, P.R. Bosch, R. Dave & L.A. Meyer (eds.), *Climate change 2007: mitigation* (pp. 323-386). Contribution of working group III to the fourth assessment report of the Intergovernmental Panel on Climate Change. Cambridge, UK: Cambridge University Press.

Katteler, H. & Roosen, J. (1989). *Vervangbaarheid van het autogebruik. Een onderzoek naar gebondenheid aan de auto.* Nijmegen, the Netherlands: Instituut voor Toegepaste Sociale wetenschappen.

Kaufmann, V. (2002). *Re-thinking mobility.* Burlington, UK: Ashgate.

Kaufmann, V. (2004). Motility: a key notion to analyse the social structure of second modernity? In: W. Bonβ, S. Kesselring and G. Vogl (eds.), *Mobility and the cosmopolitan perspective* (pp. 75-82). A workshop at the Reflexive Modernisation Research Centre. Neubiberg/Munich, Germany: Sonderforschungsbereich 536.

Kaufmann, V., Bergman, M.M. & Joye, D. (2004). Motility: mobility as capital. *International Journal of Urban and Regional Research,* 28: 745-756.

Kemp, R. & Loorbach, D. (2006). Transition management: a reflexive governance approach. In: J. Voss, R. Kemp & D. Bauknecht (eds.), *Reflexive governance* (pp.103-130). Cheltenham, UK: Edward Elgar.

Kemp, R. & Rotmans, J. (2004). Managing the transition to sustainable mobility. In: B. Elzen, F. Geels & K. Green (eds.), *System innovation and the transition to sustainability: theory, evidence and policy* (pp. 137-167). Camberley, UK: Edward Elgar Publishers.

Kemp, R., Geels, F.W. & Dudley, G. (2012). Introduction: sustainability transitions in the automobility regime and the need for a new perspective. In: F.W. Geels, R. Kemp, G. Dudley & G. Lyons (eds.), *Automobility in transition? A socio-technical analysis of sustainable transport* (pp. 3-28). London, UK: Routledge.

Kenyon, S. & Lyons, G. (2003). The value of integrated multimodal traveller information and its potential contribution to modal change. *Transportation Research Part F,* 6: 1-21.

KiM (2007). *Beleving en beeldvorming van mobiliteit.* Den Haag, the Netherlands: Kennisinstituut voor Mobiliteitsbeleid.

KiM (2010). *Mobiliteitsbalans 2010.* Den Haag, the Netherlands: Kennisinstituut voor Mobiliteitsbeleid.

Klein, L.R. and Ford, G.T. (2003). Consumer search for information in the digital age: an empirical study of prepurchase search for automobiles. *Journal for Interactive Marketing,* 17: 29-49.

Klein, N. (2000). *No logo.* London, UK: Harper Collins.

Koens, J.F. & Nijhuis, J.O. (2006). Het *digipanel over autogebruik en het kopen van een auto.* Utrecht, the Netherlands: Milieu Centraal.

KpVV (2007). *Mobiliteitsmanagement definitie, toepassingen, maatregelen en checklists.* Rotterdam, the Netherlands: Kennisplatform Verkeer en Vervoer.

Kurani, K.S. & Turrentine, T.S. (2002). Household adaptations to new personal transport options. Constraints and opportunities in household activity spaces. In: H.S. Mahmassani (ed.), *In perpetual motion. Travel behaviour research opportunities and application challenges* (pp. 43-69). Oxford, UK: Pergamon Press.

Lambert-Pandraud, R., Laurent, G. & Lapersonne, E. (2005). Repeat purchasing of new automobiles by older consumers: empirical evidence and interpretations. *Journal of Marketing*, 69: 97-113.

Lane, B. (2005). *Car buyer research report. Consumer attitudes to low carbon and fuel-efficient cars.* Bristol, UK: Ecolane Transport Consultancy.

Loeber, A. (2003). *Inbreken in het gangbare. Transitiemanagement in de praktijk: de NIDO-benadering.* Leeuwarden, the Netherlands: NIDO.

Loorbach, D. (2007). *Transition management. A new mode of governance for sustainable development.* Rotterdam, the Netherlands: Erasmus University Rotterdam.

Lury, C. (1996). *Consumer culture.* New Brunswick, New York, USA: Rutgers University Press.

Lury, C. (2004). *Brands.* The logos of the global economy. New York, USA: Routledge.

Lyons, G. & Urry, J. (2005). Travel time use in the information age. *Transportation Research Part A – Policy and Practice*, 39: 257-276.

Mackett, R.L. (2003). Why do people use their cars for short trips? *Transportation*, 30: 329-349.

Macnaghten, P. (2003). Embodying the environment in everyday life practices. *The Sociological Review*, 51: 63-84.

Makimoto, T. & Manners, D. (1997). *Digital nomad.* Hoboken, USA: Wiley.

Martens, S & Spaargaren, G. (2006). The politics of sustainable consumption: the case of the Netherlands. In: T. Jackson (ed.), *Sustainable consumption* (pp. 197-221). London, UK: Earthscan.

Maslow, A.H. (1954). *Motivation and personality.* New York, USA: Harper and Row.

Miller, D. (ed.) (2001). *Car cultures.* Oxford, UK/New York, USA: Berg.

Ministerie van Verkeer en Waterstaat en Ministerie van Volkshuisvesting Ruimtelijke Ordingen en Milieubeheer (2004). *Nota mobiliteit. Naar een betrouwbare en voorspelbare bereikbaarheid.* Den Haag, the Netherlands: Staatsuitgeverij.

Ministerie van Verkeer en Waterstaat (1990). *Tweede structuurschema verkeer en vervoer (SVV2).* Den Haag, the Netherlands: Staatsuitgeverij.

Ministerie van Verkeer en Waterstaat (1996). *Samen werken aan bereikbaarheid (SWAB).* Den Haag, the Netherlands: Staatsuitgeverij.

Ministerie van Verkeer en Waterstaat (2000). *Nationaal verkeers- en vervoersplan (NVVP).* Den Haag, the Netherlands: Staatsuitgeverij.

Ministerie van Verkeer en Waterstaat (2002). *Nota mobiliteitsmanagement.* Den Haag, the Netherlands: Staatsuitgeverij.

MNP (2006). *Milieubalans 2006.* Bilthoven, the Netherlands: Milieu- en Natuurplanbureau.

MNP (2007a). *Milieubalans 2007.* Bilthoven, the Netherlands: Milieu- en Natuurplanbureau.

MNP (2007b). *Nederland en een duurzame wereld. Armoede, klimaat en biodiversiteit. Tweede duurzaamheidsverkenning.* Bilthoven, the Netherlands: Milieu en Natuur Planbureau.

Mokhtarian, P.L. & Salomon, I. (2001). How derived is the demand for travel? Some conceptual and measurement considerations. *Transportation Research Part A*, 35: 695-719.

Mol, A.P.J. & Spaargaren, G. (2000). Ecological modernisation theory in debate: a review. *Environmental Politics*, 9: 17-50.

Mol, A.P.J. (1995). *The refinement of production. Ecological modernisation and the chemical industry.* Utrecht, the Netherlands: Van Arkel.

Mol, A.P.J. (2005). Environment in the information age. The transformative powers of environmental information. In: CGEE. *Ciência, Technologia e Sociedade. Novos Modelos de Governanca* (pp. 99-134). Brasília, Brazil: CGEE.

Mom, G., Staal, P. & Schot, J. (2002). De beschaving van het gemotoriseerde avontuur. ANWB en KNAC als wegbereiders bij de inburgering van de auto in Nederland. *Tijdschrift voor de Sociale Geschiedenis*, 3: 323-346.

Mommaas, J.T. (2000). De culturele industrie in het tijdperk van de netwerkeconomie. *Boekmancahier*, 12: 26-45.

Mont, O. & Emtairah, T. (2006). *Systemic changes for sustainable consumption and production.* Proceedings: perspectives on radical changes to sustainable consumption and production. Workshop of the sustainable consumption research exchange network. Copenhagen, Denmark: conference proceedings.

Mont, O. & Plepys, A. (2003). *Customer satisfaction: review of literature and application to the product-service systems*. Lund, Sweden: International Institute for Industrial Environmental Economics.

Munters, Q.J., Meijer E., Mommaas H., Van der Poel, H., Rosendal R. & Spaargaren, G. (eds.) (1993). *Anthony Giddens: een kennismaking met de structuratietheorie*. Wageningen, the Netherlands: Landbouwuniversiteit Wageningen.

NIDO (2002). *Informatie over duurzaamheid: een zoektocht*. Leeuwarden, the Netherlands: NIDO.

Nieuwenhuis, P. & Wells, P. (1997). *The death of motoring?* Hoboken, USA: Wiley.

Nijhuis, J.O. & Van den Burg, S. (2009). Consumer-oriented strategies for car purchases: an analysis of environmental information tools and taxation schemes in the Netherlands. In: T. Geerken & M. Borup (eds.). System innovation for sustainability 2. Case studies in sustainable consumption and production – Mobility (pp. 90-108). Sheffield, UK: Greenleaf Publishing.

Nordlund, A.M. & Garvill, J. (2003). Effects of values, problem awareness, and personal norm on willingness to reduce personal car use. *Journal of Environmental Psychology*, 23: 339-347.

Nye, M. & Hargreaves, T. (2008). *The intrinsic role of context in negotiating sustainable behaviour: a comparative study of intervention processes at home and work*. Norwich, UK: School of Environmental Sciences, University of East Anglia.

Nykvist, B. & Whitmarsh, L. (2007). *Identifying opportunities and pathways for transitions to sustainable transport in Sweden and the UK*. Matisse Working Paper. Available at: http://www.matisse-project.net/projectcomm/uploads/tx_article/Working_Paper_7_01.pdf.

OECD (1996). *Towards sustainable transportation*. Proceedings of the Vancouver Conference, 24-27 March, 1996. Vancouver, Canada: conference proceedings.

Ornetzeder, M., Hertwich, E.G., Hubacek, K., Korytarova, K. & Haas, W. (2008). The environmental effect of car-free housing: a case in Vienna. *Ecological Economics*, 65: 526-530.

Orsato, R.J. & Clegg, S.R. (2005). Radical reformism: towards critical ecological modernization. *Sustainable Development*, 13: 253-267.

Ory, D.T. & Mokhtarian, P.L. (2005). When is getting there half the fun? Modeling the liking for travel. *Transportation Research Part A*, 39: 97-123.

Otnes, P. (ed.) (1988), *The sociology of consumption. An anthology.* New Jersey, USA: Humanities Press International.

Oudshoorn, N. & Pinch, T. (2003) (eds.). *How users matter. The co-construction of users and technology*. Cambridge, USA: MIT Press.

Pallant, J. (2005). *SPSS survival manual: a step by step guide to data analysis using SPSS for Windows*. Maidenhead, UK: Open University Press.

Paredis, E. (2009). *Socio-technische systeeminnovaties en transities: van theoretische inzichten naar beleidsvertaling.* Gent, Belgium: Centrum voor Duurzame Ontwikkeling – Universiteit Gent.

PBL (2009). *Energielabels en autokeuze: effect van het energielabel op de aanschaf van nieuwe personenauto's door consumenten*. Bilthoven, the Netherlands: Planbureau voor de Leefomgeving.

PBL (2013). *Wissels omzetten. Bouwstenen voor een robuust milieubeleid voor de 21ste eeuw*. Den Haag, the Netherlands: Planbureau voor de Leefomgeving.

Peters, P. (1998). De smalle marges van de politiek. In: H. Achterhuis en B. Elzen (ed.), *Cultuur en mobiliteit* (38-63). Den Haag, the Netherlands: Staatsuitgeverij.

Peters, P. (1999). *In de praktijk. Naar een andere conceptualisering van verplaatsingen*. Anticiperend onderzoek in opdracht van het projectbureau integrale verkeers- en vervoerstudies. Maastricht, the Netherlands: Universiteit van Maastricht.

Peters, P. (2003). *De haast van Albertine. Reizen in de technologische cultuur: naar een theorie van passages*. Amsterdam, the Netherlands: Uitgeverij De Balie.

Peters, P. (2004). *Roadside wilderness. US national parks design in the 1950s and 1960s*. Paper presented at the alternative mobility futures conference session: tourist mobilities. Lancaster, UK: Lancaster University.

PODO II (2004). *Verhandelbare mobiliteitsrechten: haalbaarheid, socio-economische effectiviteit en maatschappelijk draagvlak*. Brussel, Belgium: Plan voor wetenschappelijke ondersteuning van een beleid gericht op duurzame ontwikkeling.

Pooley, C.G., Turnbull, J. & Adams, M. (2005). *A mobile century? Changes in everyday mobility in Britain in the twentieth century*. Aldershot, UK: Ashgate.

Poortinga, W., Steg, L. & Vlek, C. (2004). Values, environmental concern, and environmental behavior. A study into household energy use. *Environment and Behavior,* 36: 70-93.

Pred, A. (1981). Social reproduction and the time-geography of everyday life. *Geografiska Annaler. Series B, Human Geography:* 63, 5-22.

Putman, L & Nijhuis, J. (2006). Burger-consumenten in transities naar duurzame productie en consumptie. In: S. Van den Burg, R. Van der Ham, & J. Grin (eds.). *Beleid in transities. SWOME/KSI Marktdag 2006* (pp. 51-60). Wageningen, the Netherlands: Wageningen Universiteit.

Raad voor Verkeer en Waterstaat (2006). *Een prijs voor elke reis. Een beleidsstrategie voor CO_2-reductie in verkeer en vervoer.* Den Haag, the Netherlands: Gezamenlijk advies van de Raad voor Verkeer en Waterstaat, de VROM-raad en de Algemene Energieraad.

Rabobank (2008). *Maatschappelijk jaarverslag 2007*. Utrecht, the Netherlands: Rabobank Groep.

Rabobank (2010). *Maatschappelijk Verantwoord Ondernemen 2010*. Utrecht, the Netherlands: Rabobank Groep.

Reckwitz, A (2002b). Toward a theory of social practices. A development in culturalist theorizing. *European Journal of Social Theory*, 5: 243-263.

Reckwitz, A. (2002a). The status of the 'material' in theories of culture. From 'social structure' to 'artefacts'. *Journal for the Theory of Social Behaviour*, 32: 195-217.

Redshaw, S. (2007). Articulations of the car: the dominant articulations of racing and rally driving. *Mobilities*, 2: 121-141.

Reed, G., Story, V. & Saker, J. (2004). Information technology: changing the face of automotive retailing? *International Journal of Retail and Distribution Management,* 32: 19-32.

Rifkin, J. (2001). *The age of access. The new culture of hypercapitalism where all of life is a paid-for experience*. New York, USA: Tarcher/Putnam.

RIVM (1988). *Zorgen voor morgen*. Bilthoven, the Netherlands: Rijksinstituut voor Volksgezondheid en Milieu.

RIVM (2004). *Maatschappelijke waardering van duurzame ontwikkeling. Achtergrondrapport bij de duurzaamheidsverkenning*. Bilthoven, the Netherlands: Rijksinstituut voor Volksgezondheid en Milieu.

Rosa, H. (2003). Social acceleration: ethical and political consequences of a desynchronized high-speed society. *Constellations*, 10: 3-33.

Rotmans, J. (2003). *Transities en transitiemanagement. Sleutel voor een duurzame samenleving*. Assen, the Netherlands: Koninklijke Van Gorcum BV.

Rotmans, J. (2007). *Duurzaamheid: van onderstroom naar draaggolf. Op de rand van een doorbraak*. Rotterdam, the Netherlands: Drift, Erasmus Universiteit Rotterdam.

Rotmans, J., Kemp, R. & Van Asselt, M. (2001). More evolution than revolution. Transition management in public policy. *The Journal of Future Studies, Strategic Thinking and Policy,* 3: 15-31.

RWS-DVS (2009). *Evaluatie mobiliteitsbeïnvloedende maatregelen A6 Hollandse Brug*. Delft, the Netherlands: Rijkswaterstaat Dienst Verkeer en Scheepvaart.

Sachs, W. (1983). Are energy-intensive life-images fading? The cultural meaning of the automobile in transition. *Journal of Economic Psychology*, 3: 347-365.

Sachs, W. (1984). *For the love of the automobile. Looking back into the history of our desires*. Berkeley, USA: University of California Press.

Sahlins, M. (1976). *Culture and practical reason*. Chicago, USA: University of Chicago Press.

Schade, J. & Schlag, B. (2003). Acceptability of urban pricing strategies. *Transportation Research Part F: Traffic Psychology and Behaviour,* 6: 45-61.

Schatzki, T. (1996). *Social practices: a Wittgensteinian approach to human activity and the social*. Cambridge, UK: Cambridge University Press.

Scheurer, J. (2001). *Urban ecology, innovations in housing policy and the future of cities: towards sustainability in neighbourhood communities*. Perth, Australia: Murdoch University.

Schipper, F., Poot, G. & Engelaar, C. (1998). *What is the value of our car? Positioning brands and models based on the value systems of car buyers*. 5th international automotive marketing conference. Lausanne, Switzerland: European Society for Opinion and Marketing Research.

Schnaiberg, A., Pellow, D.N. & Weinberg, A. (2002). The treadmill of production and the environmental state. In: A.P.J. Mol & F. Buttel (eds.), *The environmental state under pressure* (15-32). Oxford, UK: Elsevier Science.

Schot, J. & De la Bruheze, A.A. (2003). The mediated design of products, consumption, and consumers in the twentieth century. In: N.T. Oudshoorn & T. Pinch (eds.), *How users matter. The co-construction of users and technology* (pp. 229-246). Cambridge, USA: MIT Press.

Schot, J., Hoogma, R. & Elzen, B. (1994). Strategies for shifting technological systems. The case of the automobile system. *Futures,* 26: 1060-1076.

Schot, J.W. & Geels, F.W. (2008). Strategic niche management and sustainable innovation journeys: Theory, findings, research agenda, and policy. *Technology Analysis & Strategic Management,* 20: 537-554.

Schwanen, T. & Dijst, M. (2003). Time windows in workers' activity patterns: empirical evidence from the Netherlands. *Transportation,* 30: 261-283.

Schwartz, S. (1977). Normative influences on altruism. *Advances in Experimental Social Psychology*, 10: 222-279.

Schwartz-Cowan, R. (1987). The consumption junction: a proposal for research strategies in the sociology of technology. In: W.E. Bijker, T.P. Hughes, & T.J. Pinch (eds.), *The social construction of technological systems: new direction in the sociology and history of technology* (253-272). Cambridge, USA: MIT Press.

SER (1994). Vervolgadvies Nationaal Milieubeleidsplan 2: verkeer en vervoer. Den Haag, the Netherlands: Sociaal-Economische Raad.

SER (2003). *Duurzaamheid vraagt om openheid. Op weg naar een duurzame consumptie*. Den Haag, the Netherlands: Sociaal-Economische Raad.

SER (2006). *Mobiliteitsmanangement*. Den Haag, the Netherlands: Sociaal-Economische Raad.

Sheller, M. & Urry, J. (2000). The city and the car. *International Journal of Urban and Regional Research*, 24:737-757.

Sheller, M. (2004). Automotive emotions. Feeling the car. *Theory, Culture and Society*, 21: 221-242.

Sheller, M. (2012). The emergence of new cultures of mobility. Stability, openings and prospects. In: F.W. Geels, R. Kemp, G. Dudley & G. Lyons (2012). *Automobility in transition? A socio-technical analysis of sustainable transport* (pp. 180-202). Lodnon, UK: Routledge.

Shove, E. & Pantzar, M. (2005). Consumer, producers and practices. Understanding the invention and reinvention of Nordic walking. *Journal of Consumer Culture*, 5: 43-64.

Shove, E. & Walker, G. (2010). CAUTION! Transitions ahead: politics, practice, and sustainable transition management. *Environment and Planning A*, 39: 763-770.

Shove, E. & Warde, A. (1998). *Inconspicuous consumption: the sociology of consumption and the environment*. Lancaster, UK: University of Lancaster.

Shove, E. (1998). *Consuming automobility. A discussion paper*. Project SceneSusTech. Dublin, Ireland: Trinity College Dublin.

Shove, E. (2002). *Rushing around: coordination, mobility and inequality*. Draft paper for the mobile network meeting, October 2002. Lancaster, UK: University of Lancaster.

Shove, E. (2003a). *Comfort, cleanliness and convenience; the social organization of normality*. Oxford, UK: Berg.

Shove, E. (2003b). Converging conventions of comfort, cleanliness and convenience. *Journal of Consumer Policy*, 26: 395-418.

Shove, E. (2003c). *Changing human behaviour and lifestyle: a challenge for sustainable consumption?* Lancaster, UK: University of Lancaster.

Shove, E., Pantzar, M. & Watson, M. (2012). *The dynamics of social practice. Everyday life and how it changes*. London, UK: Sage.

Shove, E., Watson, M. & Ingram, J. (2005). *Products and practices*. Paper submitted to the Nordic design research conference. Copenhagen, Denmark: conference proceedings.

Slater, D. (1997). *Consumer culture and modernity*. Cambridge, UK: Polity Press.

Smink, C.K., Van Koppen, C.S.A. & Spaargaren, G. (2003). Ecological modernisation theory and the changing dynamics of the European automotive industry: the case of Dutch end-of-life vehicle policies. *International Journal of Environment and Sustainable Development*, 2: 284-304.

Smith, A., Stirling, A. & Berkhout, F. (2005). The governance of sustainable socio-technical transitions. *Research Policy*, 34: 1491-1510.

Soldaat, K. (2007). *Bewonersgedrag en balansventilatie. De invloed van bewonersgedrag op de effectiviteit van balansventilatie*. Gouda, the Netherlands: Habiforum.

Spaargaren, G. & Martens, S. (2004), Globalization and the role of citizen-consumers in environmental politics. In: F. Wijen, K. Zoeteman, & J. Pieters (eds.), *A handbook of globalization and environmental policy* (pp.211-245). Cheltenham, UK: Edward Elgar.

Spaargaren, G. & Martens, S. (2005), Consumption domains in transition. The social practices approach as a tool for analyzing sustainability in everyday life. In: S. Van den Burg, G. Spaargaren & H. Waaijers. *Wetenschap met beleid, beleid met wetenschap* (pp. 100-107). Swome/GaMON Marktdag 2005. Wageningen, the Netherlands: Wageningen Universiteit.

Spaargaren, G. & Oosterveer, P. (2010). Citizen-consumers as agents of change in globalizing modernity: the case of sustainable consumption. *Sustainability*, 7: 1887-1908.

Spaargaren, G. & Van Koppen, C.S.A. (2009). Provider strategies and the greening of consumption practices. In: H. Lange, & L. Meier (eds), *The New Middle Classes. Globalising lifestyles, consumerism, and environmental concern* (pp. 81-100). Heidelberg, Germany: Springer Verlag.

Spaargaren, G. & Van Vliet, B. (2000). Lifestyles, consumption and the environment; the ecological modernisation of domestic consumption. *Environmental Politics*, 9: 50-76.

Spaargaren, G. (1997). *The ecological modernisation of production and consumption. Essays in environmental sociology.* Wageningen, the Netherlands: Wageningen University.

Spaargaren, G. (2003a). Sustainable consumption: a theoretical and environmental policy perspective. *Society and Natural Resources*, 16: 687-702.

Spaargaren, G. (2003b). Duurzaam consumeren of ecologisch burgerschap. In: H. Dagevos, H. & L. Sterrenberg (eds.), *Burgers en consumenten: tussen tweedeling en twee-eenheid* (pp. 70-84). Wageningen, the Netherlands: Wageningen Academic Publishers.

Spaargaren, G. (2005) Political consumerism for sustainable consumption practices: rethinking the commitments of citizen-consumers with environmental change. In: CGEE. *Ciência, technologia e sociedade. Novos modelos de governanca* (pp. 135-168). Brasília, Brazil: CGEE.

Spaargaren, G. (2006). *The ecological modernisation of social practices at the consumption junction.* Discussion-paper for the ISA-RC24 conference. Madison, USA: conference proceedings.

Spaargaren, G. (2011). Theories of practices: agency, technology, and culture. Exploring the relevance of practice theories for the governance of sustainable consumption practices in the new world-order. *Global Environmental Change: Human and Policy Dimensions*, 21: 813-822.

Spaargaren, G., Beckers, T., Martens, S., Bargeman, B., & Van Es, T. (2002). *Gedragspraktijken in transitie. De gedragspraktijkenbenadering getoetst in twee gevallen: duurzaam wonen en duurzame toeristische mobiliteit.* Den Haag, the Netherlands: Ministerie van VROM. Publicatiereeks milieustrategie 2002/1.

Spaargaren, G., Mommaas, H., Van den Burg, S., Maas, L., Drissen, E., Dagevos, H., Bargeman, B., Putman, L., Nijhuis, J., Verbeek, D., & Sargant, E. (2007a). *More sustainable lifestyles and consumption patterns. A theoretical perspective for the analysis of transition processes within consumption domains.* Wageningen, the Netherlands: WUR/Telos/MNP/LEI.

Spaargaren, G., Munsters, Q.J. & Hendriksen, A. (1995). *Detailhandel, consument en milieu-advisering: een onderzoek naar de milieu-voorlichtende rol van het midden- en kleinbedrijf bij de aanschaf van watergedragen verven, spaarlampen en CFK-vrije koelkasten.* Wageningen, the Netherlands: Universiteit Wageningen.

Staats, H (2003). Understanding pro-environmental attitudes and behaviour: an analysis and review of research based on the theory of planned behaviour. In: M. Bonnes, T. Lee & M. Bonaiuto (eds), *Psychological Theories for Environmental Issues* (pp. 171-202). Aldershot, UK: Ashgate.

Steg, L. & Gifford, R. (2005). Sustainable transport and quality of life. *Journal of Transport Geography*, 13: 59-69.

Steg, L. (1999). *Verspilde energie? Wat doen en laten Nederlanders voor het milieu.* Den Haag, the Netherlands: Sociaal en Cultureel Planbureau.

Steg, L. (2005). Car use: lust and must. Instrumental, symbolic and affective motives for car use. *Transportation Research Part A*, 39: 147-162.

Stern, P. (2000). Toward a coherent theory of environmentally significant behavior. *Journal of Social Issues*, 56: 407-424.

Stienstra, J. & Jansen, A. (2001). *Evaluatie energie-etiket.* Amsterdam, the Netherlands: Bureau Ferro.

Stock, M. & Duhamel, P. (2005). A practice-based approach to the conceptualisation of geographical mobility. *Belgeo*, 1-2: 59-68.

Stones, R. (2005). *Structuration theory (traditions in social theory).* New York, USA: Palgrave Macmillan.

Storey, J. (1999). *Cultural consumption and everyday life.* London, UK: Arnold.

T&E (2008). *Can you hear us? Why it is finally time for the EU to tackle the problem of noise from road and rail traffic.* Brussels, Belgium: European Federation for Transport and Environment.

T&E (2010). *How clean are Europe's cars? An analysis of carmaker progress towards EU CO_2 targets in 2009*. Brussels, Belgium: European Federation for Transport and Environment.

Teisl, M.F., Noblet, C.L. & Rubin, J. (2007). The design of an eco-marketing and labelling program for vehicles in Maine. In: U. Grote, A.K. Basu, & N.H. Chau (eds.), *New frontiers in environmental and social labeling* (pp.11-35). Heidelberg, Germany: Springer.

Thøgersen, J. &. Møller, B. (2008). Breaking car use habits: the effectiveness of a free one-month travelcard. *Transportation*, 35: 329-345.

Thøgersen, J. (2000). Psychological determinants of paying attention to eco-labels in purchase decisions: model development and multinational validation. *Journal of Consumer Policy*, 23: 285-313.

Thøgersen, J. (2002). Promoting green consumer behavior with eco-labels. In: T. Dietz, & P. Stern (eds.), *New tools for environmental protection: education, information, and voluntary measures* (pp. 83-104). Washington DC, USA: National Academic Press.

Thøgersen, J. (2005). How may consumer policy empower consumers for sustainable lifestyles? *Journal of Consumer Policy*, 28: 143-178.

Thomson, T.U., Nielsen, L.D. & Gudmundsson, H. (eds.) (2005). *Social perspectives on mobility*. Aldershot, UK: Ashgate.

Timmer, J. (1998). Chroom en charisma. Een sociologisch essay over de auto. *Mens en Maatschappij*, 73: 157-175.

TNO (2006). *Review and analysis of the reduction potential and costs of technological and other measures to reduce CO_2-emissions from passenger cars*. Delft, the Netherlands: TNO.

TNS Emnid (2004). *Internet gewinnt bei der fahrzeugvermittlung weiter an bedeutung*. Bielefeld, Germany: TNS Emnid.

TTR (2002). *The impact of sustainable transport policies on the travel behaviour of shoppers*. Lichfield, UK: Transport and Travel Research.

Urry, J. (1999). *Automobility, car culture and weightless travel. A discussion paper*. Project SceneSusTech. Dublin, Ireland: Trinity College.

Urry, J. (2000). *Sociology beyond societies. Mobilities for the twenty-first century*. London, UK: Routledge.

Urry, J. (2002). Mobility and proximity. *Sociology*, 36: 255-274.

Urry, J. (2003a). *Global complexity*. Cambridge, UK: Polity press.

Urry, J. (2003b). Social networks, travel and talk. *British Journal of Sociology*, 54: 155-175.

Urry, J. (2004). The 'system' of automobility. *Theory, Culture and Society*, 21: 25-39.

Urry, J. (2006). Inhabiting the car. *The Sociological Review*, 54: 17-31.

Urry, J. (2007). *Mobilities*. Cambridge, UK: Polity Press.

Van Beynen de Hoog, P. & Brookhuis, K. (2005). *Winkelen en vervoerwijzekeuze. Een data-analyse.*, Antwerpen, Belgium: conference proceedings of Colloquium Vervoersplanologisch Speurwerk.

Van den Brink, R.M.M. & Van Wee, B. (2001). Why has car-fleet specific fuel consumption not shown any decrease since 1990? Quantitative analysis of Dutch passenger car-fleet specific fuel consumption. *Transportation Research Part D*, 6: 75-93.

Van den Brink, R.M.M., Hoen, A., Van den Wijngaart, R.A., Geilenkirchen, G.P., Geurs, K.T., Drissen, E., & Olivier, J.G.J. (2007). *Beoordeling van milieumaatregelen in het Belastingplan 2008*. Bilthoven, the Netherlands: Milieu- en Natuurplanbureau (MNP).

Van den Burg, S. (2006). *Governance through information. Environmental monitoring from a citizen-consumer perspective*. Wageningen, the Netherlands: Wageningen University.

Van Meegeren, P. (1997). *Communicatie en maatschappelijke acceptatie van milieubeleid. Een onderzoek naar de houding ten aanzien van de 'dure afvalzak' in Barendrecht*. Wageningen, the Netherlands: Wageningen Universiteit.

Van Soest, J.P. (2007). Stoppen met gedragsbeïnvloeding. *Stromen*, 20: 6.

Van Vliet, B. (2002). *Greening the grid. The ecological modernisation of network-bound systems*. Wageningen, the Netherlands: Wageningen University.

Van Wee, B. & Dijst, M. (2002) (red.). *Verkeer en Vervoer in Hoofdlijnen*. Bussum, the Netherlands: Coutinho.

Van West (2004). *Transport and Kyoto protocol. Consumer information*. Presentation given for Gas Natural Foundation, Bilbao. Available at: http://www.fiafoundation.org/Documents/Environment/bilbao_co2_presentation_1104.ppt.

Vannini, P. (ed.) (2009). *The cultures of alternative mobilities: the routes less travelled*. Aldershot, UK: Ashgate.

Veblen, T. (1899). *The theory of the leisure class. An economic study of institutions*. New York, USA: Macmillan.

Verbeek P. & Slob, A. (eds.) (2006). *User behavior and technology design – shaping sustainable relations between consumers and technologies*. Dordrecht, the Netherlands: Springer.

Verbeek, A. (2007). *Met een djellaba aan kan je niet fietsen*. Wageningen, the Netherlands: Wageningen Universtiteit.

Verbeek, D. (2009). *Sustainable tourism mobilities: a practice approach*. Tilburg, the Netherlands: Tilburg University.

Verplanken, B. & Wood, W. (2006). Interventions to break and create consumer habits. *Journal of Public Policy and Marketing*, 12: 90-103.

Verplanken, B. (2006). Beyond frequency: habit as mental construct. *British Journal of Social Psychology*, 45: 639-656.

Vigar, G. (2000). Local 'barriers' to environmentally sustainable transport planning. *Local Environment*, 5: 19-32.

Vigar, G. (2001). Reappraising UK transport policy 1950-99: the myth of 'mono-modality' and the nature of 'paradigm shifts'. *Planning Perspectives*, 16: 269-291.

Vigar, G. (2002). *The politics of mobility. Transport, the environment and public policy*. London. UK: Spon Press.

Visser, H., Aalbers, T.G. & Vringer, K. (2007). *How Dutch citizens prioritize the social agenda. An analysis of the 2003, 2005 and 2006 Surveys*. Bilthoven, the Netherlands: Milieu- en Natuur Planbureau.

Vlek, C., Reisch, L., & Scherhorn, G. (1999). Transformation of unsustainable consumer behaviour and consumer policies. Problem analysis, solution approaches and a research agenda. In: P. Vellinga (ed.), *Industrial transformation project. Research approaches to support the industrial transformation science plan* (pp. 59-114). Amsterdam, the Netherlands: Institute for environmental studies.

Vogl, G. (2004). *Mobility between first and second modernity*. Paris, France: conference proceedings of the 4S & EASST Conference.

Vringer, K. (2005). *Analysis of the energy requirement of household consumption*. Utrecht, the Netherlands: Universiteit Utrecht.

Vringer, K. (2007). *Analysis of the energy requirement for household consumption*. Bilthoven, the Netherlands: RIVM/MNP.

Vringer, K., Aalbers, T. & Blok, K. (2007). Household energy requirements and value patterns. *Energy Policy*, 35: 553-566.

VROM-raad (2005). *Milieu en de kunst van het goede Leven. Advies voor de toekomstagenda Milieu*. Den Haag, the Netherlands: VROM-raad.

Walrave, M. & De Bie, M. (2005). *Teleworking home or close to work. Attitudes towards and experiences with homeworking, mobile working, working in satellite offices and telecentres*. Brussels, Belgium: ESF Agentschap.

Warde, A. (2005). Consumption and theories of practice. *Journal of Consumer Culture*, 5: 131-153.

WBCSD (2001). *Mobility 2001. World mobility at the end of the twentieth century and its sustainability*. World Business Council for Sustainable Development. Available at: http://www.wbcsd.org/web/projects/mobility/english_full_report.pdf.

WBCSD (2004). Mobility 2030. *Meeting the challenges to sustainability*. World Business Council for Sustainable Development. Available at: http://www.wbcsd.org/web/publications/mobility/mobility-full.pdf.

Weterings, R. (2010). *Werk in uitvoering. Ervaringen met het Nederlandse transitiebeleid*. Utrecht, the Netherlands: Compententiecentrum Transitie.

Whittles, M.J. (2003). *Urban road pricing: public and political acceptability*. Aldershot, UK: Ashgate.

Wickham, J. & Lohan, M. (1999). *The social shaping of European urban car systems*. Project SceneSusTech. Dublin, Ireland: Trinity College.

World Watch Institute (2004). *State of the world 2004. Special focus: the consumer society*. Washington D.C, USA: World Watch Institute.

Zemp, S. (2002). *From class theory to mobility styles. Applying the lifestyle concept to explain leisure travel behaviour*. Zürich, Switzerland: ETH Nr. 23/04.

Interviews

Frank Versteege, Manager Car Marketing Toyota, Raamsdonkveer
Luuk van der Schoot, Consulant Tchai International
Dick Bakker, Commercial Director Arval Leasing

Appendix A. Environmental developments in the lease market

Fleet Europe, the major information provider for European fleet owners and lease companies published an article in June 2004 to reflect on the green initiatives undertaken by leasing companies. The conclusion was that demand remained very limited. The reactions of Dutch lease companies on the possibilities of introducing an eco-label for green leasing (an exploration issued by the Ministry for Environment) were also very reserved. Some companies cautiously showed some interest and were sympathetic to the initiative but the lease companies did not see a leading role for themselves in promoting green leasing (see also CE, 2004). Three and a half years Fleet Europe re-examined the situation; the renewed conclusion was that the picture has turned much greener as there are numerous green initiatives undertaken by leasing companies (www.fleeteurope.com, 2007). Interesting to note is that the number and impact of environmental initiatives of the leasing branch differs per country; especially the Netherlands and the UK are considered to be frontrunners of environmental lease policies. Table A1 provides an overview of these green lease activities provided by Dutch lease companies; nearly all of these activities have been implemented in the last six years.

Two explanations can be given for the increase of green initiatives in the car leasing branch. First, green leasing is mostly a demand driven development. As has been explained in Chapter 3, most of the larger Western companies consider that firms have a certain social and environmental responsibility towards society. Therefore, under the heading of Corporate Social Responsibility (CSR) increasingly companies focus on green procurement and similarly demand good environmental housekeeping from their suppliers. Moving towards a sustainable car fleet is one of the measures that companies take in the light of CSR. A survey conducted among 2,000 fleet managers in Europe showed that the environment is on the top of the agenda of fleet managers (www.fleeteurope.com, 2007); similarly they expect lease companies to play an active role in meeting their environmental targets. Some of the green activities, such as Ecolease by ING Car Lease and Save Lease by Athlon (part of the Rabobank), are direct CSR spin-offs undertaken by

Table A1. Examples of green initiatives undertaken by Dutch lease companies.

Name of green initiative	Initiated by
GREENLease	Terberg Leasing
Greenplan	Leaseplan
EcoLease	ING Car Lease
SaveLease	Athlon Car Lease
ECOLease	Kroymans Autolease
Schonerleasen	MKB Lease
Natuurlijk leasen	MultiLease
Hybride Lease	Strixt Lease

the mother company. Both companies, which have thousands of lease cars in their fleet, strive to have a car fleet existing of only A, B and C labelled cars within the next few years. This is achieved by replacing current cars only with more energy-efficient cars. Green procurement of the car fleet is therefore an important demand-oriented environmental change in the lease branch.

Second, for lease companies the provision of green initiatives also makes a good business case as green car leasing significantly reduces total cost of ownerships. Clearly, energy-efficient cars use less fuel. Furthermore, the same applies for driving in accordance with the New Driving scheme, a new form of driving which optimises fuel-efficient driving, which also reduces car maintenance costs. For example, SaveLease by Athlon advertises with a 15% reduction in fuel costs.

In the Netherlands, a minor role is also played by Friends of the Earth who issued a responsible car leasing campaign. In 2005, over thirty-seven Dutch companies signed an agreement of intent with Friends of the Earth to start with a responsible car leasing policy (existing of four components: (1) incorporation of leasing prices; (2) lease cars limited to A, B and C labelled cars; (3) diesel particulate filters for new diesel cars; and (4) allowance for public transport usage). Two years later thirty-five of these companies had met, on average, with 75% of the intent.

An eye catching development is the vast increase in the environmental monitoring activities by leasing companies. This monitoring can take place at the level of the leasing company, at the level of the fleet management, or even at the level of the individual driver. Increasingly lease companies and fleet owners monitor the total CO_2-emissions of the car fleet. Monitoring of these emissions allows the setting of emission targets and helps to select specific measures to reach the targets. For example, GREENLease has a special greenscan programme which compares the (costs and CO_2-emissions of the) current car fleet of a customer with a green fleet. Greenplan, provided by Leaseplan, is a reporting and consulting service which identifies a company's fleet carbon footprint. This monitoring is complemented by advisory measures to reduce the carbon emissions and reach the set targets of a company. So far, 40% of the Leaseplan's clients make use of Greenplan services. Arval, each quarter, calculates the running costs per driver and designates a report-mark (1-10) for the driver's performance with the aim to influence the driver's behaviour. Those who excel on all of the monitored elements, as a reward, receive a new bicycle for free. Finally, ING Ecolease enables clients to check on fuel use on line.

Next to environmental monitoring, many of the leasing companies have introduced special green provision programmes; the first of these initiatives in the Netherlands was GREENLease initiated in 2005. GREENLease takes a special three-tier approach: (1) only cars with an A, B and C energy label are offered; (2) all customers receive a training in the New Driving scheme; and (3) the CO_2-emissions are compensated via climate compensation. This initiative has been taken up by several other lease companies as well. Also many lease companies now provide training days for drivers to inform lease customers of the New Driving scheme.

To conclude, in the Netherlands attention and interest of the leasing branch for environmental aspects of car leasing are quickly increasing. Lease companies through various mechanisms are undertaking environmental initiatives; in essence these companies have a facilitating and advisory role in promoting green leasing. Through monitoring leasing companies make information available about the environmental performance of actors and car fleets. Analysis of emissions and fuel consumption is made available to fleet owners and drivers and in some cases complemented with advise on the possibilities of emission reduction.

References

CE (2004). Responsible lease. De haalbaarheid voor een keurmerk voor maatschappelijk verantwoorde autolease. Delft.

www.fleeteurope.com accessed between November 2006 and February 2008.

www.milieudefensie.nl accessed in February 2008.

www.vna-lease.nl accessed in February 2008.

Appendix B. Investigating mobility practices: institutional analysis and the analysis of strategic conduct

The dual methodological strategy of the structuration theory implies that each of the four dimensions stipulated in Figure 5.3 can be analysed both from an institutional perspective and from the perspective of strategic conduct of the agent. Examples of the types of dynamics and processes which can be analysed in this respect are portrayed in Table B1. Though the dynamics and processes described in Table B1 are just an indication and not a complete overview, it does show the interplay between the different elements of mobility practices on the one hand, and the relation between institutional analysis of structural properties and the analysis of strategic conduct on the other.

Table B1. Processes and dynamics between different elements of practices.

	1 Temporal-spatial	**2 Material-symbolic**	**3 Temporal-symbolic**
Institutional perspective	Separation of activities in time and space	Collectively shared meanings of transport modes	Temporal structuration of society Social meaning of time in society
Perspective of strategic conduct of actors	Daily time-space prisms	Travel experience / travel liking	Time management
On-going processes	Time-space distanciation Time-space convergence	Merging of different transport modalities shifting meanings of automobile in society	Social Acceleration, 24 Hour Society, Slow Travel movement

	4 Temporal-material	**5 Material-spatial**	**6 Spatial-symbolic**
Institutional perspective	Transport speed & temporal flexibility/fixity of modalities	Infrastructural layout Place accessibility	Place images and place identity
Perspective of strategic conduct of actors	Travel time use	Individual accessibility	Experience of place
On-going processes	Development of faster transport modes	Car-free neighbourhoods Shared space principle Compact city planning	Interweaving of leisure and space Increasing distance of personal networks

To illustrate this table, the temporal dimension of mobility practices, and the processes and dynamics taking place within this dimension, could be analysed both from an institutional and from a human agent perspective. The temporal dimension in Giddens' framework is presented in three forms. First, there is the reversible temporality of day-to-day actions, encounters and activities. This temporal dimension is considered to be reversible because it indicates the repetitive character of everyday routinized practices (Giddens, 1984, p. 35). Second, there is the finite lifespan of human agents which undertake these activities (therefore the lifespan of the individual is referred to as irreversible time). Finally, there is the longue durée of supra-individual institutions. The longue durée of institutional time is labelled reversible time as well because of the long-term existence of institutions. This reversible time is both the condition and the outcome of practices organised in everyday life. The temporal component, when analysed from the human agent perspective, focuses upon the activities, events and practices that an individual may incorporate within the daily path. This refers to questions such as: At what time of the day are social activities undertaken? Are the mobility practices undertaken during weekday or during the weekend? For which period of time are the activities undertaken? It may also include questions concerning the lifespan which influence the possibilities of everyday mobility choices. The participation of human agents in everyday practices is strongly influenced by the choices made during the life course (for example, the choice of residence, household composition, mobility portfolio, lifestyle choice). For example, Cullen (1978) has investigates the linkages between everyday travel patterns within the context of medium and long-term choices. Clearly, there are many forms of times involved when analysing the temporal component of mobility.

Furthermore, also the analyses of processes and dynamics within the temporal dimension can be analysed from an institutional and a human agent perspective. A well-known process is the acceleration of time and the increase in the pace of life in contemporary societies and their influence on practices of mobility (see Rosa, 2003; Jensen, 2006; Shove, 2002). Rosa (2003) argues that the process of modernisation is best understood by the change in the temporal structures and a wide-ranging speed-up of society. An institutional analysis focuses on the changes in historically constructed structural properties of social systems such as nine-to-five working day[102]. While technological changes (including faster transport modes) decrease the time needed to carry out everyday actions it should entail an increase in free time and a slowdown in the pace of life (Rosa, 2003). Furthermore, the constitution of a 24-hour-society creates opportunities for undertaking and scheduling activities which were not present in the collectivist schedules of modernity (Cass *et al.*, 2003). However, the breakdown of institutionalized time-schedules also has the effect that

[102] In Rosa's analysis there are three forms of acceleration which functions as a self-enforcing triad: technological acceleration, acceleration of social change and the acceleration of the pace of life. The motor behind technological acceleration is the logic of capitalism in the sense that in the capitalist mode of production time is money. Social change is accelerated by the structural motor (based on Luhmann's system's theory) in which increasing societal complexity drives social change. The pace of life is accelerated by the cultural motor which is based on the cultural ideals of modernity. The current cultural ideal is based on the idea that life should be experienced in its fullness. Acceleration is a strategy of people to realize more available projects and plans into a certain time period. However, as acceleration simultaneously increases the options available the promise of acceleration is never fulfilled (Rosa, 2002). This last motor is therefore closely linked to the cultural perspective on escalating consumer demand which I described in chapter three (see Campbell, Rifkin).

people have increasing problems of coordinating encounters and social occasions. Organising periods of co-presence is a more demanding task in a 24-hour-society than in traditional society with fixed time-schedules (Shove, 2002). As Shove (2002) indicates the changes in the temporal dimension lead to problems of temporal coordination and social participation which has both collective as well as individual significance in the domain of mobility. So, when analysing temporal dimensions of mobility from the perspective of strategic conduct more emphasis is put on the necessary rules and resources that individuals must gather to manage their spatially dispersed social activities. In this analysis the focus shifts from the institutional perspectives of the temporal structuration of mobility towards the personal time-space prisms of daily movement and the time management of individual households coping with the individualized freedom of mobility and the necessity of scheduling their everyday lives. For example, the development of a 'slow travel' movement in which human agents pursue a lifestyle in which they regard travel time not as a cost component but rather attach an added value to a relaxed way of travelling is an interesting counter-process of the acceleration of society and the increase in travel speeds. It runs counter to the cultural trend in which time spend on mobility is an aspect which needs to be reduced as much as possible.

Summary

The possibility to be mobile forms an essential part of contemporary society and its economy. More than other consumption domains everyday mobility is inherently intertwined with the fabric of social life. The freedom to invest mobility capital for work, social life and recreation is therefore seen by the members of society as an unconditional right. The central assumption of this thesis is therefore that only when a connection can be made with the social-cultural concerns of citizen-consumers, and when the inclusion of citizen-consumers is no longer approached as an obstacle to change but rather as a source for it, transitions to sustainable mobility can be realized.

The main objective of this thesis is to contribute to the academic and political discourse and debate on transitions to sustainable development in the domain of personal mobility. By adopting a practice based approach for the study of socio-technical transitions an attempt is made to develop a novel framework to analyse, understand and influence resource-intensive mobility patterns of everyday life. The main supposition is that a focus on social practices provides a major asset to the study of innovation processes and the complex interaction between social-technical innovations and human behaviour in the domain of everyday mobility.

The following research questions are investigated in this thesis:

1. What are the characterizing elements in political and academic thought regarding the role of citizen-consumers in sustainable mobility transitions?
2. How can a practice approach help to transcend the limitations of the individualistic, voluntaristic perspective on consumer behaviour with the long-term, systemic perspective of socio-technical transitions?
3. When applied to everyday mobility routines, what can the social practices model contribute to the existing body of knowledge on transitions and system innovations?
4. By which means can environmental policy better incorporate citizen-consumers as agents of change in strategies towards sustainable mobility practices?

The shifts in the storylines of policy-makers and in the political debate on the role of citizen-consumers in mobility related problems are illustrated in Chapter 2. From a historical perspective the transport policy responses of the Dutch and UK government with respect to sustainability challenges are described. When analysing the discourse and policies in the field of mobility in these two countries one can see fluctuations between demand and supply oriented approaches. Over time mobility policies, both within and outside the Netherlands, shift in the way mobility related problems are defined, while the suggested directions for solving the problems vary accordingly. The specific policy measures implemented (such as increasing road network capacity, traffic management, mobility management) reflect how the role of mobility in society is perceived at that moment in time. Moreover, the adopted policy approaches reflect the changing perception of the role of citizen-consumers in everyday mobility practices and in environmental policy in general. The chapter shows how mobility policies at the start of the 21st century tended to be strongly supply-oriented in character, thereby lacking a clear cut and convincing orientation on citizen-consumers and their everyday life mobility behaviours. As a result, mobility policies and politics represented a bias towards infrastructural, technological and spatial strategies in the governance of mobility transitions. Citizen-consumers are often seen as actors which are not willing or able to implement structural changes in their everyday life patterns and are therefore frequently considered as barriers to the successful realisation of sustainable transitions, especially in the field of mobility. Citizen-consumer are considered to be prone to an 'attitude-action gap': there are differences in what they say and what they do, they are unwilling to change their existing behaviour and are too opportunistic to start consuming in a more sustainable manner. Focusing on behaviour change is therefore seen as ineffective considering the large-scale transformation that is required within the mobility domain. Especially in the last decades it seems as if in scientific and policy debates there is primarily faith in achieving structural behaviour change with the use of formal rules, regulations and taxations. Arguments are presented to stress why a citizen-consumer orientation in mobility is not only a fruitful but also an essential ingredient in (policies targeting) transitions towards sustainable mobility.

To understand the role of citizen-consumers in transition trajectories, in Chapter 3 various theoretical perspectives on consumption behaviour and the development of material culture are described. The first part of the chapter focuses on conspicuous consumption, as consumption choices have become very important aspects in the (reflexive) constitution of personal identities. Clearly, passenger mobility is inherently related to consumer cultures, the symbolic and affective motives for car usage, functional aspects of trips, consumer perceptions and appreciation of different modalities. Analysing (the history of) automobility from the perspective of different consumption theories provides information about the social-cultural meanings associated with the car and its uses and functions in society. By taking the changing structures of car related meanings and images into account, it becomes possible to deliver a satisfactory analysis of the diffusion of the car and its contemporary cultural functions. Images of the car as a material object related to freedom, adventure and especially speed, are reproduced over and over again through car design and advertising, in car magazines, and through social talks. However, in this chapter it is also shown how meanings associated with automobiles are not fixed. Automotive meanings develop and change over time, depending on the social and cultural context of the consumer, and on the role of important actors and organizations influencing this cultural context.

The viewpoint of conspicuous consumption is especially helpful in understanding the acquisition process of commodities. In addition, in this chapter the role of routines is described to understand the more habitual, taken-for-granted character of the vast bulk of activities of day-to-day social life.

Chapter 3 shows that practice theories contain characteristics which enable them to study the role of agency in sustainable transitions because they connect human agency with structural properties of socio-technical systems. All theories of practice have in common that they consider practices as the key unit of analysis to understanding social life. Authors of practice theories commonly agree upon the notion that practices not only consist of a specific type of behaviour, but also of objects, materials and associated meanings. Furthermore, practice theories put emphasis on the routinized nature of everyday activities. In a simplified way, one could say that practices are the taken for granted ways of doing things. This is a different way of speaking about routines when compared to the social-psychological tradition of habits and routines which focuses on personal past experiences and the automatic nature of behavioural habits. While also in the psychological tradition human agents are assumed to perform individual routine behaviour, in the sense that the activity is habitually undertaken, the activities or behaviours under study are not the collectively shared routines which form the starting point of sociological analyses of routines.

Though practices are routinized in character, in the sense that shared practices can be regarded as institutions recursively reproduced by groups of actors, this in no way means that they are inherently stable. Drawing upon the work of scholars in the field of contemporary theories of practice it was shown how co-evolution of practices, objects and systems provides meaningful insight into the ways expectations and technologies may change over the course of time.

However, what needed to be conceptually refined were the (differences in) individual actions on the one hand and the structuration of collectively shared routines which takes shape in the form of social practices, on the other hand. As practices are internally differentiated, and with individuals showing different levels of understandings, skills and attributions of meaning, there are multiple ways of doing things in practices. There is a plurality of lifestyles and activities in society and this is reflected in the multiple and diverse sets of practices human agents can participate in, in the multiple ways in which practitioners connect or bind together the elements of practices, and in the degree or level of involvement in particular practices. Therefore, to understand variations in practices, an integrated approach of 'individual action' and collectively shared routines is important. Without this consideration the development of a consumer-oriented policy, targeting specific groups with the aim to involve citizen-consumers in sustainable transitions, is hard to imagine.

The social practices model (Spaargaren, 1997) is presented as a promising practice-based framework to investigate the active involvement of citizen-consumers in transition processes to sustainable development. The social practices model considers citizen-consumers as ecologically more or less committed actors which may be increasingly involved in the process of the ecological transformation of production-consumption chains. However, to be able to participate in or potentially alter social practices, citizen-consumers need to be invited and challenged with attractive visions and ideas about sustainability, while at the same time being actively supported to get involved in transition processes. Therefore the social practices model introduces the concept of the system of provision which makes clear that the possibilities for sustainable consumption to a large extent depend on the amount and kind of socio-technical innovations available in a specific domain. Next to the system of provision the social practices model uses conceptualisations of the lifestyles of (groups of) citizen-consumers to depict variation in the reproduction of practices.

In three empirical case studies it is shown how the theoretical framework can be applied to study sustainable development in the domain of everyday mobility. The main focus of each of the three cases is on situated interactions taking place at the crossroads between modes of access and modes of provisioning in the domain of everyday mobility; that is, the ways that citizen-consumers get access to forms of mobility on the one hand and the ways in which mobility is supplied by mobility service providers to them on the other.

In the first empirical chapter (Chapter 4) the practice of purchasing a new car is investigated. The focal point is on the different environmental information tools, such as the fuel efficiency label, and the supporting taxation schemes that were implemented in the Netherlands from 2001 onwards as new consumer-oriented policy strategies.

This chapter shows how the practice of car purchasing has changed considerably over the last decade. Consumers today are increasingly knowledgeable about (environmentally relevant) vehicle-characteristics. This is the result of a strong consumer empowerment which has taken place in the last decade as (also environmental) information has become more accessible. The internet has disclosed information about the environmental impact of cars and about alternative vehicles that was until recently inaccessible to consumers. Car salesmen notice that consumers are asking more and more environmentally related questions in the showroom, and car advertisements are actively promoting the environmental aspects of new cars. A car market analysis showed that the energy efficiency of new cars in the Netherlands has improved greatly, especially after 2008, the year that new fiscal policies were implemented. While the energy efficiency of European cars is improving as a whole, the Dutch fuel efficiency is improving in an even faster rate. It is more than likely that the consumer-oriented strategies in new car purchasing, as discussed in this chapter, played a crucial role in this development.

Nevertheless, this chapter also shows that there is a clear need for help in reducing the complexity of the environmental information in the car-buying process. Much environmental information as yet is abstract and hard to understand, while the information tools have the aim that consumers are able to compare cars at the level of sustainability. In addition, the chapter shows that the meaning of the energy efficiency label at the time of research was contested by car-salesmen who felt the energy label to be too abstract to be useful as an informational tool. This chapter therefore points out the importance of the social context in which environmental information is provided. One cannot assume that consumers are empowered with information solely by increasing the amount of environmental information.

While the practice of purchasing a new car can be seen as a form of (conspicuous) consumption similar to other practices of buying and selling commodities, the next two case studies focus on (routinized) everyday life mobility patterns. In order to analyse transformation in these practices of mobility, the general theoretical framework presented by the social practices approach needed to be conceptualised and operationalized for the domain of everyday mobility. This task is taken up in Chapter 5, which focuses on the question what everyday life mobility practices actually are about and how they can be characterised. Mobility is portrayed as the result of the participation of human agents in social occasions and social activities which take place at specific moments in time at specific sites and for specific reasons and motives. Therefore the most important mobility practices we distinguish are derived from the 'social bases of everyday travel', namely: commuting, business travel, home-school travel, visiting family/friends, shopping and leisure travel. These everyday journeys allow travellers to connect the most important social activities and encounters which constitute their daily lives. In addition to defining these mobility practices, it is investigated which basic analytical characteristics are shared by these mobility practices. It is argued that mobility practices can be described as the interplay between four different constitutive dimensions: the temporal, the spatial, the material and the symbolic dimension.

In Chapter 5 also the individual and lifestyle part of the social practices approach is conceptualised and operationalized for mobility practices. We introduce and use the notions of 'motility' – a human agent's potential to be mobile – and 'mobility portfolio's' (Flamm and Kaufmann, 2006) to describe the ways that human agents get access to and appropriate the capacity for mobility. Motility involves not only all the means (mobility products and services) available to get access to the various forms of mobility, but also the skills and knowledge necessary to make use of these forms of mobility, and the assessment that comes along with them.

The second empirical case study, described in Chapter 6, focuses on the practice of commuting. In this chapter three cases of mobility management are presented in which an attempt was made to orchestrate a shift in the mobility practices of commuters. The aim of this chapter is to understand stability and change in these particular practices of everyday life mobility.

The routinization of commuting activities is one of the key mechanisms that explain their relative stability and robustness over time. Therefore, in most circumstances a trigger moment is required in order to shake up the taken for granted character of these day to day activities. Most long-term changes in consumer behaviour require that a change in the contextual situation occurs as well. The case studies in this chapter provide various examples of both negative and positive trigger moments that provide such a change in context.

Based on the case studies we can conclude that mobility management may lead to innovation in mobility practices through various means. In addition to making travel alternatives more attractive, the case studies have shown that practitioners are proactively 'confronted' with socio-technical innovations and may start to experiment with these innovations. Thus, the measures can have the effect that new ways of doing things are introduced in mobility practices. However, this chapter also shows that innovation in mobility practices does not always take shape. The chapter on car purchasing already revealed that merely increasing the amount of environmental information is no guarantee for its effective uptake and use. In a similar way, this chapter shows that merely broadening the mobility portfolio of practitioners does not guarantee the uptake or enrolment into new practices of commuting.

Theoretically the cases show that alterations in practices imply that a shift in one or more dimensions of the mobility practices has to occur. That is, socio-technical innovations need to be integrated in the specific mobility practices which require that adjustments are being made by practitioners, also to non-technical dimensions of their daily life mobility routines. A shift from car use to public transport not only implies that the mode of transport changes, it also means a shift in the spatio-temporal organization of mobility practices and their cultural framings.

The third and final case study, described in Chapter 7, presents the results of a large-scale consumer survey. The aim is to investigate how consumers assess and evaluate the various environmentally friendly alternatives on offer in the domain of everyday mobility.

An important outcome of the quantitative analysis is that environmental change in the domain of mobility is supported by a surprisingly large part of the Dutch citizen-consumers. Environmental issues are recognized and acknowledged as problematic and almost a majority of respondents feels at least partly co-responsible for dealing with these issues. More importantly, the current system of petrol-based automobility is seen as unfavourable and a systemic change to other modes of transport or fuels is supported by most. This survey supports the outcome presented in the chapter on new car purchasing that in a relatively short time-period in the domain of mobility possibilities for sustainable change have emerged.

In addition, it is examined how the receptiveness of citizen-consumers for socio-technical innovations in the domain of mobility is also dependent on the specific system of mobility provision and its particular provider strategies. With the assessment of provider strategies it is shown how these strategies may target barriers for the integration of socio-technical innovations in mobility practices.

Finally, by using the concept of mobility portfolios, individual variety in the conduct of social practices in the domain of mobility is analysed in an innovative detail. It was investigated whether or not citizen-consumers with a 'greener' mobility portfolio are more receptive to sustainable mobility alternatives. The survey showed how mobility portfolios are indeed an important factor in the receptiveness of citizen-consumers to sustainable mobility alternatives such as multi-modal travelling. Citizen-consumers with a larger mobility potential – i.e. more access to mobility products and services, and with a positive evaluation of these products and services – are more likely to conduct their mobility practices in a more sustainable way.

The concluding Chapter 8 makes the balance by describing both the theoretical and empirical contributions of this thesis to the study of sustainable transitions in the domain of mobility. An important scientific insight is that each of the empirical case studies has shown the important role of contextual factors in understanding and influencing mobility behaviour of designated (lifestyle) groups of consumers. Furthermore, theoretically interesting is that motility and mobility portfolios provide the means to analyse mobility behaviour between the level of the individual and lifestyle on the one hand and the social practices and collective conventions on the other. By showing how citizen-consumers have different levels of understandings, competences and access to mobility options, and more importantly, how these have an impact on the way citizen-consumers participate in practices, a relevant contribution has been made to theories of practice. Next to describing the scientific contributions, in this chapter four policy recommendations are presented to strengthen the crucial role of citizen-consumers in transitions to sustainable mobility.

About the author

Jorrit Nijhuis (1977, the Netherlands) was educated at Utrecht University where, after two years of human geography, he started a master's programme in environmental sciences. After an internship at Alterra, he graduated on the topic of landscape experience: individual differences in the aesthetic evaluation of urban natural landscapes. After his graduation he worked for one and a half year in public healthcare as a supervisor for people with intellectual disabilities.

In 2005 he started working as one of the four PhD students of the Contrast research programme at the Environmental Policy Group of Wageningen University. Contrast (consumption transitions for sustainability) was co-funded by GaMON (gamma-onderzoek milieu, omgeving, natuur) and KSI (knowledge network on system innovations and transitions). The aim of the research programme was to develop knowledge and methodologies to involve citizen-consumer groups in transitions towards sustainable development on a number of policy-relevant domains of consumption.

In 2009 he continued his professional career at Rijkswaterstaat, the national agency for road and water management of the Ministry of Infrastructure and Environment. At Rijkswaterstaat he works as an advisor on freight traffic and mobility management. He is an expert on 'Toekan', a methodology aimed to investigate the potential for travel alternatives and mobility management strategies during road construction works. In addition, he is the secretary of the national meetings on mobility management within the Ministry. In 2013 he was appointed as a member of the International Program Committee of the European Platform on Mobility Management. Currently, he is also involved in the monitoring and evaluation of the Programme Better Utilization, a public-private investment programme containing over 270 local measures (ranging from infrastructure construction, to mobility management and traffic management) aimed to improve accessibility.

Printed in the United States
by Baker & Taylor Publisher Services